Landslides in Forests around the World: Causes and Mitigation

Landslides in Forests around the World: Causes and Mitigation

Editors

Haijia Wen
Weile Li
Chong Xu
Hiromu Daimaru

MDPI • Basel • Beijing • Wuhan • Barcelona • Belgrade • Manchester • Tokyo • Cluj • Tianjin

Editors
Haijia Wen
School of Civil Engineering
Chongqing University
Chongqing
China

Weile Li
State Key Laboratory of
Geohazard Prevention and
Geoenvironment Protection
Chengdu University of
Technology
Chengdu
China

Chong Xu
Geological Hazards Research
Center
National Institute of Natural
Hazards
Beijing
China

Hiromu Daimaru
Laboratory of Watershed
Environmental Studies
Ishikawa Prefectural
University
Nonoichi
Japan

Editorial Office
MDPI
St. Alban-Anlage 66
4052 Basel, Switzerland

This is a reprint of articles from the Special Issue published online in the open access journal *Forests* (ISSN 1999-4907) (available at: www.mdpi.com/journal/forests/special_issues/forests_landslides).

For citation purposes, cite each article independently as indicated on the article page online and as indicated below:

LastName, A.A.; LastName, B.B.; LastName, C.C. Article Title. *Journal Name* **Year**, *Volume Number*, Page Range.

ISBN 978-3-0365-7507-0 (Hbk)
ISBN 978-3-0365-7506-3 (PDF)

Contents

About the Editors

Haijia Wen

Professor Haijia Wen obtained his MS and Ph.D. degrees from Chongqing University of China. Now, he is a full professor of the School of Civil Engineering at Chongqing University. He is also the Corresponding Member of TC309, ISSMGE, a Member of Geological Engineering Teaching Steering Committee of the Ministry of Education, PRC.; a Member of Geology, Mining and Environment Special Teaching Steering Committee of Chongqing; the vice chairman of Risk and Insurance Research Branch, CCES; the Standing director of Geological Education Research Branch, GSC; and a Member of the Landslide and Debris Flow Prevention Special Committee, China Society of Soil and Water Conservation. Prof. Wen's major research interests cover geotechnical risks, geological disasters, and data science in civil engineering. He has published over 160 journal or conference papers and 17 Patents, and has received approximately 10 awards, including the science and technology invention award or the science and technology progress award from Chongqing City, the Ministry of Education, and the Rock Mechanics and Engineering Society of China.

Weile Li

Professor Weile Li was born in Anqing, Anhui Province, China, in 1982. He received his B.S. degree in surveying and mapping engineering, his M.S. degree in GIS and RS, and his Ph.D. degree in engineering geology from Chengdu University of Technology, China, in 2005, 2008, and 2019, respectively. From 2008 to 2014, he was a Research Assistant with the State Key Laboratory of Geohazard Prevention and Geoenvironment Protection (SKLGP), Chengdu University of Technology. From 2015 to 2019, he has been an assistant professor with the SKLGP. Since 2021, he has been a full professor with the SKLGP. His research interests include landslide detection, monitoring and risk assessment, and earthquake-induced landslide rapid mapping and prediction. He is the author of 9 books and more than 150 papers on the detection and monitoring of landslides.

Chong Xu

Professor Chong Xu is the Director of the Committee on Earthquake Hazard Chain, Seismological Society of China, the Director of the Key Laboratory of Compound and Chained Natural Hazards Dynamics, Ministry of Emergency Management of China, the Director of the Geological Hazards Research Center, National Institute of Natural Hazards, and Research Professor in National Institute of Natural Hazards, Ministry of Emergency Management of China. He obtained his Ph.D. from the Institute of Geology and Geophysics, Chinese Academy of Sciences, in 2010. His research interests include earthquake-triggered landslides, rainfall-induced landslides, and geological hazards using GIS, remote sensing, and statistical analysis techniques. By virtue of the study of earthquake-triggered landslides over 10 years, he has obtained a series of scientific research achievements of this subject, including database construction and the criteria of earthquake-triggered landslides, a quantitative description of landslide distribution, addressing issues of earthquake geology, landslide volume modeling, and hazard assessment. He has published over 300 papers with over 11000 citations and an h-index of 55 from Google Scholar. He was selected for inclusion in Elsevier's "China's highly cited scholars" from 2020 to present day. His research achievements, mainly on landslides triggered by recent large earthquakes, have been extended to about 50 research institutes and universities worldwide and have been adopted directly by more than 100 research teams.

Hiromu Daimaru

Professor Hiromu Daimaru received his Ph.D. from Shinshu University. Now, he is a full professor at the Laboratory of Watershed Environmental Studies at Ishikawa Prefectural University. After studying natural geography at Hokkaido University, he conducted studies on mountain geoecology and the disaster science of mountain areas at the FFPRI (Forestry and Forest Products Research Institute) from 1989 to 2021. Prof. Daimaru's major research interests cover the geoecology of mountainous areas, landslide geomorphology, and the visualization and classification of topographic information using spatial information technology.

Preface to "Landslides in Forests around the World: Causes and Mitigation"

Landslides are a common natural disaster in forested mountainous regions. Because of the increased frequency of extreme rainfall events resulting from global warming, the risk of landslides in forested areas remains high. Forest landslide research has always been of great interest to scholars. Key areas of research include understanding the interaction between forests and mountain landslides, assessing the hazard and risk of landslides, and developing rainfall-induced landslide warning systems in forested areas. In recent years, the development of technologies such as artificial intelligence, remote sensing (including optical remote sensing, InSAR, LiDAR, and others), and big data has led to rapid advancements in the detection, monitoring, warning, and risk assessment of forest landslides. These developments provide reliable solutions for effectively and quickly responding to potential landslide prevention and mitigation measures in forested areas. To showcase the latest advancements in this field, we organized 14 papers to form this reprint, named "Landslides in Forests around the World: Causes and Mitigation". This reprint covers various topics, such as the impact mechanism of tree roots on landslide stability, landslide deformation monitoring, landslide disaster prevention and control engineering technology, the automatic identification of regional landslides, susceptibility and hazard assessment, and rainfall-induced mass landslide warning. We are delighted to see that the articles featured in this reprint have contributed new knowledge to the field of landslides in forests, provided innovative ideas for future development, and opened up new opportunities for collaboration and innovation. We extend our gratitude to the staff of *Forests* for their tremendous support of our Editor Team, as well as their assistance with the review, revision, and verification process. We also thank all authors for their insightful contributions to this reprint and all reviewers for their valuable comments and suggestions.

Haijia Wen, Weile Li, Chong Xu, and Hiromu Daimaru
Editors

forests

Editorial

Landslides in Forests around the World: Causes and Mitigation

Haijia Wen [1], Weile Li [2], Chong Xu [3,4,*] and Hiromu Daimaru [5]

1 School of Civil Engineering, Chongqing University, Chongqing 400045, China; jhw@cqu.edu.cn
2 State Key Laboratory of Geohazard Prevention and Geoenvironment Protection, Chengdu University of Technology, Chengdu 610059, China; whylwl01@163.com
3 National Institute of Natural Hazards, Ministry of Emergency Management of China, Beijing 100085, China
4 Key Laboratory of Compound and Chained Natural Hazards Dynamics, Ministry of Emergency Management, Beijing 100085, China
5 Tama Forest Science Garden, Forestry and Forest Products Research Institute, Tokyo 305-8687, Japan; hiromudaimaru@gmail.com
* Correspondence: xc11111111@126.com

1. Introduction

Landslides are a common natural disaster in forested mountainous regions. Because of the increased frequency of extreme rainfall events resulting from global warming, the risk of landslides in forested areas remains high. Forest landslide research has always been of great interest to scholars. Key areas of research include understanding the interaction between forests and mountain landslides, assessing the hazard and risk of landslides, and developing rainfall-induced landslide warning systems in forested areas. In recent years, the development of technologies such as artificial intelligence, remote sensing (including optical remote sensing, InSAR, LiDAR, and others), and big data has led to rapid advancements in the detection, monitoring, warning, and risk assessment of forest landslides. These developments provide reliable solutions for effectively and quickly responding to potential landslide prevention and mitigation measures in forested areas.

To showcase the latest advancements in this field, a Special Issue named "Landslides in Forests around the World: Causes and Mitigation" has been organized. The issue comprises 14 papers covering various topics such as the impact mechanism of tree roots on landslide stability, landslide deformation monitoring, landslide disaster prevention and control engineering technology, automatic identification of regional landslides, susceptibility and hazard assessment, and rainfall-induced mass landslide warning.

Citation: Wen, H.; Li, W.; Xu, C.; Daimaru, H. Landslides in Forests around the World: Causes and Mitigation. *Forests* **2023**, *14*, 629. https://doi.org/10.3390/f14030629

Received: 14 March 2023
Accepted: 19 March 2023
Published: 20 March 2023

2. Mechanism, Monitoring, and Control of Large-Scale Landslides in Forested Areas

To gain a better understanding of the role that forests play in preventing landslides, Okada et al. conducted a study on the impact of Japanese cedar roots on the occurrence of shallow landslides [1]. They continuously monitored changes in root strength over the course of several years and calculated the safety factor of the slope every five years. The study revealed that root strength is at its lowest after approximately 10 years, and the risk of landslide increases between 5 and 15 years. These findings provide valuable insight for selecting the most appropriate time for forest operations and mitigating the occurrence of landslide disasters. Another study conducted by P. Li et al. investigated the effects of root distribution and stress path on the shear strength of root–soil composites using the consolidated undrained (CU) triaxial test method [2]. The results of this study contribute to a better understanding of the impact of tree roots on slope stability and provide insight into prevention and mitigation strategies for forest landslides.

In October 21, 2017, heavy rainfall triggered a significant landslide in Guang'an Village, Wuxi County, Chongqing City, China. To investigate the stability of the landslide accumulation, K. Zhang et al. analyzed the long-term deformation process of the accumulation after sliding using time series InSAR technology and LiDAR data [3]. Their study provides valuable insight into preventing the accumulation from sliding again and forming secondary disasters caused by landslides.

To gain a deeper understanding of the mechanism behind large-scale landslides, Guo et al. analyzed the formation process and mechanism of the Nuugaatsiaq landslide in Greenland in 2017 [4]. This was achieved through examining the similarity and amplitude characteristics of multi-temporal seismic waveforms. Their findings suggest that the landslide was likely caused by a series of small earthquakes that accelerated and migrated. This study provides a credible case study report for analyzing the nucleation process of large landslides based on seismic waveform data. In another study, Huang et al. investigated the macroscopic friction fluctuation of the spherical particle system and breccia particles [5]. Their results showed that the friction fluctuation of the spherical particle system is determined by the friction of the particle surface and the normal stress of the system. Similarly, for breccia particles, the macroscopic friction fluctuation is determined by particle surface friction, system normal stress, and shear velocity.

In light of the successful interception of a landslide near the entrance of a railway tunnel using a flexible barrier to ensure the safety of railway operations, Zhao et al. conducted an analysis of the interaction between the landslide debris and the barrier [6]. Their analysis was based on the numerical simulation method, and the results were validated by field measurements. The study analyzed the impact of landslide debris on the flexible barrier under different impact energy and velocity conditions. The findings suggest that such a barrier can resist landslide debris with a maximum energy roughly four times that of resisting rockfall. These results provide valuable support for the prevention and control of shallow landslide disasters in mountainous regions

3. Automatic Identification and Susceptibility Analysis of Regional Landslides

The automatic identification of landslides plays a vital role in mass landslide inventorying. Pang et al. focused on the automatic identification of coseismic landslides using the example of the Mw6.7 earthquake that occurred in Hokkaido, Japan, on September 6, 2018 [7]. They mapped 4000 landslide label samples and developed a convolution neural network model based on the YOLOv3 algorithm. The results demonstrated that the model is efficient, with high recognition accuracy and speedy processing. This research provides a reference method for rapidly extracting mass landslides.

The majority of papers in this special issue focus on landslide hazard and susceptibility assessment, highlighting its significance in research. L. Li et al. conducted their study in Hechuan District, Chongqing, China, utilizing the random forest method to assess landslide susceptibility based on 754 landslides and the associated impact factors [8]. Furthermore, by considering natural, social, and ecological factors in the region, they evaluated the suitability of construction land in mountainous areas. This research may support the spatial layout of urban land and critical infrastructure in mountainous areas. To compare the shallow machine learning method with the deep learning method for landslide hazard and susceptibility assessments, Xu et al. conducted experiments using logistic regression, random forest, deep fully connected neural network, and long short-term memory neural networks in the Three Gorges reservoir area of China [9]. The results showed that the random forest model performs comparably to deep learning models while being more efficient, making it a valuable reference for selecting a landslide susceptibility assessment model. W. Zhang et al. investigated how zoning based on qualitative information affects landslide susceptibility assessment with a machine learning model in Yunyang County, Chongqing City, China [10]. The results showed that considering zoning in the evaluation process leads to better accuracy than not taking zoning into account. The combination of qualitative analysis and a quantitative model demonstrates a promising application prospect in landslide susceptibility assessment. To assess the impact of forests on landslide susceptibility, Y. Zhang et al. conducted a study in Bijie City, Guizhou Province, China to establish a model for evaluating the contribution of forests at different levels to landslide susceptibility [11]. They also examined the impact of various forest indicators on landslides. The findings indicated that shrub forests have a stronger protection ability against landslides, while natural forests have a greater inhibitory effect compared with artificial forests. Additionally, middle-aged and near-mature forests showed better landsliding control ability. This work provides valuable insights into the impact of different forest types on landslides.

4. Rainfall-Triggered Landslide Warning

Early warning systems for landslides triggered by rainfall in forested areas have proven effective in mitigating landslide disasters. Liu et al. focused on Fujian Province, China, and developed six machine learning models for predicting landslide occurrences based on geological and meteorological data, as well as other factors affecting landslides [12]. The results showed that the random forest model performed the best, providing a reference for future landslide early warning systems. During 9–11 August 2019, heavy rainfall resulted in numerous landslides in Ningguo City, Anhui Province, China. Cui et al. utilized satellite images to identify 414 rainfall-induced landslides and constructed a distribution map of rainfall-induced landslides [13]. They also developed a probability model for landslide occurrence triggered by rainfall and calculated the absolute probability of landslides under different rainfall conditions. This study provides support for landslide assessment and early warning systems under various rainfall scenarios. Taking 1520 landslide events in Fengjie County, Chongqing, China in 2016 as samples, Sun et al. developed a landslide susceptibility model based on the random forest model and a landslide early warning model based on precipitation [14]. The results showed that the precipitation early warning model combined with landslide susceptibility, early effective precipitation, and daily precipitation threshold has a high early warning ability. The results can provide scientific and technological support for rainfall-triggered landslide disaster prevention and mitigation triggered by local rainfall.

5. Conclusions and Prospects

We are delighted to see that the articles featured in this Special Issue have contributed new knowledge to the field of landslides in forests, provided innovative ideas for future development, and opened up new opportunities for collaboration and innovation. We extend our gratitude to the staff of *Forests* for their tremendous support to our Guest Editor Team in organizing this Special Issue, as well as their assistance with the review, revision, and verification process. We also thank all authors for their insightful contributions to this Special Issue and all reviewers for their valuable comments and suggestions on the submitted manuscripts.

Funding: This study was supported by the National Key Research and Development Program of China (2021YFB3901205) and the Natural Science Foundation of Chongqing (CSTB2022NSCQ-MSX0594).

Conflicts of Interest: The authors declare no conflict of interest.

References

1. Okada, Y.; Cai, F.; Kurokawa, U. Changes in slope stability over the growth and decay of Japanese cedar tree roots. *Forests* **2023**, *14*, 256. [CrossRef]
2. Li, P.; Xiao, X.; Wu, L.; Li, X.; Zhang, H.; Zhou, J. Study on the shear strength of root-soil composite and root reinforcement mechanism. *Forests* **2022**, *13*, 898. [CrossRef]
3. Zhang, K.; Gong, F.; Li, L.; Ng, A.H.-M.; Liu, P. Mapping the long-term evolution of the post-event deformation of the Guang'an Village Landslide, Chongqing, China using multibaseline InSAR techniques. *Forests* **2022**, *13*, 887. [CrossRef]
4. Guo, Z.; Hou, X.; Gao, D.; Liu, J. Nucleation process of the 2017 Nuugaatsiaq, Greenland Landslide. *Forests* **2022**, *14*, 2. [CrossRef]
5. Huang, Y.; Wang, Y.; Wang, S. Effect of particle form and surface friction on macroscopic shear flow friction in particle flow system. *Forests* **2022**, *13*, 1107. [CrossRef]
6. Zhao, L.; Zhang, L.; Yu, Z.; Qi, X.; Xu, H.; Zhang, Y. A case study on the energy capacity of a flexible rockfall barrier in resisting landslide debris. *Forests* **2022**, *13*, 1384. [CrossRef]

7. Pang, D.; Liu, G.; He, J.; Li, W.; Fu, R. Automatic remote sensing identification of co-seismic landslides using deep learning methods. *Forests* **2022**, *13*, 1213. [CrossRef]
8. Li, L.; Chen, X.; Zhang, J.; Sun, D.; Liu, R. Landslide susceptibility-oriented suitability evaluation of construction land in mountainous areas. *Forests* **2022**, *13*, 1621. [CrossRef]
9. Xu, S.; Song, Y.; Hao, X. A comparative study of shallow machine learning models and deep learning models for landslide susceptibility assessment based on imbalanced data. *Forests* **2022**, *13*, 1908. [CrossRef]
10. Zhang, W.; Liu, S.; Wang, L.; Samui, P.; Chwała, M.; He, Y. Landslide susceptibility research combining qualitative analysis and quantitative evaluation: A case study of Yunyang County in Chongqing, China. *Forests* **2022**, *13*, 1055. [CrossRef]
11. Zhang, Y.; Shen, C.; Zhou, S.; Luo, X. Analysis of the influence of forests on landslides in the Bijie Area of Guizhou. *Forests* **2022**, *13*, 1136. [CrossRef]
12. Liu, Y.; Huang, J.; Xiao, R.; Ma, S.; Zhou, P. Research on a regional landslide early-warning model based on machine learning—A case study of Fujian Province, China. *Forests* **2022**, *13*, 2182. [CrossRef]
13. Cui, Y.; Jin, J.; Huang, Q.; Yuan, K.; Xu, C. A data-driven model for spatial shallow landslide probability of occurrence due to a typhoon in Ningguo City, Anhui Province, China. *Forests* **2022**, *13*, 732. [CrossRef]
14. Sun, D.; Gu, Q.; Wen, H.; Shi, S.; Mi, C.; Zhang, F. A hybrid landslide warning model coupling susceptibility zoning and precipitation. *Forests* **2022**, *13*, 827. [CrossRef]

forests

MDPI

Article

Changes in Slope Stability over the Growth and Decay of Japanese Cedar Tree Roots

Yasuhiko Okada [1,*], Fei Cai [2] and Ushio Kurokawa [3]

[1] Forestry and Forest Products Research Institute, Forest Research and Management Organization, 1 Matsunosato, Tsukuba 305-8687, Ibaraki, Japan
[2] Department of Environmental Engineering Science, Gunma University, Tenjin 1-5-1, Kiryu 376-8515, Gunma, Japan
[3] Kyushu Research Center, Forestry and Forest Products Research Institute, Forest Research and Management Organization, Kurokami 4-11-16, Chuo, Kumamoto 860-0862, Kumamoto, Japan
* Correspondence: okada10@affrc.go.jp

Abstract: In Japan, repeated driftwood landslide disasters have become a major issue; thus, studies are required to better understand forest function to implement appropriate forest management and prevent such disasters. We investigated the effect of Japanese cedar tree roots on shallow landslide initiation. To incorporate the effect of roots on the two side-flanks of the shallow landslide, we propose a new slope-stability analysis method in which the sliding block is simplified as a three-prism model. The root reinforcement was approximated by the sum of the root pullout forces over a unit area, incorporating changes in the root strength with the growth of the trees after planting and the decay of the stumps after cutting. The reinforced root strength after the stump-cutting decreased linearly with time, with no strength remaining at 9 years. In contrast, the reinforced root strength of the new plants increased according to a logistic curve with time; thus, the root strength increased only slightly up to 9 years after planting, and the minimum total reinforced root strength was observed at this time. The safety factor of the slopes in a forest basin in Ibaraki Prefecture was calculated using the proposed three-prism method at intervals of 5 years on a 1-metre-resolution digital elevation model. The number of unstable grids peaked at 10 years, and a higher risk of slope instability was observed at 5–15 years. Therefore, implementing forest operations for lowering slope instability during this period should be important to prevent landslide disasters.

Keywords: reinforcement of roots; slope stability; growth and decay; Japanese cedar trees; shallow landslides

Citation: Okada, Y.; Cai, F.; Kurokawa, U. Changes in Slope Stability over the Growth and Decay of Japanese Cedar Tree Roots. *Forests* **2023**, *14*, 256. https://doi.org/10.3390/f14020256

Academic Editors: Haijia Wen, Weile Li, Chong Xu and Hiromu Daimaru

Received: 14 December 2022
Revised: 21 January 2023
Accepted: 22 January 2023
Published: 29 January 2023

1. Introduction

In Japan, driftwood landslide disasters are a major concern, owing to the regional development of residences in mountainous areas and an increase in the number of severe rainfall events, probably influenced by climate change [1]. In the last quarter century, debris flows mobilised from shallow landslides or complex earth slides–earth flows [2] have hit some residential areas, extending to alluvial fans in many places [3–5]. Notably, in Hiroshima Prefecture, debris flows have severely damaged areas within a short period, causing 31 and 74 deaths in 1999 and 2014, respectively [6–9].

In July 2018, a large amount of driftwood was observed in landslides induced by heavy rainfall events over the western part of the main island and Shikoku Island [10]. The relationship between forests and landslide disasters has received substantial attention from researchers, engineers, and administrative officers. In particular, the differences in landslide susceptibility between natural and artificial forests and between coniferous and broadleaf forests have been examined [11,12]. After an extended period of afforestation from around the 1950s to the 1970s, the trees have grown large and are ready for harvest [13]; therefore, the strategies for controlling and managing forests must be carefully considered. Trees are hazardous once they are engulfed and become driftwood, but they reinforce slopes through networking root systems within the subsoil that prevent shallow landslides. In some cases, trees can capture flows of driftwood and debris to decrease downslope damage [14,15]. Regarding the multifunctional role of forests, the public has been aware of their disaster prevention function against landslides and erosion for 40 years and now understands that they help prevent climate change by absorbing greenhouse gases [16]. As of 31 March 2021, the Forestry Agency of the Ministry of Agriculture, Forestry and Fisheries has designated areas in the national forest as forest reserves (1.3×10^5 km^2) [17]; these forest areas are expected to prevent soil run-off (2.6×10^4 km^2) and landslides (6.0×10^2 km^2), thereby acting as countermeasures against forest disasters.

The ability of tree roots to prevent shallow landslides has been investigated intensively. For example, the mechanical behaviour of roots crossing a sliding surface during shearing has been closely examined, and reinforced root strength is known to act as the root cohesion component (c_r) in the Mohr–Coulomb failure equation [18–22]. In addition, root pullout experiments have revealed that the summation of root pullout strength over a unit area acts as c_r [23,24]. Accordingly, the root bundle model was proposed, in which the root length, maximum tensile force, and average Young's modulus were approximated using a function of the root diameter [25]. The effects of the root distribution, root angles across the sliding surface, and root diameter have also been examined in triaxial compression tests [26]. Large-scale, in situ direct shear experiments have been conducted, in which specimens (about 1 m in both width and length) holding a tree stump were excavated for direct shearing [27–29], with one such experiment conducted on a steep slope of 32° [30].

Analyses of slope stability have been conducted considering c_r [31–33]. In this regard, the two-dimensional (2-D) infinite slope model is commonly employed, given its simplicity, i.e., the reinforced strength of roots is given only over the bottom surface of an oblique quadrangular prism; therefore, only the vertical roots penetrating to the bottom of a sliding surface contribute to reinforcing the slope stability. Considering the importance of lateral roots, slope models that enable the evaluation of their effects have been developed. The effects of lateral roots have been considered in the 2-D infinite slope model [34]; however, only one vertical surface of the oblique quadrangular prism at the upslope side in the longitudinal direction was considered, and the effects of lateral roots on the vertical surface at the downslope side and two side-flanks of the prism were not examined. Regarding the transverse direction, Ueno [35] investigated the effect of widths on landslide initiation using an infinite slope model in the longitudinal direction but with a transverse trapezium section; however, the transverse slope of the side-flanks of the trapezium section was assumed to be constant and equal to 45°, and c_r was not considered.

To evaluate the effects of lateral roots across the side-flanks, a three-dimensional (3-D) slope model should be used. Indeed, 3-D slope stability has commonly been analysed to calculate the safety factors related to relatively large, deep-seated landslides for which the whole sliding surface of the landslide mass could be approximated before the final failure [36–39]. General and rigorous limit equilibrium methods have been proposed for performing a 3-D slope-stability analysis [40,41], and recent advances in such analyses were reviewed by Kumar [42]. However, for most shallow landslides, the length is much greater than the transverse width, and the side-flanks of the sliding surface cannot be determined before the failure initiation, making it difficult to apply a 3-D analysis method to shallow landslides. Thus, changes in the slope stability associated with shallow landslides in forest basins have been investigated using digital elevation models with the 2-D infinite slope model [21,43–45]. In such studies, the main focus was the changes in triggering conditions, such as rainfall infiltration, determined through saturated and unsaturated seepage flow analysis, which is investigated because rainfall-induced shallow landslides are frequent disasters.

The focus of the present study was the stability changes in forest slopes over the years as an important factor in the induction of shallow landslides. The forest growth over time improves the ability of a forest to prevent shallow landslides, but a forest loses this ability after a certain level of cutting is conducted. Accordingly, the yearly changes in the ability of a forest to prevent shallow landslides are complex. Because yearly changes affect the induction of shallow landslides, a slope stability evaluation should include the growth and decay of the trees. Japanese cedars are the most planted trees in Japan, occupying about 18% of all Japanese forests, as of 31 March 2018 [46]; hence, Japanese cedars were examined in the present study. We propose a method for assessing slope stability using three prisms to analyse the effects of the lateral roots. The proposed method was applied to a forest basin in Ibaraki Prefecture, northeast of Tokyo, Japan, for which the topography was represented by a 1-metre-resolution digital elevation model. The change in the slope stability of the forest basin was evaluated over a 50-year period after the stump cutting and immediate planting of Japanese cedar trees.

2. Materials and Methods

2.1. Root Pullout Experiments

A series of pullout experiments was conducted to investigate the reinforced strength of Japanese cedar roots, i.e., the root cohesion component (c_r) in the Mohr–Coulomb failure criterion. Half-sphere-shaped trenches with a diameter of 0.3–0.5 m centred in the test tree trunk were excavated below the trunks. A pullout experiment was conducted on each root exposed on the trench surface following diameter measurements. In the pullout experiment, the roots were pinched with a pliers-like tool and pulled manually to measure the maximum pullout force, using a force gauge (Digital Force Gauge, ZTS-1000N, IMADA Limited, Aichi, Japan) connected to the pliers-like tool. The diameter of the pulled-out roots was 2–40 mm. Although broken and unbroken roots were observed during the experiment, we did not distinguish the measured maximum forces of the broken roots from those without ruptures. We selected uncut tree trunks to investigate live roots, and those cut 1, 2, 4, 7, and 8 years ago to examine changes following decay. At least two trunks were examined for each condition.

2.2. Roots on the Sliding Surface

To calculate c_r, the number and diameter of roots on a potential sliding surface must be obtained. Although tree root distributions are generally not simple, we used the root distribution model for Japanese cedar trees proposed by Abe [22], who excavated a layer of soil below the test trees to a depth of 10 cm and measured the number and diameter of all the exposed roots; this excavation process was repeated at 10 cm intervals until no roots were found. Abe surveyed four locations in Japan, one of which was in Ibaraki Prefecture [22], where the changes in the slope stability in a forest basin were investigated in the present study.

Abe [22] proposed a root volume depth-wise distribution of 10 cm, as follows:

$$V(z) = V_r \int_{z_c-10}^{z_c} f(z)dz, \tag{1}$$

$$f(z) = \frac{mz^{m-1}}{\alpha_0} \times \mathrm{EXP}\left[\frac{-z^m}{\alpha_0}\right], \tag{2}$$

$$m = \frac{2.0}{\log Z_{max} - \log(0.3522 \times Z_{\max} - 10.7990)}, \tag{3}$$

$$\alpha_0 = (0.3522 \times Z_{\max} - 10.7990)^m, \tag{4}$$

where $V(z)$ is the root volume in a 10-centimetre-thick layer (cm^3), V_r is the total root volume (cm^3), z is the depth (cm), $Z_{\max}$ is the maximum root depth (cm), and z_c is the depth parameter varying by 10 cm (z_c = 10, 20, ..., $z_{c\max}$, e.g., in Equation (1), at z_c = 10, the root volume in the soil layer 0–10 cm deep would be calculated, whereas that at z_c = 20 would be calculated in the soil layer 10–20 cm deep, and so on; $z_{c\max}$ is the upper limit of a 10-centimetre-deep class containing $Z_{\max}$, e.g., at $Z_{\max}$ = 97 cm; $z_{c\max}$ would be 100 cm);

$f(z)$ is a Weibull probability density function; and m and α_0 are the Weibull parameters. The total root volume, V_r, was calculated using an equation for the total weight [47]:

$$\log W_r = -0.3085 + 0.8216 \times \log(DBH^2 \times H), \\ V_r = W_r / G_s, \quad (5)$$

where H is the height of the tree (cm), DBH is the diameter at breast height (cm), W_r is the total weight of the roots (gf), and G_s is the unit weight of the roots, set at 0.973 gf/cm^3 in the present study [48].

The soil was divided into three zones: shallow, middle, and deep. The roots were then classified into 0.5-centimetre-wide classes, and the relative frequencies were calculated for the shallow, middle, and deep zones before they were expressed using the diameter class:

$$Y_t(D_i) = 6.43 \times D_i^{-1.53}, \\ Y_m(D_i) = 4.73 \times D_i^{-1.86}, \\ Y_b(D_i) = 6.04 \times D_i^{-1.81}, \quad (6)$$

where D_i is the class mark for the 0.5-centimetre-wide diameter of roots (D_i = 0.25, 0.75, 1.25 cm, and so on), and $Y_t(D_i)$, $Y_m(D_i)$, and $Y_b(D_i)$ are the relative frequencies in the topsoil, middle, and bottom zones, respectively. In addition, the average volume of a root in each class [$V_a(D_i)$] was measured and expressed using the diameter class:

$$V_a(D_i) = 7.81 \times D_i^{2.14} \quad (7)$$

Based on these equations, we calculated the number of roots in each diameter class (0.5-centimetre-wide) on a potential sliding surface in a certain 10-centimetre-thick soil layer after incorporating DBH, H, and Z_{max} as the input parameters. The reinforced strength of the roots was estimated by summing the pullout strengths over a unit area, according to the calculated number of roots in each diameter class on a potential sliding surface [23,24,49], a method that has been widely used in other studies [32,50–53]. We calculated the reinforced strength of the roots in a soil layer at depths of >30 cm (i.e., 30–40 cm, 40–50 cm, etc.), which were labelled as 35, 45, 55, 65, 75, 85, and 95 cm deep.

2.3. Slope-Stability Analysis

The reinforced strength of the roots (i.e., c_r) should be considered when evaluating slope stability. The 2-D infinite slope model is usually employed, where the reinforced strength of the roots is assigned on the bottom surface. However, a 2-D infinite slope model does not consider the reinforcement of the roots on the two transverse sides of the sliding block, which ought to increase the safety factor of shallow landslide blocks, to some extent, especially for sliding blocks with smaller widths. Thus, a 3-D slope model should be used to consider the reinforcement effect of the roots across the flanks on the safety factor of the sliding blocks. Because the length of shallow landslides is usually much larger than the transverse width, we assumed an infinite slope in the longitudinal direction to maintain the simplicity of our slope-stability analysis. Thus, we propose a three-prism model consisting of an oblique quadrangular prism sandwiched between two oblique triangular prisms (Figure 1) to evaluate the reinforcement effect of the roots on the flanks of shallow landslides. In this model, a constant reinforced root strength is assigned to the bottom surfaces of the central prism and the two side triangular prisms. The variable transverse slope, β, is determined as the safety factor of the three-prism model, which takes its lowest value under unsaturated conditions without a water table in the soil. The results of the three-prism method were compared with those of Hovland's method [36] and simplified Janbu's method [37] because these have generally been employed in Japan for 3-D examinations. In this fundamental analysis of the grid calculation on a 1-metre-resolution digital elevation model, the soil depth, D, and longitudinal slope, α, were kept constant at 0.7 m and 34.5°, respectively. The wet soil unit weight was 11.12 kN/m^3, the saturated soil unit weight was 15.09 kN/m^3, the angle of internal friction was 34.5°, and the soil cohesion was 0 kPa. These parameter values were determined according to the results of soil tests performed on soil samples taken from a scarp, underlain by biotite gneiss, in a forest basin in Ibaraki Prefecture. These soil parameters were later incorporated in the grid calculations on a digital elevation model obtained via airborne laser surveys, in which four laser points per m^2 were measured. Both the c_r and the total width of the three-prism model at the ground surface were changed to investigate their influence on the slope stability.

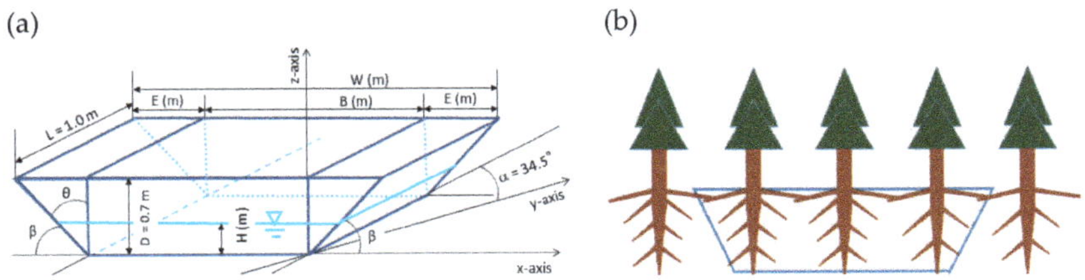

Figure 1. Schematic of the three-prism model used for slope-stability analysis. (**a**) Three prisms. (**b**) Reinforcement by roots on a potential sliding surface.

For the three-prism model shown in Figure 1, the safety factor can be expressed by Equation (8) for Hovland's method (F_{h}) and Equation (9) for the 3-D simplified Janbu's method (F_{j}):

$$F_{\text{h}} = \frac{\sum_{i=1}^{3}\left[(W_i - U_i)/J_i\cdot\tan\phi' + c_r\cdot A_i\right]}{\sum_{i=1}^{3} W_i\cdot\sin\alpha}, \tag{8}$$

$$F_{\text{j}} = \frac{\sum_{i=1}^{3}\left[\left(W_i - U_i\cdot\cos^2\alpha\right)\cdot\tan\phi' + c_r\cdot A_i/J_i\right]/(\cos\alpha\cdot m_{\alpha i})}{\sum_{i=1}^{3} W_i\cdot\tan\alpha}, \tag{9}$$

$$\begin{gathered} W_i = 0.5\cdot H^2\cdot\tan\theta\cdot L\cdot\cos\alpha\cdot\gamma_{\text{sat}} + 0.5\cdot\left(D^2 - H^2\right)\cdot\tan\theta\cdot L\cdot\cos\alpha\cdot\gamma_{\text{t}} + q_i\ (i = 1,\ 3), \\ W_2 = B\cdot H\cdot L\cdot\cos\alpha\cdot\gamma_{\text{sat}} + B\cdot(D - H)\cdot L\cdot\cos\alpha\cdot\gamma_{\text{t}} + q_2, \end{gathered} \tag{10}$$

$$U_i = 0.5\cdot H^2\cdot\tan\theta\cdot L\cdot\cos\alpha\cdot\gamma_{\text{w}}\ (i = 1,\ 3);\ U_2 = B\cdot H\cdot L\cdot\cos\alpha\cdot\gamma_{\text{w}}, \tag{11}$$

$$A_i = D\cdot L\sqrt{1 - \sin^2\alpha + 1/\tan^2\beta}\ (i = 1,\ 3);\ A_2 = B\cdot L, \tag{12}$$

$$J_i = \sqrt{1 + \tan^2\alpha + \tan^2\beta}\ (i = 1,\ 3);\ J_2 = 1/\cos\alpha, \tag{13}$$

$$m_{\alpha i} = 1/J_i + \sin\alpha\cdot\tan\phi'/F_{\text{j}}\ (i = 1,\ 2,\ 3), \tag{14}$$

where W_1, W_3, and W_2 are the soil weights of the two oblique triangular quadrangular prisms and the central oblique quadrangular prism, respectively; U_1, U_3, and U_2 are the total water pressures on the bottom of the two oblique triangular quadrangular prisms and oblique quadrangular prism, respectively; q_1, q_3, and q_2 are the tree surcharges on the two oblique triangular quadrangular prisms and oblique quadrangular prism, respectively; A_1, A_3, and A_2 are the bottom areas of the two oblique triangular quadrangular prisms and oblique quadrangular prism, respectively, on which a constant reinforced root strength is mobilised; ϕ' is the effective angle of internal friction; c_r is the root cohesion component; γ_{t} is the unit weight of wet soil (11.12 kN/m^3); γ_{sat} is the unit weight of saturated soil (15.09 kN/m^3); and γ_{w} is the unit weight of water (9.81 kN/m^3). As shown in Figure 1, α and β are the longitudinal and transverse slopes, respectively, θ is the complementary angle of β (= 90° − β), and E, B, H, L, and D define the sizes of the sliding block.

However, both Hovland's method and the 3-D simplified Janbu's method have some weaknesses when applied to the three-prism model, as described in Section 3.3. To resolve these issues, we proposed a slope-stability analysis method applying the three-prism model as follows: the self-weight component was calculated using the self-weight of the oblique triangular prisms to act as the oblique quadrangular prism with the same volume; hence, the self-weight component was obtained through the one assumed oblique quadrangular prism. Meanwhile, the reinforced root strength acted on the flank-surfaces of the oblique triangular prisms. The safety factor of this proposed three-prism method is given by Equation (15):

$$F_{3p} = \frac{\sum_{i=1}^{3}\left[(W_i - U_i)\cdot \cos\alpha \cdot \tan\phi' + c_r \cdot A_i\right]}{\sum_{i=1}^{3} W_i \cdot \sin\alpha}. \tag{15}$$

Using the forest basin in Ibaraki Prefecture as a case study, the slope stability was examined using the three-prism method (described in detail in Section 3.3), assuming that the whole basin was covered with 50-year-old Japanese cedar trees. The changes in slope stability on a 1-metre-resolution digital elevation model (Figure 2) were analysed, as 50-year-old trees were clear-cut and new seedlings were immediately planted. The assumed changes in the number, height, and diameter at breast height of Japanese cedar trees over 50 years (the second rank of the site index) were determined using the yield tables of Japanese cedar trees, developed for regions in and around Ibaraki Prefecture [54]. To perform the slope-stability analysis, the soil depth at each grid in the target area must be obtained. Ideally, several simplified cone penetration tests should be conducted to obtain the soil depth at N_c = 10 (N_c indicates the number of attempts required for a 5 kg weight dropped from a height of 0.50 m to drive the cone 0.10 m into the soil; it is generally assumed that slopes are subject to failure under certain triggering conditions involving N_c values of <10) [55]. However, conducting such tests in large areas is difficult, owing to time and cost limitations. In the last decade, some soil depth estimations have been generated using machine learning [56–59]; however, soil depth data will likely be acquired using this method in the near future. In the present study, we instead used a soil depth estimation method in which the slope and transverse angle on each grid were incorporated in the following equations [60]:

$$D = 0.9581 \times EXP[-0.0036 \times D_{br}] \times D_{br}, \tag{16}$$

$$D_{br} = -8.30 \times \alpha - 1.29 \times \delta + 681, \tag{17}$$

where D is the soil depth (cm), D_{br} is the depth above the bedrock surface of N_c = 40 (cm), α is the slope (degrees), and δ is the transverse angle (degrees). The transverse angle was obtained following the survey guide for investigations on forest disasters [61], as follows:

A circle (with C set at the centre point) was inscribed in a square grid on a contour map.

Two intersection points of the contour line passing through point C and the inscribed circle (S_1 and S_2) were obtained.

The transverse angle, S_1CS_2, was measured on the downslope side (for the upslope direction, the angle is smaller on a valley topography and larger on a ridge).

Notably, some grids had soil depths of <30 cm in the area. For these grids, the reinforced strength of roots (the root cohesion component) calculated on a potential sliding surface at 35 cm was instead applied. For each grid with a slope and soil depth, the three-prism slope model was applied to evaluate the slope stability by calculating the safety factor. The stability analysis was conducted under fully saturated conditions with two different tree weight conditions, i.e., with and without the aboveground weight of trees.

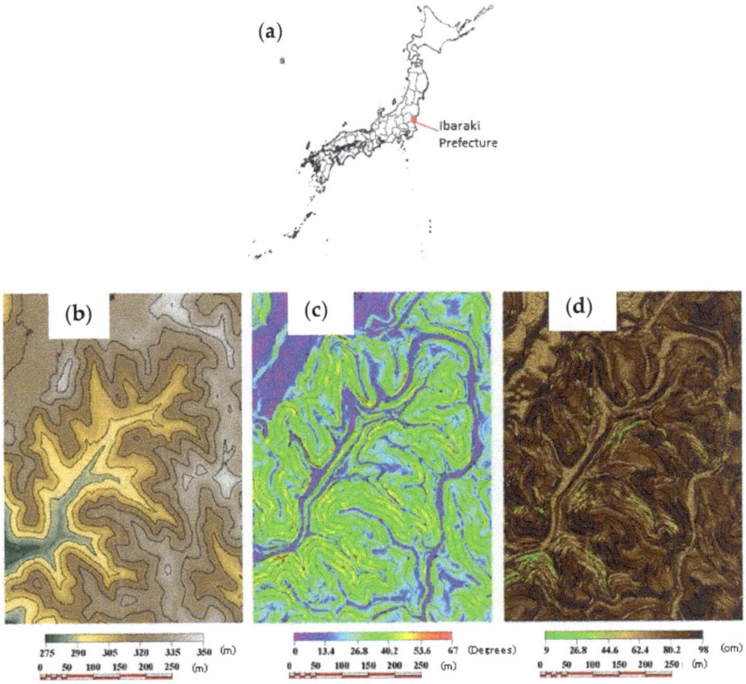

Figure 2. Topography maps of the studied forest basin. (**a**) Location of Ibaraki Prefecture. (**b**) Elevation, (**c**) slope (calculated by TNTmips (Version 2019, Redmond, DC, USA)), and (**d**) soil depth.

3. Results

3.1. Pullout Strength of Roots

The obtained maximum pullout strength of the roots is shown in Figure 3 in relation to the diameter at a point where the root was pinched with the pliers-like tool. The maximum pullout strength of the live roots of uncut trunks exhibited the exponent function of the diameter, as reported in a previous study [24]. Notably, the maximum pullout strength of the roots of cut trunks had similar characteristics, exhibiting the exponent function of the diameter, regardless of the time after cutting. The roots of trunks cut 8 years previously were vulnerable to breakage; thus, we could not conduct pullout experiments on roots with diameters of about <20 mm because they broke when pinched. Given the same diameter, the maximum pullout strength was the highest for the living roots of uncut trunks and decreased as more time elapsed after the trunks were cut.

3.2. Root Strength on the Sliding Surface at Different Depths

In shallow landslides, a sliding surface forms within a soil layer above the bedrock. The number of tree roots in each diameter class across the sliding surface at different soil depths was calculated using the root distribution model [22] and is summarized for Japanese cedar stands at different ages (10–50 years) in Table A1 in Appendix A.

The reinforced strengths of tree roots, calculated as the summation of the maximum pullout forces of the roots across a unit area sliding surface, are shown in relation to the elapsed time after new planting in Figure 4a. At each depth of the sliding surface, the reinforced strength had increased markedly by about 25 years after planting, after which it remained relatively constant. Although some scatter was observed, the reinforced strength of the tree roots was approximated using a logistic function of the elapsed time after planting. Along the curve, the reinforced strength of the roots gradually increased up to about 5 years, followed by a rapid increase, with the maximum acceleration of strength at about 7 years. However, from around 13 years, the rate of reinforcement appeared to decelerate.

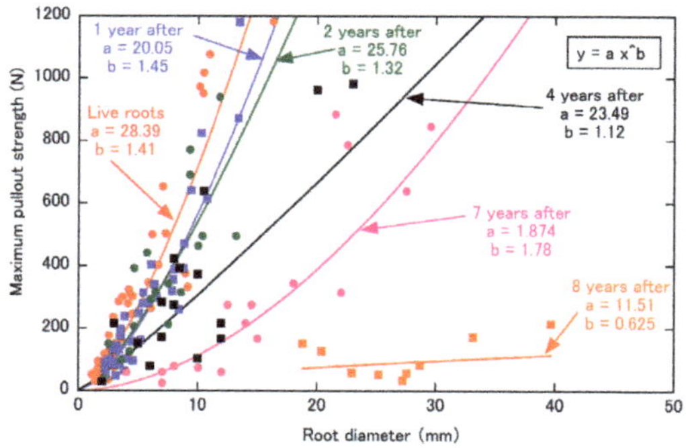

Figure 3. Maximum pullout strengths of the Japanese cedar roots of live trees and stumps cut at different times (labelled on the graph).

(a)

(b)

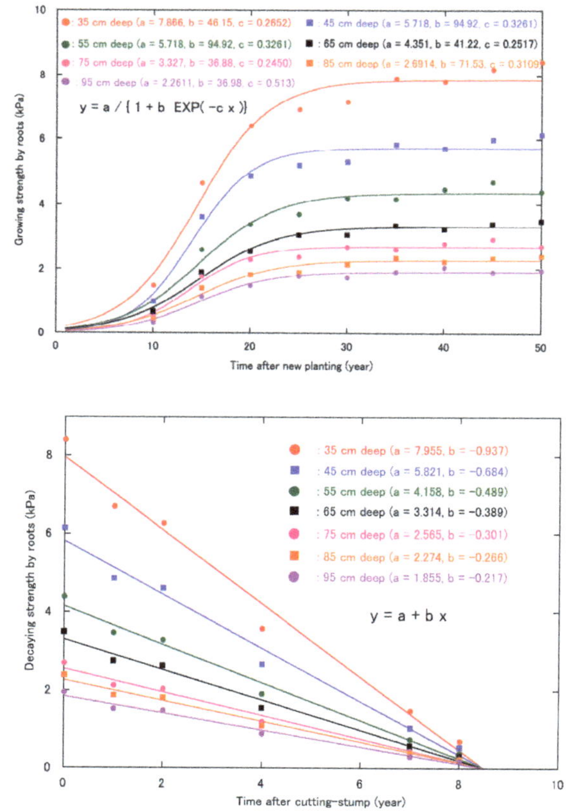

Figure 4. Changes in the reinforced strength of Japanese cedar roots over time. (**a**) Increased strength caused by roots growing after new planning. (**b**) Decreased strength caused by roots decaying after stump-cutting.

After clear-cutting, the roots of trees begin to decay, reducing the reinforced strength. Given that the trees were cut at the age of 50 years, the decreased reinforced strength is shown according to the elapsed time after cutting in Figure 4b. The reinforced strength of the roots linearly decreased with the elapsed time after cutting, and no strength appeared to remain at around 9 years.

3.3. Changes in Slope Stability in the Three-Prism Model

The changes in the risk of shallow landslides were examined using the proposed three-prism model for slope-stability analysis. Figure 5 shows the relationship between the transverse slope, β, of the side-oblique triangular prisms and the prism width, and that between the safety factor, F_h, and the prism width, both determined using Hovland's method [36]. Both the β and F_h decreased when the c_r was of <0.8 kPa, with smaller prism widths. This was likely because, in Hovland's method, the inter-prism force is approximated to zero, and the total balance of the force and the moment of force are not satisfied. In this regard, previous studies have revealed that smaller safety factors are obtained using Hovland's method for 3-D analyses of sandy soils [62–64].

(**a**)

(**b**)

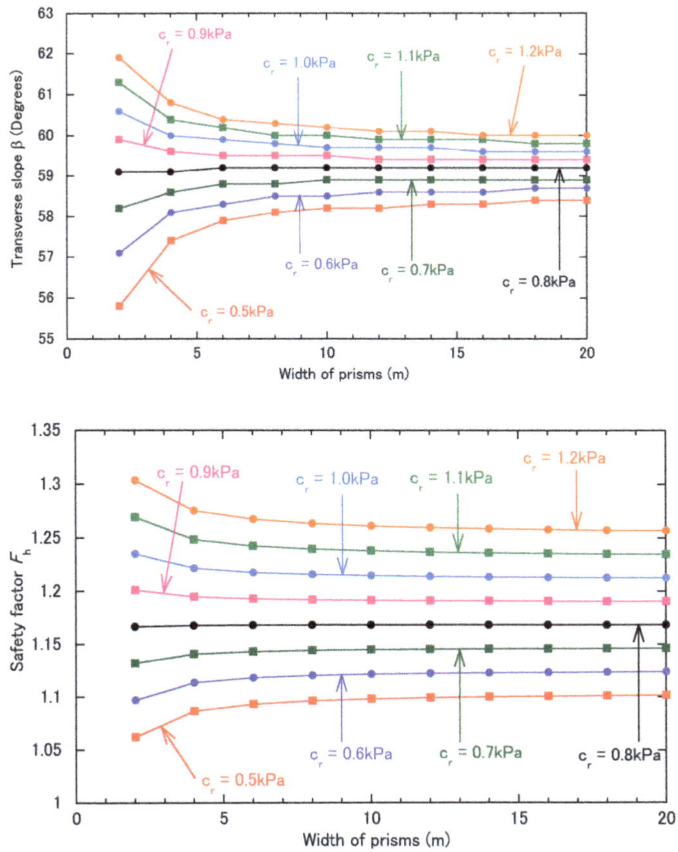

Figure 5. Results of slope-stability analysis obtained using Hovland's method. Changes in (**a**) the transverse slope (β) and (**b**) safety factor, according to prism width.

The safety factor determined using the proposed three-prism method (F_{3p}) was compared with those determined using Hovland's method and the 3-D simplified Janbu's method (Figure 6). Notably, the 3-D simplified Janbu's method [37–39] obtained the smallest safety factor when the side-oblique triangular prisms disappeared (i.e., when the transverse angle β was 90°), because the method does not consider the shear strength on the vertical sides in the 3-D slope-stability analysis [65]. Thus, in the calculation for the 3-D simplified Janbu's method, we also used the values of the transverse slope obtained using the proposed three-prism method.

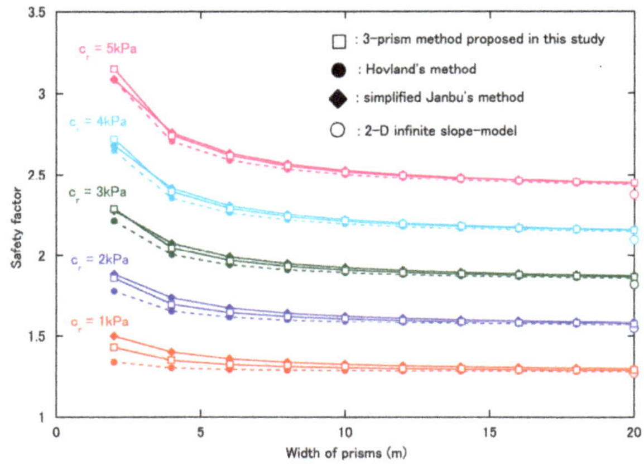

Figure 6. Comparison of safety factors according to prism width evaluated using different analysis methods.

The safety factors obtained using Hovland's method were generally lower than those obtained using the proposed three-prism method or the 3-D simplified Janbu's method when the prism widths were about <10 m. However, the safety factors of the proposed method and the 3-D simplified Janbu's method were generally consistent over the range of prism widths. Notably, the safety factors of the three methods converged when the prism width was about >15 m. For reference, the results of the simplest 2-D infinite slope model method are also shown in Figure 6. In this method, the prism width did not affect the safety factor (expediently plotted at a prism width of 20 m). At a block width of 20 m, a small difference was observed in the safety factors between the proposed three-prism method and the 2-D infinite slope model method, in which the difference was up to about 0.07 when the reinforced root strength was 5 kPa.

The widths of most shallow landslides are about 10–15 m according to Kosugi [66] and 5–20 m according to Okimura [67]. According to these findings and avoiding the overestimation of tree root effects, the shallow landslide risk was demonstratively evaluated on a 1-metre-resolution digital elevation model over the forest basin using the newly proposed three-prism method with a block width of 20 m. For each grid, the longitudinal slope, α, and soil depth, D, were the inputs, whereas β was estimated. For this estimation, the value of β with a minimum 3-D safety factor was calculated for various values of α (15.0°, 25.0°, 30.0°, 34.5°, 35.0°, 40.0°, and 45.0°) and D (0.3–1.3 m at intervals of 0.1 m). A multiple regression analysis indicated that β can be estimated using the following regression equation, where α and D are the two independent variables:

$$\beta = 3.538 \times 10^{-3}\alpha^2 - 3.757 \times 10^{-2}\alpha D + 1.930 \times 10^{-2}\alpha + 3.397 \times D + 59.70. \quad (18)$$

Because the interpretation of 3-D illustrations can be difficult, changes in β versus α with a constant D and β versus D with a constant α are shown in Figure 7, in which the curves are plotted using the regression Equation (18).

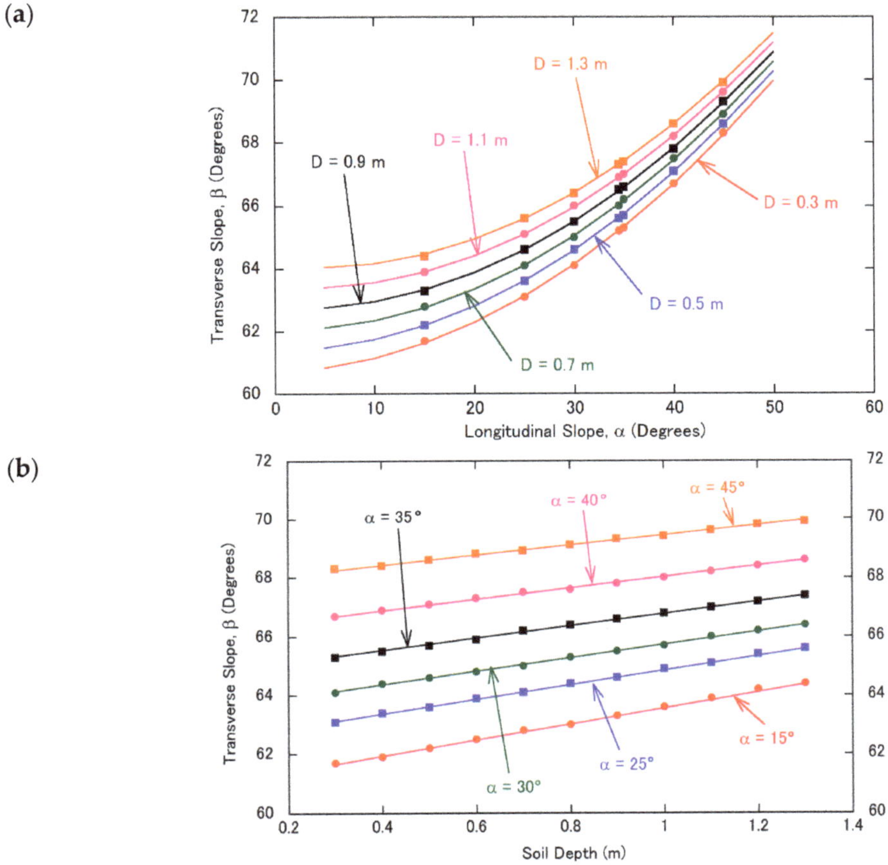

Figure 7. Changes in the transverse slope (β) according to (**a**) the longitudinal slope (α) and (**b**) soil depth (D).

4. Discussion

The reinforced root strength is expected to increase and decrease after new planting and stump cutting, respectively. Given that whole stands of Japanese cedar trees are harvested at the age of 50 years, and that seedlings are newly planted at the same time, the total reinforced root strength should be the sum of that of the decaying roots of the old stumps and the growing roots of the new stumps. Figure 8 shows the change in the total reinforced root strength at different depths according to the elapsed time up to 50 years. In the early stage, the decrease in the reinforced strength of decaying roots plays a more important role than that of the increase in the reinforced strength of growing roots, resulting in a rapid decrease in the total reinforced root strength, with a minimum strength observed at about 9 years. Thus, even with prompt planting just after clear-cutting, a considerable loss of reinforced root strength can occur, reducing the total reinforced root strength to a fraction of its maximum value. However, total root strength increases from 9 years to about 25 years, after which it remains almost constant up to 50 years.

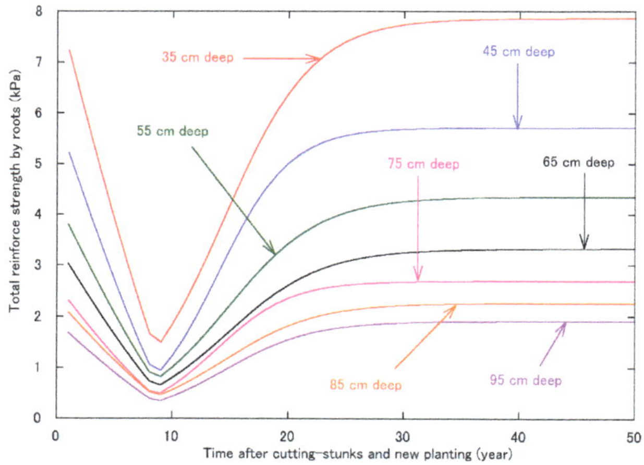

Figure 8. Changes in the total reinforced root strength of Japanese cedar according to the time after stump-cutting and new planting.

Uprooting experiments have revealed changes in the strength of tree stumps [68], which have been discussed in relation to the effects of forests on shallow landslides in Japan [69]. In such experiments, the cut stumps of Japanese cedar trees were towed horizontally using a wire rope attached to a hand winch, and the mobilised maximum pullout forces when the cut stumps were removed from the ground were evaluated in relation to the stem volume. Because the sum of the uprooting strength of the decaying stumps and growing stumps was lowest at about 10 years after clear-cutting and new planting, these previous results were considered to explain why shallow landslides occur mostly during 10–15 years after clear-cutting and new planting [68]. Notably, these previous results are similar to the present findings. However, in these previous studies, the observed decrease in the uprooting strength of the cut stumps was not linear but rather negatively exponential according to the time after cutting. Regarding the decrease of strength under decay, 30% of the initial strength that remained at 10 years after cutting was finally lost at about 38 years. In addition, the rate of acceleration of the total strength after 10 years increased more slowly and gradually, continuing to increase up to 50 years. The minimum total strength at 10 years was about 60% of the initial strength, which was a lower reduction relative to that observed in our study. These results likely arose because the strength from the uprooting experiments was mostly related to the stump itself, rather than the roots in relatively deeper soil layers in which a probable sliding surface could form. Furthermore, Tsukamoto [70] indicated that the results of uprooting experiments explained the resistance strength against the uprooting of stumps but did not clarify how that strength was related to the resistance against sliding as a key factor of shallow landslide prevention. Therefore, a merit of our study is that the reinforced root strength on a sliding surface was determined yearly as a factor for preventing shallow landslides.

The newly proposed method for slope-stability analysis using three prisms (with a central quadrangular prism sandwiched between side triangular prisms) was applied to a 1-metre-resolution digital elevation model in a forest basin in Ibaraki Prefecture (Figure 2). Specifically, a three-prism model with a width of 20 m was assumed for each grid on the map to calculate the safety factor, and the effects of Japanese cedar forest dynamics were examined at intervals of 5 years up to 50 years. The results of the slope-stability analysis, considering the aboveground weight of the trees, is shown in Figure 9 for three different times: (a) just after clear-cutting, when the roots of cut trunks are not rotten, (b) 10 years after clear-cutting and new planting, when the reinforced root strength is almost at its lowest value, and (c) 20 years after clear-cutting and new planting, when the reinforced root strength has recovered to a certain degree. As shown in Figure 9, the distribution of red grids, indicating a safety factor of <1.0, just after clear-cutting (a), is similar to that o20 years after clear-cutting and immediate new planting (c), whereas markedly more red grids appear 10 years after clear-cutting and new planting (b), i.e., when the total reinforced root strength has almost reached its lowest value.

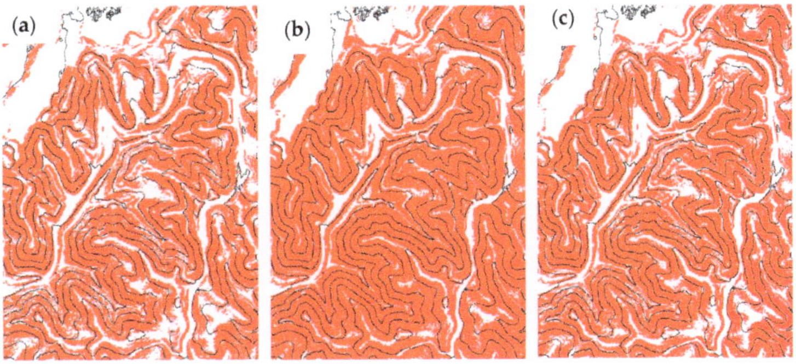

Figure 9. Changes in unstable grids calculated for the forest basin with 284,204 grids in total under fully saturated conditions with the weight of stems (**a**) just after clear-cutting (number of unstable grids: 163,720), (**b**) 10 years after clear-cutting and new planting (number of unstable grids: 216,606), and (**c**) 20 years after new planting (number of unstable grids: 178,157).

In Figure 10, the number of grids with a safety factor of <1.0 is shown according to the elapsed time after both clear-cutting and new planting. We found little difference between the safety factors with and without the aboveground weight of trees. The aboveground weight of trees is usually obtained by measuring the stem volume directly in situ; however, new methods incorporating high-density point-cloud data obtained with airborne or terrestrial laser scanners have been developed to construct 3-D tree models [71,72]. Nevertheless, as the aboveground weight of trees exerted little effect on the safety factor, we ignored these data in our stability analysis. Notably, in our stability analysis, the effect of the aboveground weight was investigated using the same ground water table conditions, i.e., the rainfall interception effect of trees, which diminishes the rainfall input [73], and the swaying of trees in the wind were not considered. These factors affect water table formation and groundwater infiltration and are driving forces in the initiation of sliding. We found that unstable grids rapidly increased for the first 5 years, peaked at 10 years, and then rapidly decreased from 10 to 20 years. After 30 years, the number of unstable grids remained almost constant.

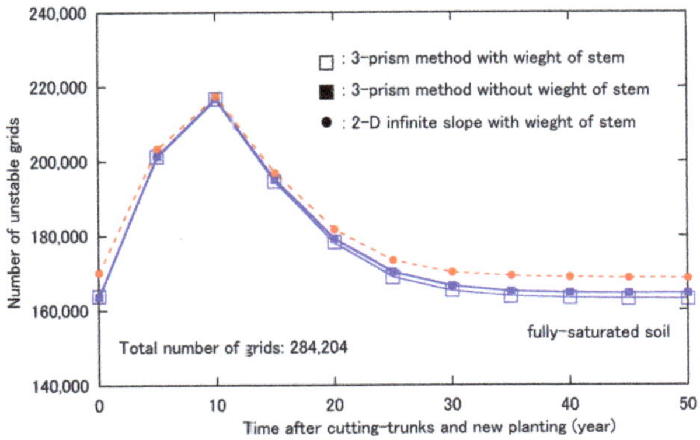

Figure 10. Changes in the number of unstable grids according to slope-stability analysis over time (after Japanese cedar stump-cutting and new planting up to 50 years).

The results of this study indicate that the reinforced strength of Japanese cedar tree roots decreases considerably between 5 and 15 years after clear-cutting and new planting, resulting in a rapid increase in slope instability. Thus, implementing forestry practices for lowering slope instability during this period should be considered important for preventing shallow landslides.

5. Conclusions

By approximating the sum of the pullout forces of roots over a unit area on a possible sliding surface, the changes in the reinforcement of roots during the growth of trees after planting and decay of stumps after cutting were simulated. Our results indicated that the root reinforcement decreased linearly after clear-cutting, with the minimum resistance strength against shallow landslides observed at around 9 years. Subsequently, by about 25 years, the root reinforcement increased to a relatively constant value. We proposed a three-prism method to examine the effect of lateral roots across the side-flanks of shallow landslides on slope stability. Using this method on a 1-metre-resolution digital elevation model, a large number of unstable slopes were observed between 5 and 15 years after clear-cutting and new planting, indicating that appropriate forestry operations should be implemented during this period to prevent landslide disasters in forests.

Author Contributions: All the authors contributed to the study's conception and design. Material preparation, data collection, and data analysis were performed by Y.O., F.C. and U.K. The first draft of the manuscript was written by Y.O. and all the authors commented on previous versions of the manuscript. All authors have read and agreed to the published version of the manuscript.

Funding: This research work was funded by the Agriculture, Forestry and Fisheries Research Council (MAFF Commissioned project study, Grant Number JPJ009840) and the Japan Society for the Promotion of Science (Grant-in-Aid for Scientific Research (c), Grant Number JP21K04601).

Data Availability Statement: Not applicable.

Acknowledgments: The authors should like to thank ABE Kazutoki of Nihon University for his valuable suggestions and advice. Our sincere appreciation goes to TSUBOYAMA Yoshio and KUBOTA Tayoko of the Forest Research and Management Organization for providing the DEM data. We are indebted to FURUYA Naoyuki and MURAKAMI Wataru of the Forest Research and Management Organization for the useful discussions. The experiments were conducted in national forests, and we express our sincere gratitude to ISHIGURI Hideto of the Forestry Agency.

Conflicts of Interest: The authors declare no conflict of interest.

Appendix A

Table A1. Number of roots in each root-diameter class on a sliding surface at different depths and stand ages.

Depth (cm)	Stand Age (Years)	Root-Diameter Class Value (cm)								
		0.25	0.75	1.25	1.75	2.25	2.75	3.25	3.75	4.25
	10	20.0	2.6	1.0						
	15	47.0	6.1	2.4	1.3	0.8				
	20	85.5	11.1	4.3	2.3	1.4	1.0			
	25	116.1	15.0	5.8	3.1	2.0	1.3	1.0		
35	30	135.7	17.6	6.8	3.6	2.3	1.6	1.2	0.9	
	35	173.2	22.4	8.7	4.6	2.9	2.0	1.5	1.1	
	40	179.9	23.3	9.0	4.8	3.0	2.1	1.5	1.2	0.9
	45	205.6	26.6	10.3	5.5	3.5	2.4	1.7	1.3	1.1
	50	225.5	29.2	11.3	6.0	3.8	2.6	1.9	1.5	1.2

Table A1. *Cont.*

Depth (cm)	Stand Age (Years)	Root-Diameter Class Value (cm)								
		0.25	0.75	1.25	1.75	2.25	2.75	3.25	3.75	4.25
	10	13.5	1.7	0.7						
	15	42.0	5.4	2.1	1.1					
	20	72.6	9.4	3.6	1.9	1.2				
	25	95.3	12.3	4.8	2.6	1.6	1.1			
45	30	108.5	14.1	5.4	2.9	1.8	1.3	0.9		
	35	138.4	17.9	6.9	3.7	2.3	1.6	1.2		
	40	140.9	18.3	7.1	3.8	2.4	1.6	1.2	0.9	
	45	161.1	20.9	8.1	4.3	2.7	1.9	1.4	1.0	
	50	176.7	22.9	8.9	4.7	3.0	2.0	1.5	1.1	
	10	9.6	1.2	0.5						
	15	30.0	3.9	1.5	0.8					
	20	50.5	6.5	2.5	1.4	0.8				
	25	68.0	8.8	3.4	1.8	1.1	0.8			
55	30	94.3	12.2	4.7	2.5	1.6	1.1			
	35	98.9	12.8	5.0	2.6	1.7	1.1	0.8		
	40	119.3	15.5	6.0	3.2	2.0	1.4	1.0		
	45	136.4	17.7	6.8	3.7	2.3	1.6	1.2		
	50	126.2	16.4	6.3	3.4	2.1	1.5	1.1	0.8	
	10	11.6	1.6							
	15	21.0	2.9	1.1	0.6					
	20	35.9	4.9	1.9	1.1	0.7				
	25	59.3	8.1	3.2	1.8	1.1				
65	30	64.6	8.8	3.5	1.9	1.2	0.8			
	35	82.4	11.3	4.5	2.4	1.5	1.1			
	40	81.1	11.1	4.4	2.4	1.5	1.1	0.8		
	45	92.8	12.7	5.0	2.7	1.7	1.2	0.9		
	50	101.7	13.9	5.5	3.0	1.9	1.3	1.0		
	10	9.0	1.2							
	15	23.7	3.2	1.3						
	20	37.3	5.1	2.0	1.1					
	25	45.9	6.3	2.5	1.4	0.9				
75	30	63.6	8.7	3.5	1.9	1.2				
	35	63.8	8.7	3.5	1.9	1.2	0.8			
	40	77.0	10.5	4.2	2.3	1.4	1.0			
	45	88.0	12.0	4.8	2.6	1.6	1.1			
	50	78.7	10.8	4.3	2.3	1.5	1.0	0.8		
	10	7.1	1.0							
	15	18.9	2.6	1.0						
	20	29.7	4.1	1.6	0.9					
	25	35.6	5.0	2.0	1.1	0.7				
85	30	50.7	6.9	2.8	1.5	0.9				
	35	64.6	8.9	3.5	1.9	1.2				
	40	61.3	8.4	3.3	1.8	1.1	0.8			
	45	70.1	9.6	3.8	2.1	1.3	0.9			
	50	76.9	10.5	4.2	2.3	1.4	1.0			
	10	5.8	0.8							
	15	15.4	2.1	0.8						
	20	24.2	3.3	1.3	0.7					
	25	40.0	5.5	2.2	1.2					
95	30	41.3	5.7	2.2	1.2	0.8				
	35	52.7	7.2	2.9	1.6	1.0				
	40	63.6	8.7	3.5	1.9	1.2				
	45	57.2	7.8	3.1	1.7	1.1	0.7			
	50	62.7	8.6	3.4	1.9	1.2	0.8			

References

1. Miyasaka, T.; Kawase, H.; Nakaegawa, T.; Imada, Y.; Takayabu, I. Future projections of heavy precipitation in Kanto and associated weather patterns using large ensemble high-resolution simulations. *SOLA* **2020**, *16*, 125–131. [CrossRef]
2. Cruden, D.M.; Varnes, D.J. Landslide types and processes. In *Landslides, Investigation and Mitigation: Special Report 247*; Turner, A.K., Schuster, R.L., Eds.; Transportation Research Board, National Research Council: Washington, DC, USA, 1996; pp. 501–516.
3. Okada, Y.; Ochiai, H.; Okamoto, T.; Sassa, K.; Fukuoka, H.; Igwe, O. A complex earth slide–earth flow induction by the heavy rainfall in July 2006, Okaya City, Nagano Prefecture, Japan. *Landslides* **2007**, *4*, 197–203. [CrossRef]
4. Duan, W.; He, B.; Takara, K.; Luo, P.; Nover, D.; Yamashiki, Y.; Huang, W. Anomalous atmospheric events leading to Kyushu's flash floods, July 11–14, 2012. *Nat. Hazards* **2014**, *73*, 1255–1267. [CrossRef]
5. Yang, H.; Wang, F.; Miyajima, M. Investigation of shallow landslides triggered by heavy rainfall during Typhoon Wipha (2013), Izu Oshima Island, Japan. *Geoenviron. Disasters.* **2015**, *2*, 15. [CrossRef]
6. Wang, G.; Sassa, K.; Fukuoka, H. Downslope volume enlargement of a debris slide–debris flow in the 1999 Hiroshima, Japan, rainstorm. *Eng. Geol.* **2003**, *69*, 309–330. [CrossRef]
7. Wang, F.; Wu, Y.H.; Yang, H.; Tanida, Y.; Kamei, A. Preliminary investigation of the 20 August 2014 debris flows triggered by a severe rainstorm in Hiroshima City, Japan. *Geoenviron. Disasters.* **2015**, *2*, 17. [CrossRef]
8. Chigira, M.; Ling, S.; Matsushi, Y. Landslide disaster induced by the 2017 northern Kyushu rainstorm. *Disaster Prev. Res. Inst. Ann. A* **2018**, *61a*, 28–35.
9. Moriya, H.; Nihei, H.; Hasebe, Y.; Mine, K.; Sushimoto, K.; Yano, S.; Watanabe, Y.; Fukuda, N. Study on driftwood disaster in Kita River, Fukuoka Prefecture, due to 2017 northern Kyushu flood. *Annu. J. Hydraul. Eng.* **2018**, *74*, I1195–I1200. [CrossRef]
10. Mori, S.; Ono, K. Landslide disasters in Ehime Prefecture resulting from the July 2018 heavy rain event in Japan. *Soils Found* **2019**, *59*, 2396–2409. [CrossRef]
11. Akiya, K. Shallow landslides and forests. Landslide prevention and slope stability. In *Landslide Prevention and Slope Stability*; Sougou Doboku Kenkyusho. Co., Ltd.: Tokyo, Japan, 1979; Volume 3, pp. 43–52.
12. Zhang, Y.; Shen, C.; Zhou, S.; Luo, X. Analysis of the influence of forests on landslides in the Bijie area of Guizhou. *Forests* **2022**, *13*, 1136. [CrossRef]
13. Annual Report on Forest and Forestry in Japan 2022. Available online: http://www.rinya.maff.go.jp/j/kikaku/hakusyo/r3hakusyo/attach/pdf/zenbun-26.pdf (accessed on 28 October 2022).
14. Okada, Y.; Konishi, C. Geophysical features of shallow landslides induced by the 2015 Kanto-Tohoku heavy rain in Kanuma city, Tochigi Prefecture, Japan. *Landslides* **2019**, *16*, 2469–2483. [CrossRef]
15. Okada, Y. Measuring critical turning moment of the Japanese cedar (Cryptomeria japonica) in situ. *J. For. Res.* **2019**, *24*, 162–167. [CrossRef]
16. Outline of Public Opinion Survey on Forests and Lives 2019. Available online: https://survey.gov-online.go.jp/r01/r01-shinrin/gairyaku.pdf (accessed on 28 October 2022).
17. The Type and Area of Forest Reserve. Available online: https://www.rinya.maff.go.jp/j/tisan/tisan/con_2_2_1.html (accessed on 28 October 2022).
18. Waldron, L.J. The shear resistance of root-permeated homogeneous and stratified soil. *Soil Sci. Soc. Am. J.* **1977**, *41*, 843–849. [CrossRef]
19. Waldron, L.J.; Dakessian, S. Soil reinforcement by roots: Calculation of increased soil shear resistance from root properties. *Soil Sci.* **1981**, *132*, 427–435. [CrossRef]
20. Waldron, L.J.; Dakessian, S.; Nemson, J.A. Shear resistance enhancement of 1.22-meter diameter soil cross sections by pine and alfalfa roots. *Soil Sci. Soc. Am. J.* **1983**, *47*, 9–14. [CrossRef]
21. Abe, K.; Ziemer, R.R. Effect of tree roots on a shear zone: Modeling reinforced shear stress. *Can. J. Forest. Res.* **1991**, *21*, 1012–1019. [CrossRef]
22. Abe, K. A method for evaluating the effect of tree roots on preventing shallow-seated landslides. *Bull. For. Forest. Prod. Res. Inst.* **1997**, *373*, 105–181.
23. Wu, T.H.; McKinnell, W.P.; Swanston, D.N. Strength of tree roots and landslides on Prince of Wales Island, Alaska. *Can. Geotech. J.* **1979**, *16*, 19–33. [CrossRef]
24. Kitahara, H. Effect of tree root systems on slope stability. *Water Sci.* **2010**, *53*, 11–37.
25. Schwarz, M.; Giadrossich, F.; Cohen, D. Modeling root reinforcement using a root-failure Weibull survival function. *Hydrol. Earth Syst. Sci.* **2013**, *17*, 4367–4377. [CrossRef]
26. Li, P.; Xiao, X.; Wu, L.; Li, X.; Zhang, H.; Zhou, J. Study on the shear strength of root-soil composite and root reinforcement mechanism. *Forests* **2022**, *13*, 898. [CrossRef]
27. Wu, T.H.; Beal, P.E.; Lan, C. In-situ shear test of soil-root systems. *J. Geotech. Eng.* **1988**, *114*, 1376–1394. [CrossRef]
28. Abe, K.; Kurokawa, U.; Takeuchi, Y. Method for evaluating thinning influence on a forest's stability to prevent shallow landslides. *J. Jpn. Landslide Soc.* **2004**, *41*, 9–19. [CrossRef]
29. Sato, H.; Otani, K.; Kanbara, T.; Torita, H. The comparison between the resistance of tree roots to shallow landslide by in-situ direct shear test and the pullout resistance of tree roots. *J. Soil Erosion Control Eng.* **2013**, *66*, 15–20. [CrossRef]
30. Okada, Y.; Kurokawa, U. An experimental examination on a reinforcement of shear resistance due to tree-roots of a Japanese cypress. *Kanto J. Forest Res.* **2012**, *63*, 141–144.

31. Okimura, T.; Ichikawa, R. A prediction method for surface failures by movements of infiltrated water in a surface soil layer. *J. Nat. Disaster Sci.* **1985**, *7*, 41–51.
32. Okada, Y.; Kurokawa, U. Examining tree roots on shearing resistance in shallow landslides triggered by heavy rainfall in Shobara in 2010. *J. For. Res.* **2015**, *20*, 230–235. [CrossRef]
33. Torii, N.; Masumoto, S.; Nonami, S. Proposal of infinite stability analysis considering effects of additional cohesion from roots. *Mem. Constr. Eng. Res. Inst. Found.* **2021**, *63*, 75–85.
34. Kagamihara, S.; Todo, C.; Okazaki, K.; Nishiwaki, H.; Kataoka, S.; Shibuya, S. Case study on evaluating the stability of natural slope by considering the effects of forest roots. *Mem. Constr. Eng. Res. Inst. Found* **2018**, *60*, 131–142.
35. Ueno, S. Influence of geological factors on configuration and scale of landslides. *J. Jpn. Landslide Soc.* **2001**, *38*, 105–114. [CrossRef]
36. Hovland, H.J. Three-dimensional slope stability analysis method. *J. Geotech. Eng. Div.* **1977**, *103*, 971–986. [CrossRef]
37. Ugai, K. Three-dimensional slope stability analysis by simplified Janbu method. *J. Jpn. Landslide Soc.* **1987**, *24*, 8–14. [CrossRef] [PubMed]
38. Hungr, O. An extension of Bishop's simplified method of slope stability analysis to three dimensions. *Géotechnique* **1987**, *37*, 113–117. [CrossRef]
39. Hungr, O.; Salgado, F.M.; Byrne, P.M. Evaluation of a three-dimensional method of slope stability analysis. *Can. Geotech. J.* **1989**, *26*, 679–686. [CrossRef]
40. Lam, L.; Fredlund, D.G. A general limit equilibrium model for three-dimensional slope stability analysis. *Can. Geotech. J.* **1993**, *30*, 905–919. [CrossRef]
41. Zhou, X.P.; Cheng, H. Analysis of stability of three-dimensional slopes using the rigorous limit equilibrium method. *Eng. Geol.* **2013**, *160*, 21–33. [CrossRef]
42. Kumar, S.; Choudhary, S.S.; Burman, A. Recent advances in 3D slope stability analysis: A detailed review. *Model Earth Syst. Environ.* 2022. [CrossRef]
43. Roering, J.J.; Schmidt, K.M.; Stock, J.D.; Dietrich, W.E.; Montgomery, D.R. Shallow landsliding, root reinforcement, and the spatial distribution of trees in the Oregon Coast Range. *Can. Geotech. J.* **2003**, *40*, 237–253. [CrossRef]
44. Tosi, M. Root tensile strength relationships and their slope stability implications of three shrub species in the Northern Apennines (Italy). *Geomorphology* **2007**, *87*, 268–283. [CrossRef]
45. Ni, J.J.; Leung, A.K.; Ng, C.W.W.; Shao, W. Modelling hydro-mechanical reinforcements of plants to slope stability. *Comput. Geotech.* **2018**, *95*, 99–109. [CrossRef]
46. Area Tables of Japanese Cedar Forests and Japanese Cypress Forests in Each Prefecture. Available online: https://www.rinya.maff.go.jp/j/sin_riyou/kafun/pdf/sugihinoki_menseki.pdf (accessed on 28 October 2022).
47. Karizumi, N. Root biomass. In *Primary Productivity of Japanese Forests—Productivity of Terrestrial Communities*; Shidei, T., Kira, T., Eds.; University of Tokyo Press: Tokyo, Japan, 1977; Volume 16, pp. 45–52.
48. Yazawa, K. Heterogeneity in specific gravity, tree rings, and water content. *Mater. Trans.* **1963**, *12*, 678–682. [CrossRef]
49. Wu, T.H. Root reinforcement of soil: Review of analytical models, test results, and applications to design. *Can. Geotech. J.* **2013**, *50*, 259–274. [CrossRef]
50. Danjon, F.; Barker, D.H.; Drexhage, M.; Stokes, A. Using three-dimensional plant root architecture in models of shallow-slope stability. *Ann. Bot.* **2008**, *101*, 1281–1293. [CrossRef]
51. Genet, M.; Kokutse, N.; Stokes, A.; Fourcaud, T.; Cai, X.; Ji, J.; Mickovski, S. Root reinforcement in plantations of *Cryptomeria japonica* D. Don: Effect of tree age and stand structure on slope stability. *Forest. Ecol. Manag.* **2008**, *256*, 1517–1526. [CrossRef]
52. Abdi, E.; Majnounian, B.; Genet, M.; Rahimi, H. Quantifying the effects of root reinforcement of Persian ironwood (*Parrotia persica*) on slope stability; a case study: Hillslope of Hyrcanian forests, northern Iran. *Ecol. Eng.* **2010**, *36*, 1409–1416. [CrossRef]
53. Schwarz, M.; Preti, F.; Giadrossich, F.; Lehmann, P.; Or, D. Quantifying the role of vegetation in slope stability: A case study in Tuscany (Italy). *Ecol. Eng.* **2010**, *36*, 285–291. [CrossRef]
54. Forestry Agency. *Yield Tables of the Japanese Cedar Developed for Regions in and around Ibaraki Prefecture*, 1st ed.; Tokyo Forestry Bureau: Tokyo, Japan, 1959; pp. 5–6.
55. Osanai, N. Site prediction of shallow landslides by means of simplified cone penetration tests. *Found Eng. Equip Mon.* **2007**, *35*, 19–22.
56. Pelletier, J.; Broxton, P.; Hazenberg, P.; Zeng, X.; Troch, P.; Niu, G.; Williams, Z.; Brunke, M.; Gochis, D. A gridded global data set of soil, intact regolith, and sedimentary deposit thicknesses for regional and global land surface modelling. *J. Adv. Model Earth Syst.* **2015**, *8*, 41–65. [CrossRef]
57. Scarpone, C.; Schmidt, M.G.; Bulmer, C.E.; Knudby, A. Modelling soil thickness in the critical zone for Southern British Columbia. *Geoderma* **2016**, *282*, 59–69. [CrossRef]
58. Chen, S.; Mulder, V.L.; Martin, M.P.; Walter, C.; Lacoste, M.; Richer-de-Forges, A.C.; Saby, N.P.A.; Loiseau, T.; Hu, B.; Arrouays, D. Probability mapping of soil thickness by random survival forest at a national scale. *Geoderma* **2019**, *344*, 184–194. [CrossRef]
59. Malone, B.; Searle, R. Improvements to the Australian national soil thickness map using an integrated data mining approach. *Geoderma* **2020**, *377*, 114579. [CrossRef]
60. Abe, K.; Kurokawa, U.; Watanabe, E.; Kubota, H.; Chang, H. A study on soil depth distribution in high elevation slopes (II). *Kanto J. Forest. Res.* **2004**, *55*, 267–270.

61. Survey Guide for Investigations on Forests Disasters. Available online: https://www.rinya.maff.go.jp/j/saigai/saigai/attach/pdf/con_1-3.pdf (accessed on 13 November 2002).
62. Azzouz, A.S.; Baligh, M.M. Discussion of three-dimensional slope stability analysis method. *J. Geotech. Eng. Div.* **1978**, *104*, 1206–1208. [CrossRef]
63. Ugai, K.; Wakai, A.; Cai, F. Examination of preciseness of 3D slope stability analysis methods using several idealized sliding masses: Comparisons of Hovland method and 3D simplified Janbu method. *J. Jpn. Landslide Soc.* **2005**, *42*, 63–68. [CrossRef]
64. Jiang, J.-C.; Yamagami, T. Three-dimensional limit equilibrium slope stability analysis: Simplified methods vs rigorous methods. *J. Jpn. Landslide Soc.* **2005**, *42*, 129–135. [CrossRef]
65. Stark, T.D.; Eid, H.T. Performance of three-dimensional slope stability methods in practice. *J. Geotech. Geoenviron. Eng.* **1998**, *124*, 1049–1060. [CrossRef]
66. Kosugi, K. Evaluation of slope failure vulnerability based on antecedent precipitation index. *Tech. Note Natl. Res. Inst. Earth Sci. Disaster Prev.* **2016**, *405*, 75–82.
67. Okimura, T.; Torii, N.; Nakagawa, W.; Haraguchi, K.; Kasahara, T.; Yamauchi, M.; Sagane, T.; Ito, M. Some problems and measures appeared during the construction of the real-time hazard system for slope disasters at a heavy rainfall (5). *Mem. Constr. Eng. Res. Inst. Found.* **2014**, *56*, 45–66.
68. Kitamura, Y.; Namba, S. Tree roots upon landslide prevention presumed through the uprooting test. *Bull. For. Forest. Prod. Res. Inst.* **1981**, *313*, 175–208.
69. Shuin, Y. Effects of vegetation on shallow landslide prevention. *Erosion. Control. Eng.* **2002**, *55*, 71–78.
70. Tsukamoto, Y. Evaluation of the effect of tree roots on slope stability. *Bull. Exp. For. Tokyo Univ. Agric. Technol.* **1987**, *23*, 65–124.
71. Awaya, Y.; Takahashi, T. Evaluating the differences in modelling biophysical attributes between deciduous broadleaved and evergreen conifer forests using low-density small-footprint LiDAR data. *Remote Sens.* **2017**, *9*, 572. [CrossRef]
72. Kumazaki, R. Application of 3D tree modelling using point cloud data by terrestrial laser scanner. *J. Jpn. Inst. Landsc. Archit.* **2021**, *84*, 527–530. [CrossRef]
73. Sheng, H.; Cai, T. Influence of rainfall on canopy interception in mixed broad-leaved—Korean pine forest in Xiaoxing'an Mountains, Northeastern China. *Forests* **2019**, *10*, 248. [CrossRef]

forests

MDPI

Article

Study on the Shear Strength of Root-Soil Composite and Root Reinforcement Mechanism

Pengcheng Li [1], Xuepei Xiao [2,*], Lizhou Wu [3], Xu Li [4], Hong Zhang [3,*] and Jianting Zhou [3]

1 School of Environment and Civil Engineering, Chengdu University of Technology, Chengdu 610059, China; lipengcheng970507@163.com
2 Sichuan Highway Planning, Survey, Design and Research Institute Ltd., Chengdu 610041, China
3 State Key Laboratory of Mountain Bridge and Tunnel Engineering, Chongqing Jiaotong University, Chongqing 400074, China; lzwu@cqjtu.edu.cn (L.W.); jtzhou@cqjtu.edu.cn (J.Z.)
4 Key Laboratory of Urban Underground Engineering, Ministry of Education, Beijing Jiaotong University, Beijing 100044, China; cexuli2012@163.com
* Correspondence: xiaoxuepei@schdri.com (X.X.); hongzhang@cqjtu.edu.cn (H.Z.)

Abstract: This study investigates the effects of root distributions and stress paths on the shear strength of root-soil composites using a consolidated-undrained (CU) triaxial test. On the basis of the limit equilibrium, two root reinforcement coefficients (n and m) are proposed for characterizing the effects of shear strength parameters on the principal stress considering different root distribution angles and root diameters. Then, n and m are introduced into the conventional limit equilibrium equation to develop a new limit equilibrium equation for root-soil composites. The results demonstrate that the root distribution angles (α) and root diameters (d) affect the shear strength of the root-soil composites. Under a consolidated-undrained condition, the effective cohesion (c'_{rs}) of the rooted soil is high and decreases in the order of 90°, 0°, 30° and 60°. For the same root distribution angle, c'_{rs} increases with the increasing root diameter. Meanwhile, the effective internal friction angle (φ'_{rs}) changes slightly. The failure principal stress of the root-soil composites is positively correlated with n and m. Furthermore, the deformation of the samples indicates that the run-through rate of α = 90° and α = 0° are both 0. Meanwhile, the lateral deformation rate declines from 17.0% for α = 60° to 10.9% for α = 90°.

Keywords: root-soil composite; root reinforcement coefficient; shear strength parameters; root distribution angle; root diameter

Citation: Li, P.; Xiao, X.; Wu, L.; Li, X.; Zhang, H.; Zhou, J. Study on the Shear Strength of Root-Soil Composite and Root Reinforcement Mechanism. *Forests* **2022**, *13*, 898. https://doi.org/10.3390/f13060898

Academic Editors: Haijia Wen, Weile Li, Chong Xu and Hiromu Daimaru

Received: 8 May 2022
Accepted: 6 June 2022
Published: 9 June 2022

1. Introduction

Landsliding is a geological disaster which occurs around the world and poses a great threat to people's lives and property [1–7]. Plant roots play an important role in stabilizing shallow slopes [8,9]. In recent years, vegetation has been widely used in ecological engineering to prevent soil erosion and stabilize slopes effectively [10–14].

Two mechanisms by which vegetation affects slope stability are recognized: (a) the mechanical reinforcement of the soil by plant roots and (b) the improvement of the hydrological conditions of slopes [15–18]. Plant transpiration affects soil matric suction [19,20], while plant roots absorb water and reduce the pore water pressure [21,22], and the mechanical reinforcement of plant roots has been proposed [23]. The most important contribution of roots is their ability to increase the shear strength of soils. Strong frictional properties of the root-soil interface improve the stability of the root-soil composites [24]. Additionally, the reinforcement effect of roots is represented in the increment of the shear strength of root-soil composites compared with that of unreinforced soil [25]. Many field and laboratory tests have been conducted and mechanical models of root-soil composites, including the Wu-Waldron model, modified Wu-Waldron model, root bundle model, and fiber bundle model, have been developed [26–31]. Rahardjo et al. [32], and Satyanaga and Rahardjo [33] investigated the effect of plant roots on slope stability under unsaturated conditions using undisturbed soil with and without roots. Triaxial tests were conducted to investigate the mechanical properties of reinforced and unreinforced soils. Zhang et al. [25] carried out a series of consolidated drained triaxial tests to examine the influence of roots on the shear strength of soils. Zhou and Wang [34] further showed that the shear strength of root-soil composites is higher than that of unreinforced soil.

At present, most researchers studied the effect of roots on the shear strength of rooted soils mainly based on the root content, root distribution pattern (usually horizontal and vertical), and moisture content [35–38]. Unfortunately, except for the influence of horizontal and vertical distribution patterns of roots on the strength of rooted soils, the reinforcement effect of other root distribution angles on root-soil composites has rarely been mentioned. Plant roots freely grow at an angle in the soils and the diameters of roots constantly change. Furthermore, the changes in the shear strength influence the failure principal stress of rooted soils, but the limit equilibrium equation for the failure principal stress of rooted soils has not been developed. Therefore, it is necessary to analyze the influence of root distribution angles and root diameters on the shear strength of root-soil composites. Moreover, the limit equilibrium equation of root-soil composites needs to be developed to investigate the impact of shear strength parameters on the failure principal stress of root-soil composites.

This study aims to examine the effect of root distribution angles (α = 0°, 30°, 60°, and 90°) and root diameters (d = 0.5 mm, 1.0 mm, and 2.0 mm) on the shear strength of soils. Based on the conventional limit equilibrium equation, two root reinforcement coefficients (n and m) are proposed to describe the influence of the variations of an effective internal friction angle and effective cohesion on the failure principal stress of root-soil composites, respectively. A new limit equilibrium equation of root-soil composites is then developed. Additionally, the deformation of soil samples is examined to demonstrate the constraint effect of roots on soils. The results provide a new insight for the analysis of the shear strength of rooted soils and provide a reference for choosing suitable slope protection plants.

2. Materials and Experimental Methods

2.1. Experimental Materials

The soils were obtained from Beijing, China (about 40°28′ N, 115°58′ E). Sieving method was used in the particle analysis test. The particle size distribution of the tested soils is shown in Figure 1 and Table 1. The soil is well-graded sand with clay and gravel (SW-SC) according to the unified soil classification system (USCS) [39] and its basic physical properties [40,41] are shown in Table 2.

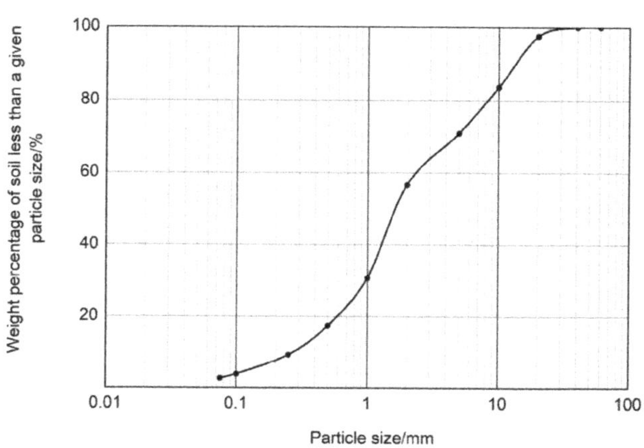

Figure 1. Particle size distribution curve of the tested soils.

Table 1. Particle size distribution of the tested soils.

Particle Size/mm	>10	10–5	5–2	2–1	1–0.5	0.5–0.25	0.25–0.075	$\leq$0.075
Percentage/%	16.3	12.8	14.2	25.97	13.3	8.23	6.5	2.7

Table 2. Physical properties of the tested soils.

Soil Dry Density (kg/m^3)	Water Content (%)	Plastic Limit (%)	Liquid Limit (%)	Plasticity Index	Specific Gravity	Unified Soil Classification System, USCS
1800	16.5	15.2	30.7	15.5	2.73	Well-graded sand with clay and gravel (SW-SC)

Pyracantha fortuneana (Maxim.) Li, an evergreen shrub, is a commonly used plant for soil and water conservation and was chosen as the experimental root system. Fifty *Pyracantha fortuneana* plants (about 80–90 cm in height) were wholly excavated and the root diameters and distribution angles were counted. A total of 200 roots were measured using vernier caliper. The root diameters ranged from 0.5 mm to 4.5 mm and were mainly distributed within 0.5 mm to 2.0 mm. Moreover, roots were widely distributed in the depth of 0.5 m below the surface and the distribution angles of roots were 0°, 30°, 60°, and 90°. Root architecture of *Pyracantha fortuneana* was horizontal [42]. In the tests, the root lengths were 30.0 and 60.0 mm and the diameters were 0.5, 1.0, and 2.0 mm, respectively. In Figure 2a–c, twelve roots with a length of 30.0 mm were set in three layers in a sample, and four roots were evenly placed in each layer. In Figure 2d, six roots with 60.0 mm length were evenly distributed in the sample. The root content of all the samples was 0.15%.

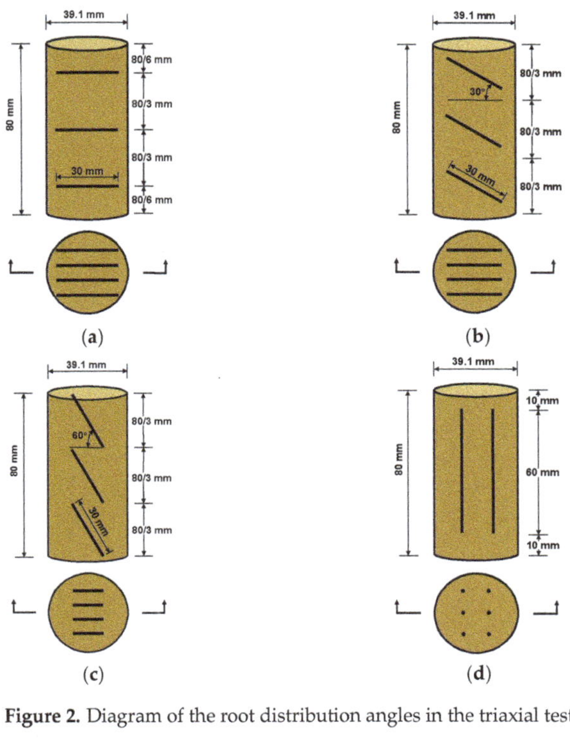

Figure 2. Diagram of the root distribution angles in the triaxial test: (**a**) 0°, (**b**) 30°, (**c**) 60°, (**d**) 90°.

2.2. Experimental Methods

Below the surface depth of 0.5 m, the enhancement effect of roots on the soil strength will be greatly weakened with the decrease of the number of roots. Therefore, the confining pressures (50 kPa, 100 kPa, and 150 kPa) were chosen to effectively evaluate the influence of roots on the soil strength. To study the influence of root distribution angles on soil strength, four root distribution angles (0°, 30°, 60°, and 90°, respectively) were employed and the root distribution patterns are shown in Figure 2.

Slope soils have generally reached a consolidated state under natural conditions. Slope instability caused by soil erosion is usually represented by the rapid increase of pore water pressure within a short time, resulting in a great decrease in soil strength under undrained condition [36]. Therefore, the consolidation undrained (CU) triaxial test was used to investigate the stress-strain response of soils using the TSZ strain-controlled automatic triaxial apparatus (control precision is ±1%). The soil samples are saturated by vacuum saturation method. The dry density and moisture content of the soils are 1800 kg/m^3 and 16.5%, respectively. The samples with 16.5% moisture content were prepared using three-layer compaction method. The samples of the root-soil composites and the unreinforced soils were 39.1 mm in diameter and 80.0 mm in height.

The range of shearing rate of TSZ was 0.002–4 mm/min ± 10%. The shearing rate of the triaxial tests was chosen as 0.1 mm/min and the stress was recorded every 0.1% increase of axial strain. When the strain reached 20%, the test ended. The mean values of the test results of effective stress were taken as the final values of deviator stress. The maximum deviator stress was taken as the failure stress. However, when the peak value of deviator stress was not recorded, the deviator stress corresponding to 15% strain was taken as the failure stress.

As for unreinforced soils, root-soil composites conform to the Mohr-Coulomb theory [27]. The shear strength of unreinforced soils and root-soil composites is rewritten as [43]:

$$\tau_f = c' + (\sigma - u)\tan\varphi' = c' + \sigma'\tan\varphi' \quad (1)$$

where τ_f is the shear strength (kPa); σ represents the normal stress (kPa); u is the pore water pressure (kPa); σ' is the effective stress (kPa); and c' and φ' represent the effective cohesion (kPa) and the effective internal friction angle (°), respectively.

3. Limit Equilibrium Equation of Root-Soil Composites

Previous studies showed that plant roots can reinforce soils [44] and shear strength parameters of root-soil composites are different from those of unreinforced soils. The limit equilibrium equations of unreinforced soil and root-soil composite are:

$$\sigma_1 = \sigma_3 K_p + 2c'_{us}\sqrt{K_p} \quad (2)$$

$$\sigma_{1l} = \sigma_3 K_{pl} + 2c'_{rs}\sqrt{K_{pl}} \quad (3)$$

where σ_1 and σ_{1l} represent the failure principal stress (kPa) of the unreinforced soil and root-soil composite under confining pressure σ_3, respectively; $K_p = \tan^2(45^\circ + \varphi'_{us}/2)$ is the passive earth pressure coefficient of the unreinforced soil; $K_{pl} = \tan^2(45^\circ + \varphi'_{rs}/2)$ is the passive earth pressure coefficient of the root-soil composite; and c'_{us}, φ'_{us}, c'_{rs}, and φ'_{rs} are the effective cohesion (kPa) and effective internal frictional angle (°) of the unreinforced soil and root-soil composites, respectively.

Under the same confining pressure (σ_3), the deviator of the failure principal stress of the unreinforced soil and root-soil composite is given by:

$$\Delta\sigma_{1l} = \sigma_{1l} - \sigma_1 \quad (4)$$

The limit equilibrium equation of the root-soil composite can also be written using the generalized equivalent confining pressure [37]:

$$\sigma_{1l} = (\sigma_3 + \Delta\sigma_{3g})K_p + 2c'_{us}\sqrt{K_p} = \sigma_1 + \Delta\sigma_{3g}K_p \tag{5}$$

where $\Delta\sigma_{3g}$ is the generalized equivalent confining pressure (kPa), which means the difference of confining pressure between unreinforced soil and root-soil composite samples under the same shear strength.

The generalized equivalent confining pressure is written as:

$$\Delta\sigma_{3g} = \frac{\sigma_{1l} - \sigma_1}{K_p} = \frac{\Delta\sigma_{1l}}{K_p} \tag{6}$$

The variations of shear strength parameters in Equation (5) are reflected in the generalized equivalent confining pressure while the variations of the shear strength parameters of Equation (3) are directly reflected in the limit equilibrium equation. Equations (3) and (5) are, therefore, equivalent.

Under the same confining pressure (σ_3), Equations (3) and (5) can be substituted into Equation (4) to obtain the deviator of the failure principal stress of the root-soil composite and the unreinforced soil. $\Delta\sigma_{1l}$ is given by:

$$\Delta\sigma_{1l} = \sigma_3\left(K_{pl} - K_p\right) + 2c'_{us}\left(\sqrt{K_{pl}} - \sqrt{K_p}\right) + 2(c'_{rs} - c'_{us})\sqrt{K_{pl}} \tag{7}$$

By substituting Equation (7) into Equation (6), we can gain a new expression of $\Delta\sigma_{3g}$, which is written as:

$$\Delta\sigma_{3g} = \frac{\sigma_3\left(K_{pl} - K_p\right) + 2c'_{us}\left(\sqrt{K_{pl}} - \sqrt{K_p}\right) + 2(c'_{rs} - c'_{us})\left(\sqrt{K_{pl}} - \sqrt{K_p}\right)}{K_p} \tag{8}$$

Substituting Equation (8) into Equation (5), the failure principal stress of the root-soil composite is obtained as:

$$\sigma_{1l} = \sigma_3 K_p \frac{K_{pl}}{K_p} + 2c'_{us}\sqrt{K_p}\left(\sqrt{\frac{K_{pl}}{K_p}} + \frac{c'_{rs} - c'_{us}}{c'_{us}}\sqrt{\frac{K_{pl}}{K_p}}\right) \tag{9}$$

The parameters in Equation (9) are written as:

$$n = \frac{K_{pl}}{K_p} \tag{10}$$

$$m = 1 + \left(\frac{c'_{rs} - c'_{us}}{c'_{us}}\right) \tag{11}$$

where n and m are the root reinforcement coefficients of the soil, which represent the variation of the internal friction angle and cohesion of the rooted soils, respectively.

Equation (9) is, therefore, rewritten as:

$$\sigma_{1l} = n\sigma_3 K_p + 2c'_{us} m\sqrt{n}\sqrt{K_p} \tag{12}$$

Equation (12) is the same as Equation (2) except that the root reinforcement coefficients are introduced into the limit equilibrium equation of the root-soil composite.

4. Results

4.1. Shear Strength Parameters of Root-Soil Composites

Table 3 lists the effects of the root distribution angles and diameters on the soil shear strength parameters. Under the condition of d = 0.5 mm, the variations of c'_r and φ'_r were different from each other. Compared with the unreinforced soil, the effective cohesions of the root-soil composites slightly increased for α = 30° and α = 60°, while that of α = 0° and α = 90° significantly increased. Moreover, c'_r respectively increased by 11.7% (α = 60°), 22.2% (α = 30°), 32.6% (α = 0°), and 66.4% (α = 90°). Meanwhile, the variations of φ'_r were slight under all root distribution angles. Additionally, the effective internal friction angle of the root-soil composites was lower than that of the unreinforced soil. φ_r' increased from 26.07° (α = 0°) to 26.66° (α = 60°), and then decreased to 26.22° (α = 90°).

Table 3. Strength parameters of unreinforced soils and root-soil composites.

Strength Parameter	WR *	Control Conditions											
		0.5 mm Root Diameter				1.0 mm Root Diameter				2.0 mm Root Diameter			
		0°	30°	60°	90°	0°	30°	60°	90°	0°	30°	60°	90°
c' (kPa)	30.17	40.00	36.88	33.70	50.21	45.43	42.14	38.99	60.27	51.92	48.64	43.03	74.15
$\Delta c'$ (kPa)	-	9.83	6.71	3.53	20.04	15.26	11.97	8.82	30.10	21.75	18.47	12.86	43.98
φ' (°)	26.69	26.07	26.12	26.66	26.22	26.96	26.69	26.93	26.77	26.87	27.07	27.26	26.85

* Without root.

Under the condition of d = 1.0 mm, the effective cohesion of the root-soil composites increased in the order of 60°, 30°, 0°, and 90° distribution angles and was greater than that of the unreinforced soil. c_r' increased by 50.6% for α = 0° and 99.8% for α = 90°. What is more, compared with the unreinforced soil, the effective internal friction angle of the root-soil composites slightly increased.

Under the condition of d = 2.0 mm, the effective cohesion of the root-soil composites increased greatly. The maximum increase of effective cohesion was 145.8% for α = 90°. c_r' respectively increased by 72.1% (α = 0°), 61.2% (α = 30°), and 42.6% (α = 60°). Compared with the unreinforced soil, φ_r' increased for all the root distribution angles. Although φ_r' increased, the maximum increase among all root distribution angles was only 2.1%.

According to the results, the root distribution angle and root diameter significantly influence the shear strength of soils. The effects of roots on the soil shear strength are mainly represented in the increase of the effective cohesion. Furthermore, roots have slight effects on the effective internal friction angle. The effective cohesion of the root-soil composites increases with the increase of the root diameter. With the increase of the root distribution angle, the effective cohesion of the root-soil composites first decreases, and then increases.

4.2. Root Reinforcement Coefficients

According to Equations (10) and (11), the root reinforcement coefficients (n and m) of the root-soil composites under different root distribution angles and diameters were obtained from the shear strength parameters of the root-soil composites (Table 3). The coefficients are listed in Table 4.

Table 4. Root reinforcement coefficients of unreinforced soils and root-soil composites.

Root Reinforcement Coefficient	WR *	Control Condition											
		0.5 mm Root Diameter				1.0 mm Root Diameter				2.0 mm Root Diameter			
		0°	30°	60°	90°	0°	30°	60°	90°	0°	30°	60°	90°
n	1.000	0.976	0.978	0.999	0.982	1.011	1.000	1.010	1.003	1.007	1.015	1.022	1.006
m	1.000	1.326	1.222	1.117	1.664	1.506	1.397	1.292	1.998	1.721	1.612	1.426	2.458

* Without root.

In Table 4, both of the root reinforcement coefficients of the unreinforced soil were 1.000 and the root reinforcement coefficients of the root-soil composites were different from each other.

The parameter n was less than 1.000 for the 0.5-mm-diameter root in the root-soil composites. Under the condition of the 1.0-mm-diameter root, n was more than 1.000 for the root distribution angles of 0°, 60°, and 90°, while n = 1.000 for the root distribution angle of 30°. Under the 2.0-mm-diameter root condition, the values of n were all more than 1.000 for the different root distribution angles. Regardless of diameter, m exceeded 1.000 for all root-soil composites and decreased in the order of 90°, 0°, 30°, and 60° of root distribution angles (α). Moreover, the maximum value of n was 1.022 and that of m was 2.458.

n represents the ratio of the passive earth pressure of the root-soil composite to that of the unreinforced soil. m describes the ratio of the effective cohesion of the root-soil composite to that of the unreinforced soil. Additionally, m increases with the increase of the root diameter. With the increase in the root distribution angle, m first decreases and then increases. However, the variations of n are slight.

4.3. Deformation Characteristics of the Soil Samples

The root had a significant impact on the sample deformation evolutions. Taking 0.5-mm-diameter root composites under a confining pressure of 50 kPa as an example, the deformation of these rooted soils was shown in Figure 3.

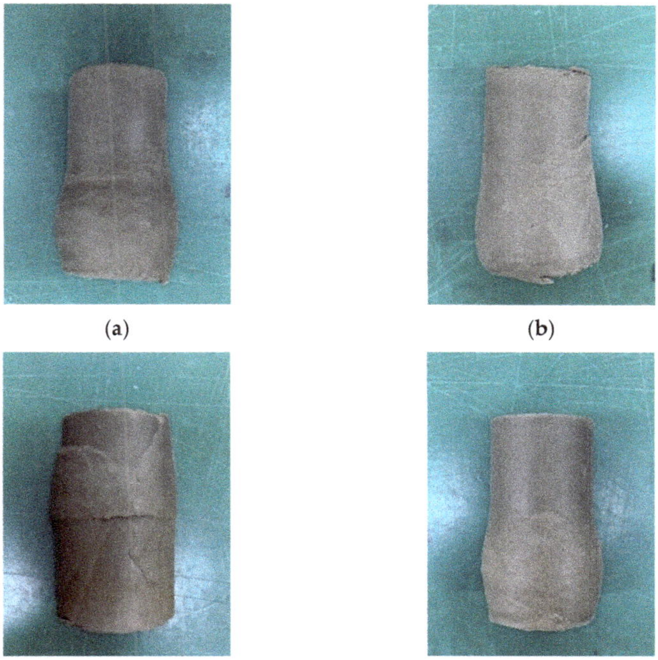

(a) (b) (c) (d)

Figure 3. Deformation characteristics of root-soil composites under the 50-kPa confining pressure and 0.5-mm root diameter: (**a**) $\alpha = 0°$, (**b**) $\alpha = 30°$, (**c**) $\alpha = 60°$, and (**d**) $\alpha = 90°$.

Figure 3 shows the obvious axial and lateral deformation of all samples. In Figure 3a,d, there is no clear failure surface in the root-soil composites with obvious lateral deformation for the root distribution angle of 0° and 90°. Meanwhile, the lateral deformation rate was 13.1% for $\alpha = 0°$ and 10.9% for $\alpha = 90°$. In Figure 3b, the failure surface did not run through the whole sample with the run-through rate of 48.7% when $\alpha = 30°$. Additionally, the main lateral deformation rate was 14.9%. Figure 3c shows an obvious failure surface of the 60°-distribution-angle composite with a 79.5% run-through rate and lateral swelling (17.0% lateral deformation rate) which appeared in the middle of the sample. However, compared with the unreinforced soil, the development of the failure surface for the rooted soil samples was greatly restrained.

The addition of roots can significantly affect the deformation characteristics of the root-soil composites. Owing to the interaction between the roots and the soil particles, the roots can bear partial shear stress, and restrain the lateral deformation of the rooted soils. With the addition of the roots, the strength and stiffness of the root-soil composites increase.

5. Discussion

5.1. Effects of Root Distribution Angles

Table 3 shows that the shear strength of the root-soil composites was higher than that of the unreinforced soil, which is consistent with many previous studies [27,38,45,46]. Among the four root distribution angles (0°, 30°, 60°, and 90°), the shear strength of the rooted soils decreased in the order of 90°, 0°, 30°, and 60°. The shear strength of rooted soils increased by 2.5 times (α = 90°), 1.7 times (α = 0°), 1.6 times (α = 30°), and 1.4 times (α = 60°). When α = 90°, the roots had the strongest effect on the shear strength and were most conductive to resist the axial pressure of the root-soil composite. Thus, the shear strength of the rooted soils can be improved by the addition of roots.

According to the Mohr-Coulomb theory, the failure angle is $\alpha_f = 45^\circ + \varphi'/2$. When φ' = 26.69° for the unreinforced soil, α_f = 58.35° can be obtained. Therefore, α = 60° was closest to the failure angle of the soils (α_f). Meanwhile, only a small part of the roots passed through the failure surface of the soils and the soils were most prone to failure and the shear strength was the lowest. This finding is consistent with the results of Meng et al. [47]. The closer the root distribution angle is to the failure angle, the less the number of roots passing through the shear failure surface. When the root-soil composites are stressed, the tensile strength of the roots cannot be fully exerted. Therefore, the shear strength of the root-soil composites for α = 60° cannot be effectively enhanced by roots. When the root is vertically distributed, the roots can run through the failure surface of the soils, which can greatly improve the shear strength of the root-soil composites. Additionally, the lateral deformation of the root-soil composites is smaller than that of the unreinforced soils [48]. The constraint of the roots on the soil deformation depends on the tensile strength of the roots. The run-through rate and the lateral deformation rate of the root-soil composites increase in the order of 90°, 0°, 30°, and 60° root distribution angles. Moreover, the larger the run-through rate and lateral deformation rate of the soils are, the smaller the restraint and the shear strength of the soils are, and the easier the soils fail.

5.2. Effects of Root Diameter on the Root-Soil Composites

In Table 3, under the same root distribution angle, the shear strength of the root-soil composites increased with the increasing root diameter. When d = 0.5 mm, the shear strength of the root-soil composites increased by the smallest amount, which is related to the soil bonding failure and the small area of contact between roots and soils. This result agrees with the previous studies [47,49]. Additionally, the shear strength of rooted soil was enhanced by 1.7 times (d = 0.5 mm), 2.0 times (d = 1.0 mm), and 2.5 times (d = 2.0 mm).

The larger the diameter of roots, the greater the shear strength of root-soil composites and the stronger the effect of roots on the shear strength. Large diameter roots can increase the root-soil contact area, and a large area of contact between the roots and soils means an increase in the frictional strength between the root-soil interface [50]. Therefore, the shear strength of the rooted soils is enhanced. Additionally, the larger the root diameter, the greater the force required for root fracture [51]. Thus, the large diameter roots are not easy to fail and the soils can be greatly reinforced. Moreover, the surface of large diameter roots is rougher than that of small diameter roots. With the increase in the root diameter, the soils around the roots are subjected to the root radial force and become relatively dense, which leads to greater friction between the root-soil interface [52]. Meanwhile, the binding force between the soil particles around the roots increases. Thus, the shear strength of the root-soil composites increases.

5.3. Variations of the Root Reinforcement Coefficients

Equation (2) is equivalent to Equation (12) when both n and m are 1.000. The result shows that Equation (12) is a generalized limit equilibrium equation for all kinds of soils. The limit equilibrium equation for the unreinforced soil is only a particular case.

In Figures 4 and 5, the variation in m is consistent with that of c', and changes of n are similar to that of φ'. Variations in root reinforcement coefficients directly describe the effects of the root distribution angle and root diameter on the shear strength parameters of the root-soil composites. In Table 3, the main contribution of roots to soil shear strength was to increase the effective cohesion. The internal friction angle is mainly affected by the soil particle structure [53]. However, compared with soil mass, the root content (0.15%) of the root-soil composites is small. Therefore, the effective internal frictional angle of the rooted soil had a slight change [36,54].

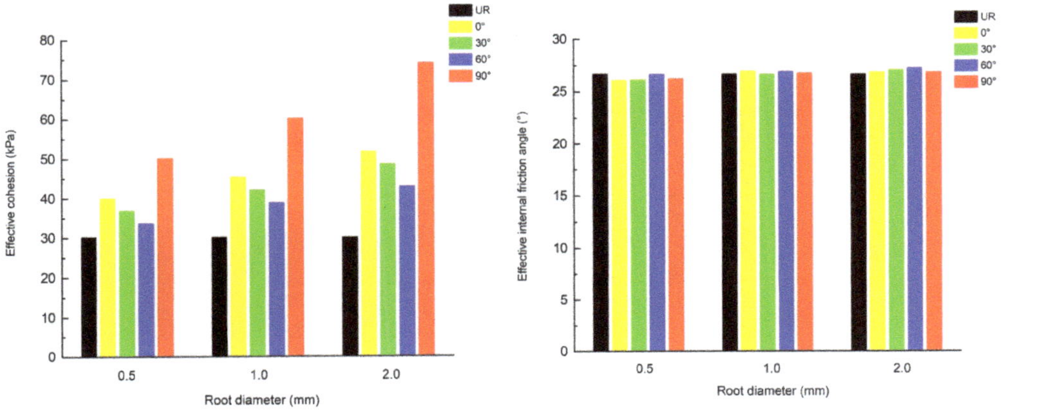

(a) **(b)**

Figure 4. Variations of soil shear strength parameters under different conditions: (**a**) c' and (**b**) φ'.

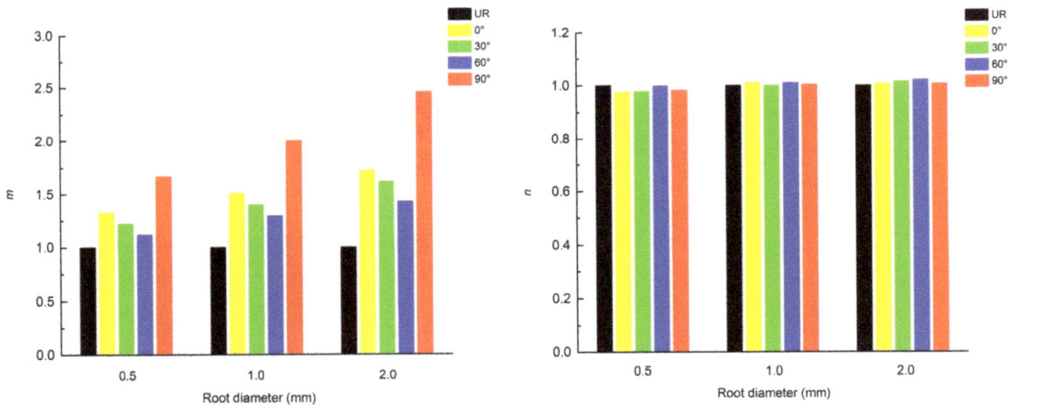

(a) **(b)**

Figure 5. Variations of the soil root reinforcement coefficient under different conditions: (**a**) m, and (**b**) n.

When the increments of the effective cohesion and the effective internal frictional angle ($\Delta c'$ and $\Delta\varphi'$) were both greater than zero, n and m were greater than 1.000. When $\Delta c'$ was greater than zero, m was more than 1.000. While $\Delta\varphi'$ was less than zero, n was less than 1.000. Furthermore, n and m were not generally equal to 1.000.

There are two different conditions, as shown in Figure 6. τ and τ_l denote the shear strength of the unreinforced soil and the root-soil composites, respectively. Considering the root distribution angles and the root diameters, there are several shear strength envelope relationships of the unreinforced soil and the root-soil composites. In Figure 6a, when $m = 1.426$ and $n = 1.002$, meaning that $\Delta c' > 0$ and $\Delta\varphi' > 0$, the shear strength envelopes of the root-soil composites are commonly larger than the shear strength envelope of the unreinforced soil. Figure 6a presents the envelopes of the shear strength of $\alpha = 60°$ under $d = 2.0$ mm. Figure 6b presents the envelopes for a 0.5-mm-diameter-root composite with $\alpha = 0°$. In Figure 6b, when $m = 1.326$ and $n = 0.978$ (i.e., $\Delta c' > 0$ and $\Delta\varphi' < 0$), the shear strength envelope of the root-soil composites is smaller than that of the unreinforced soil. Additionally, Figure 6 demonstrates that the changes of the root reinforcement coefficients affect the failure principal stress of the root-soil composites. The failure principal stress of the root-soil composites decreases with decreasing n and m.

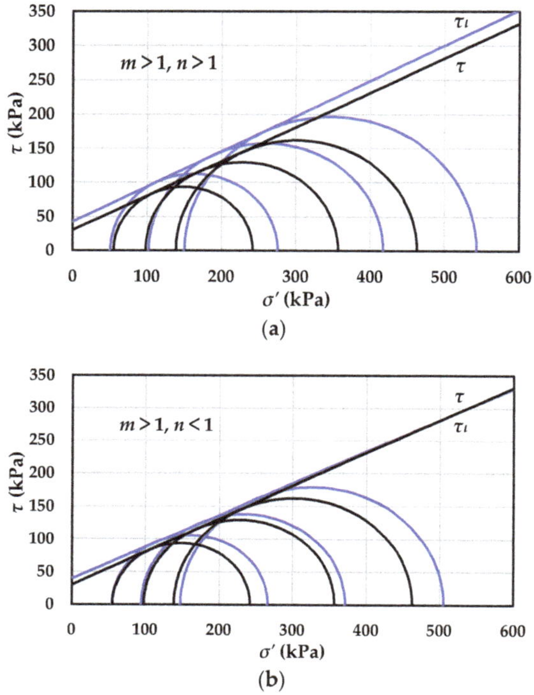

Figure 6. Shear strength envelope diagrams for unreinforced soil and root-soil composites: (**a**) $m = 1.426$, $n = 1.002$, (**b**) $m = 1.326$, $n = 0.978$.

5.4. Root Reinforcement Mechanism

The interaction between roots and soils can increase the mechanical properties and shear strength of soils, thus increasing the stability of slopes [36]. The exertion of the root tensile strength depends on the friction between the root-soil interface. Therefore, enhancing the friction between the root-soil interface is key to improve the shear strength of the root-soil composites.

Compared with soils, plant roots have a strong tensile strength, which can offset a part of the shear stress. Under loading, the deformation of the root-soil composites causes the relative displacement and mutual dislocation between roots and soils, which can generate the friction between the root-soil interface. The tensile property of plant roots can be combined with the compressive property of soils through the friction between the root-soil interface [52]. Roots can effectively improve the shear strength of soils, and enhance the capacity of soils to resist the shear failure. On the one hand, the increase of the root diameter can enlarge the root-soil contact area, root surface roughness, and the compactness of soils around roots, which is beneficial to improve the friction on the root-soil interface. On the other hand, the root distribution angle affects the number of plant roots passing through the shear failure surface of soils, thus impacting the enhancement of roots on the shear strength of rooted soils. The more the number of roots passing through the shear failure surface, the stronger the transmission and dispersion effect of roots on the soil stress, and the greater the shear strength of the rooted soils.

In general, the greater the shear strength of the rooted soils, the higher the slope stability [55]. According to the results of this study, the shear strength of the rooted soils increases with an increasing root diameter, and the vertical root distribution pattern can greatly enhance the shear strength of root-soil composites. Thus, plants with a larger root diameter and mainly 90° root distribution angle are suggested to reinforce slopes.

6. Conclusions

The following conclusions are drawn from the results of the study.

1. Root distribution angles and root diameters affect the shear strengths of the root-soil composites and the shear strength of rooted soils is enhanced by 1.1–2.5 times. The shear strength of the root-soil composites becomes high in the order of 60°, 30°, 0°, and 90° distribution angles, and decreases in the order of 2.0 mm, 1.0 mm, and 0.5 mm root diameters. Moreover, roots mainly affect the effective cohesion of the soils. However, the effective internal friction angle of the rooted soils changes slightly.
2. The run-through rate and the lateral deformation of the root-soil composites increase in the order of 90°, 0°, 30°, and 60° root distribution angles (α), and the run-through rate of $\alpha = 90°$ and $\alpha = 0°$ are both 0. Meanwhile, the lateral deformation rate declines from 17.0% for $\alpha = 60°$ to 10.9% for $\alpha = 90°$. Roots can effectively restrain the deformation of the root-soil composites.
3. Two root reinforcement coefficients n and m were proposed to develop the limit equilibrium equation of the root-soil composites. n and m can be calculated by φ' and c'. The limit equilibrium of the unreinforced soil is equivalent to that of the root-soil composite when both n and m are 1.000, which means that the limit equilibrium equation of the unreinforced soil is only a particular case. Therefore, the limit equilibrium equation of the root-soil composites has a wide applicability. Additionally, n and m represent the effects of root distribution angles and root diameters on the failure principal stress of the root-soil composites, respectively. The failure principal stress of the root-soil composites is positively correlated with n and m.

Author Contributions: P.L. and L.W. conceived and designed the experiments; L.W. and X.L. guided the experiment; P.L. performed the experiments and analyzed the data; P.L., L.W., X.L., H.Z. and J.Z. wrote the manuscript; P.L., X.X., L.W., X.L., H.Z. and J.Z. contributed critically to improve the manuscript. All authors have read and agreed to the published version of the manuscript.

Funding: This research was funded by the National Natural Science Foundations of China (Grant No. 41790432, Project leader: Peng Cui).

Institutional Review Board Statement: Not applicable.

Informed Consent Statement: Not applicable.

Data Availability Statement: Not applicable.

Conflicts of Interest: The authors declare no conflict of interest.

Notation

τ_f	Shear strength
σ	Normal stress
u	Pore water pressure
σ'	Effective stress
c'	Effective cohesion
φ'	Effective internal friction angle
σ_3	Confining pressure
σ_1	Failure principal stress of the unreinforced soil
σ_{1l}	Failure principal stress of the root-soil composite

K_p	Passive earth pressure coefficient of the unreinforced soil
K_{pl}	Passive earth pressure coefficient of the root-soil composite
c'_{us}	Effective cohesion of the unreinforced soil
φ'_{us}	Effective internal frictional angle of the unreinforced soil
c'_{rs}	Effective cohesion of the root-soil composite
φ'_{rs}	Effective internal frictional angle of the root-soil composite
$\Delta\sigma_{1l}$	Deviator of the failure principal stress of the unreinforced soil and root-soil composite
$\Delta\sigma_{3g}$	Generalized equivalent confining pressure
n, m	Root reinforcement coefficients of the soil

References

1. Kim, D.H.; Gratchev, I.; Balasubramaniam, A. Determination of joint roughness coefficient (JRC) for slope stability analysis: A case study from the Gold Coast area, Australia. *Landslides* **2013**, *10*, 657–664. [CrossRef]
2. Wu, L.Z.; Selvadurai, A.P.S.; Zhang, L.M.; Huang, R.Q.; Huang, J.S. Poro-mechanical coupling influences on potential for rainfall-induced shallow landslides in unsaturated soils. *Adv. Water Resour.* **2016**, *98*, 114–121. [CrossRef]
3. Lin, F.; Wu, L.Z.; Huang, R.Q.; Zhang, H. Formation and characteristics of the Xiaoba landslide in Fuquan, Guizhou, China. *Landslides* **2018**, *15*, 669–681. [CrossRef]
4. Wu, L.Z.; Li, S.H.; Huang, R.Q.; Xu, Q. A new grey prediction model and its application to predicting landslide displacement. *Appl. Soft. Comput.* **2020**, *95*, 106543. [CrossRef]
5. Li, S.H.; Luo, X.H.; Wu, L.Z. A new method for calculating failure probability of landslide based on ANN and convex set model. *Landslides* **2021**, *18*, 2855–2867. [CrossRef]
6. Zhu, S.R.; Wu, L.Z.; Huang, J.S. Application of an improved P(m)-SOR iteration method for flow in partially saturated soils. *Comput. Geosci.* **2022**, *26*, 131–145. [CrossRef]
7. Zhu, S.R.; Wu, L.Z.; Song, X.L. An improved matrix split-iteration method for analyzing underground water flow. *Eng. Comput.* **2022**, Online. [CrossRef]
8. Montagnoli, A.; Terzaghi, M.; Magatti, G.; Scippa, G.S.; Chiatante, D. Conversion from coppice to high stand increase soil erosion in steep forestland of European beech. *Reforesta* **2016**, *2*, 60–75. [CrossRef]
9. Zegeye, A.D.; Langendoen, E.J.; Tilahun, S.A.; Mekuria, W.; Poesen, J.; Steenhuis, T.S. Root reinforcement to soils provided by common Ethiopian highland plants for gully erosion control. *Ecohydrology* **2018**, *11*, 1940–1951. [CrossRef]
10. Mitsch, W.J. Ecological engineering—The 7-year itch. *Ecol. Eng.* **1998**, *10*, 119–138. [CrossRef]
11. Osman, N.; Barakbah, S.S. Parameters to predict slope stability—Soil water and root profiles. *Ecol. Eng.* **2006**, *28*, 90–95. [CrossRef]
12. Wu, T.H. Root reinforcement of soil: Review of analytical models, test results, and applications to design. *Can. Geotech. J.* **2013**, *50*, 259–274. [CrossRef]
13. Ai, X.Y.; Wang, L.; Xu, D.P.; Rong, J.J.; Ai, S.H.; Liu, S.; Li, C.L.; Ai, Y.W. Stability of artificial soil aggregates for cut slope restoration: A case study from the subalpine zone of southwest China. *Soil Till. Res.* **2021**, *209*, 104934. [CrossRef]
14. Li, S.H.; Cui, P.; Cheng, P.; Wu, L.Z. Modified Green–Ampt Model considering vegetation root effect and redistribution characteristics for slope stability analysis. *Water Resour. Manag.* **2022**, *36*, 2395–2410. [CrossRef]
15. McGuire, L.A.; Rengers, F.K.; Kean, J.W.; Coe, J.A.; Mirus, B.B.; Baum, R.L.; Godt, J.W. Elucidating the role of vegetation in the initiation of rainfall-induced shallow landslides: Insights from an extreme rainfall event in the Colorado Front Range. *Geophys. Res. Lett.* **2016**, *43*, 9084–9092. [CrossRef]
16. Gonzalez-Ollauri, A.; Mickovski, S.B. Plant-soil reinforcement response under different soil hydrological regimes. *Geoderma* **2017**, *285*, 141–150. [CrossRef]
17. Feng, S.; Liu, H.W.; Ng, C.W.W. Analytical analysis of the mechanical and hydrological effects of vegetation on shallow slope stability. *Comput. Geotech.* **2020**, *118*, 103335. [CrossRef]
18. Löbmann, M.T.; Geitner, C.; Wellstein, C.; Zerbe, S. The influence of herbaceous vegetation on slope stability-A review. *Earth-Sci. Rev.* **2020**, *209*, 103328. [CrossRef]
19. Zhu, H.; Zhang, L.M.; Garg, A. Investigating plant transpiration-induced soil suction affected by root morphology and root depth. *Comput. Geotech.* **2018**, *103*, 26–31. [CrossRef]
20. Yildiz, A.; Graf, F.; Rickli, C.; Springman, S.M. Assessment of plant-induced suction and its effects on the shear strength of rooted soils. *Geotech. Eng.* **2019**, *172*, 507–519. [CrossRef]
21. Liu, H.W.; Feng, S.; Ng, C.W.W. Analytical analysis of hydraulic effect of vegetation on shallow slope stability with different root architectures. *Comput. Geotech.* **2016**, *80*, 115–120. [CrossRef]
22. Ng, C.W.W.; Zhang, Q.; Ni, J.J.; Li, Z.Y. A new three-dimensional theoretical model for analysing the stability of vegetated slopes with different root architectures and planting patterns. *Comput. Geotech.* **2021**, *130*, 103912. [CrossRef]
23. Su, X.M.; Zhou, Z.C.; Liu, J.E.; Cao, L.G.; Liu, J.Y.; Wang, P.P. Estimating slope stability by the root reinforcement mechanism o Artemisia sacrorum on the Loess Plateau of China. *Ecol. Model* **2021**, *444*, 109473. [CrossRef]
24. Jiang, K.Y.; Chen, L.H.; Gai, X.G.; Yang, W.J. Relationship between tensile properties and microstructures of three different broadleaf tree roots in North China. *Trans. Chin. Soc. Agric. Eng.* **2013**, *29*, 115–123. (In Chinese)

25. Zhang, C.B.; Chen, L.H.; Liu, Y.P.; Ji, X.D.; Liu, X.P. Triaxial compression test of soil-root composites to evaluate influence of roots on soil shear strength. *Ecol. Eng.* **2010**, *36*, 19–26. [CrossRef]
26. Wu, T.H. *Investigation of Landslides on Prince of Wales Island, Alaska*; Geotechnical Engineering Report 5; Department of Civil Engineering, Ohio State University: Columbus, OH, USA, 1976.
27. Waldron, L.J. The shear resistance of root-permeated homogeneous and stratified soil. *Soil Sci. Soc. Am. J.* **1977**, *41*, 843–849. [CrossRef]
28. Wu, T.H.; McKinell, W.P.; Swanston, D.N. Strength of tree roots and landslides on Prince of Wales Island, Alaska. *Can. Geotech. J.* **1979**, *16*, 19–33. [CrossRef]
29. Gray, D.H.; Sotir, R.B. Biotechnical and soil bioengineering slope stabilization: A practical guide for erosion control. *Soil Sci.* **1998**, *163*, 83–85. [CrossRef]
30. Pollen, N.; Simon, A. Estimating the mechanical effects of riparian vegetation on stream bank stability using a fiber bundle model. *Water Resour. Res.* **2005**, *41*, 226–244. [CrossRef]
31. Schwarz, M.; Preti, F.; Giadrossich, F.; Lehmann, P.; Or, D. Quantifying the role of vegetation in slope stability: A case study in Tuscany (Italy). *Ecol. Eng.* **2010**, *36*, 285–291. [CrossRef]
32. Satyanaga, A.; Rahardjo, H. Stability of unsaturated soil slopes covered with Melastoma malabathricum in Singapore. *Proc. Inst. Civ. Eng.-Geotech. Eng.* **2019**, *172*, 530–540. [CrossRef]
33. Rahardjo, H.; Satyanaga, A.; Wang, C.L.; Wong, J.L.H.; Lim, V.H. Effects of unsaturated properties on stability of slope covered with Caesalpinia Crista in Singapore. *Environ. Geotech.* **2020**, *7*, 393–403. [CrossRef]
34. Zhou, Y.Y.; Wang, X.M. Mesomechanics characteristics of soil reinforcement by plant roots. *Bull. Eng. Geol. Environ.* **2019**, *78*, 3719–3728. [CrossRef]
35. Zhang, C.B.; Chen, L.H.; Jiang, J. Vertical root distribution and root cohesion of typical tree species on the Loess Plateau, China. *J. Arid. Land* **2014**, *6*, 601–611. [CrossRef]
36. Lian, B.Q.; Peng, J.B.; Zhan, H.B.; Wang, X.G. Mechanical response of root-reinforced loess with various water contents. *Soil Till. Res.* **2019**, *193*, 85–94. [CrossRef]
37. Guo, P.; Xia, Z.; Liu, Q.; Xiao, H.; Gao, F.; Zhang, L.; Li, M.Y.; Yang, Y.S.; Xu, W.N. The mechanism of the plant roots' soil-reinforcement based on generalized equivalent confining pressure. *PeerJ* **2020**, *8*, e10064. [CrossRef]
38. Liu, Y.B.; Hu, X.S.; Yu, D.M.; Zhu, H.L.; Li, G.R. Influence of the roots of mixed-planting species on the shear strength of saline loess soil. *J. Mt. Sci.* **2021**, *18*, 806–818. [CrossRef]
39. *ASTM (D4287-11)*; Standard Practice for Classification of Soils for Engineering Purpose (Unified Soil Classification System). ASTM International: West Conshohocken, PA, USA, 2006.
40. *ASTM. (D854-02)*; Standard Test Methods for Specific Gravity of Soil Solids by Water Pycnometer. Annual Book of ASTM Standards. ASTM International: West Conshohocken, PA, USA, 2002.
41. *ASTM. (D4318-00)*; Standard Test Methods for Liquid Limit, Plastic Limit and Plasticity Index of Soils. Annual Book of ASTM Standards. ASTM International: West Conshohocken, PA, USA, 2000.
42. Li, Y.P.; Wang, Y.Q.; Ma, C.; Zhang, H.L.; Wang, Y.J.; Song, S.S.; Zhu, J.Q. Influence of the spatial layout of plant roots on slope stability. *Ecol. Eng.* **2016**, *91*, 477–486. [CrossRef]
43. Gyssels, G.; Poesen, J.; Bochet, E.; Li, Y. Impact of plant roots on the resistance of soils to erosion by water: A review. *Prog. Phys. Geogr.* **2005**, *29*, 189–217. [CrossRef]
44. Alam, S.; Banjara, A.; Wang, J.; Patterson, W.B.; Baral, S. Novel approach in sampling and tensile strength evaluation of roots to enhance soil for preventing erosion. *Open J. Soil Sci.* **2018**, *8*, 330–349. [CrossRef]
45. Cazzuffi, D.; Crippa, E. Shear strength behaviour of cohesive soils reinforced with vegetation. In Proceedings of the 16th International Conference on Soil Mechanics and Geotechnical Engineering: Geotechnology in Harmony with the Global Environment, Osaka, Japan, 12 September 2005.
46. Fan, C.C.; Su, F. Role of roots in the shear strength of root-reinforced soils with high moisture content. *Ecol. Eng.* **2008**, *33*, 157–166. [CrossRef]
47. Meng, S.Y.; Zhao, G.Q.; Yang, Y.Y. Impact of plant root morphology on rooted-soil shear resistance using triaxial testing. *Adv. Civ. Eng.* **2020**, *2020*, 8825828. [CrossRef]
48. Tan, H.M.; Chen, F.M.; Chen, J.; Gao, Y.F. Direct shear tests of shear strength of soils reinforced by geomats and plant roots. *Geotext Geomembr.* **2019**, *47*, 780–791. [CrossRef]
49. Schmidt, K.M.; Roering, J.J.; Stock, J.D.; Dietrich, W.E.; Montgomery, D.R.; Schaub, T. The variability of root cohesion as an influence on shallow landslide susceptibility in the Oregon Coast Range. *Can. Geotech. J.* **2001**, *38*, 995–1024. [CrossRef]
50. Ji, X.D.; Cong, X.; Dai, X.Q.; Zhang, A.; Chen, L.H. Studying the mechanical properties of the soil-root interface using the pullout test method. *J. Mt. Sci.* **2018**, *15*, 882–893. [CrossRef]
51. Vergani, C.; Chiaradia, E.A.; Bischetti, G.B. Variability in the tensile resistance of roots in Alpine forest tree species. *Ecol. Eng.* **2012**, *46*, 43–56. [CrossRef]
52. Su, L.J.; Hu, B.L.; Xie, Q.J.; Yu, F.W.; Zhang, C.L. Experimental and theoretical study of mechanical properties of root-soil interface for slope protection. *J. Mt. Sci.* **2020**, *17*, 2784–2795. [CrossRef]
53. Rasti, A.; Adarmanabadi, H.R.; Pineda, M.; Reinikainen, J. Evaluating the effect of soil particle characterization on internal friction angle. *Am. J. Eng. Appl. Sci.* **2021**, *14*, 129–138. [CrossRef]

54. Normaniza, O.; Faisal, H.A.; Barakbah, S.S. Engineering properties of Leucaena leucocephala for prevention of slope failure. *Ecol. Eng.* **2008**, *32*, 215–221. [CrossRef]
55. Hu, X.S.; Brierley, G.; Zhu, H.L.; Li, G.R.; Fu, J.T.; Mao, X.Q.; Yu, Q.Q.; Qiao, N. An exploratory analysis of vegetation strategies to reduce shallow landslide activity on loess hillslopes, Northeast Qinghai-Tibet Plateau, China. *J. Mt. Sci.* **2013**, *10*, 668–686. [CrossRef]

forests

Article

Mapping the Long-Term Evolution of the Post-Event Deformation of the Guang'an Village Landslide, Chongqing, China Using Multibaseline InSAR Techniques

Kui Zhang [1,*], Faming Gong [1], Li Li [2], Alex Hay-Man Ng [3,4] and Pengfei Liu [5]

1 School of Microelectronics and Communication Engineering, Chongqing University, Chongqing 400044, China; 20134576@cqu.edu.cn
2 Chongqing Natural Resources Safety Dispatch Center, Chongqing 401147, China; llywhu@whu.edu.cn
3 School of Civil and Transportation Engineering, Guangdong University of Technology, Guangzhou 510006, China; hayman.ng@gdut.edu.cn
4 Key Laboratory for City Cluster Environmental Safety and Green Development of the Ministry of Education, Guangdong University of Technology, Guangzhou 510006, China
5 Chongqing Institute of Geological Environment Monitoring, Chongqing 401121, China; 2018019@ctbu.edu.cn
* Correspondence: zk@cqu.edu.cn

Abstract: On 21 October 2017, days of heavy rainfall triggered a landslide in Guang'an Village, Wuxi County, Chongqing, China. According to the field investigation after the incident, there is still a massive accumulation body, which could possibly reactivate the landslide. In this study, to explore the long-term evolution of the deformation after the initial Guang'an Village Landslide, a time-series InSAR technique (TS-InSAR) was applied to the 128 ascending Sentinel-1A datasets spanning from October 2017 to March 2022. A new approach is proposed to enhance the conventional TS-InSAR method by integrating LiDAR data into the TS-InSAR process chain. The spatial–temporal evolution of post-event deformation over the Guang'an Village Landslide is analyzed based on the time-series results. It is found that the post-event deformation can be divided into three main stages: the post-failure stage, the post-failure and reactivation stage, and the reactivation stage. It is also suggested that, although the study area is currently under the reactivation stage, there are two active deformation zones that may become the origin of a secondary landslide triggered by heavy rainfall in the future. Moreover, the nearby Yaodunzi landslide might also play an important role in the generation and reactivation of a secondary Guang'an Village Landslide. Therefore, continuous monitoring for post-event deformation of the Guang'an Village Landslide is important for early warning of a secondary landslide in the near future.

Keywords: multibaseline InSAR techniques; Guang'an Village Landslide; InSAR; time-series analysis; post-event deformation mapping

Citation: Zhang, K.; Gong, F.; Li, L.; Ng, A.H.-M.; Liu, P. Mapping the Long-Term Evolution of the Post-Event Deformation of the Guang'an Village Landslide, Chongqing, China Using Multibaseline InSAR Techniques. *Forests* **2022**, *13*, 887. https://doi.org/10.3390/f13060887

Academic Editors: Chong Xu, Haijia Wen, Weile Li and Hiromu Daimaru

Received: 9 May 2022
Accepted: 5 June 2022
Published: 7 June 2022

Publisher's Note: MDPI stays neutral with regard to jurisdictional claims in published maps and institutional affiliations.

1. Introduction

The Guang'an Village Landslide occurred on 21 October 2017 and was triggered by continuous rainfall in Wuxi County, Chongqing between September and October of the same year [1]. The disaster caused nine missing people and serious damage to the roadway and infrastructure nearby. Over 1000 people were affected by the disaster. According to the field investigation after the incident, the size of the whole landslide was approximately 1500 m in length and 1300 m in width, with a total volume of up to 3.3×10^8 m^3. According to Wang et al. [1], there is still a massive accumulation body at the middle section of the landslide, which could form a secondary landslide and cause a serious threat to the residents nearby. In addition, the potential secondary landslide could block the major river and form a barrier lake, which could lead to a serious flood disaster. Therefore, continuous field investigation and monitoring over the affected area of the Guang'an Village Landslide are essential for early warning of the secondary landslide. However, the elevation of the

Guang'an Village Landslide region is over 1500 m, with a slope angle of 45–60° at the middle and lower part of the region. As a result, it is very challenging to conduct an on-site field investigation to evaluate the stability of the middle section of the landslide and to obtain the comprehensive deformation phenomenon over the area. Satellite interferometric synthetic aperture radar (InSAR), an active remote sensing technique, can hence play an important role in monitoring the deformation process of the sliding area at the middle section of the landslide and providing important information for assessing slope stability.

The InSAR technique, with all-day and all-weather capability, can measure regional scale deformations of the ground surface with relatively high resolution and accuracy [2–6]. Time-series InSAR (TS-InSAR), an extension of the InSAR technique, utilizes multi-temporal SAR imagery to enhance the measurement capability of the conventional InSAR technique. Typical time-series InSAR algorithms include PSInSAR™ [7], SBAS [8,9], SqueeSAR™ [10], CSI [11], CPT [12,13], GEOS-PSI [14], GEOS-ATSA [15,16], IPTA [17], π-RATE [18,19], PSP [20], StaMPS [21,22], STUN [23], and TCPInSAR [24,25]. The development of TS-InSAR allows researchers to monitoring the ground surface deformation of a slow-moving landslide, which is extremely useful for landslide investigation over remote and inaccessible regions. TS-InSAR has been successfully applied in many landslide investigation studies, including landslide detection and identification [26–30], precursory deformation detection [31,32], pre- and post-failure analysis [33,34], landslide process mechanism inversion [35], and landslide inventory, susceptibility, and hazard assessment [36,37].

The main objectives of this work are to (1) to investigate the spatial-temporal evolution of the post-event deformation over the Guang'an Village Landslide area, (2) to analyze the failure mechanism of the post-event deformation, and (3) to identify the possible origin of the potential secondary landslide. Following the idea of the TS-InSAR technique, one hundred and twenty-eight Sentinel-1A TOPS SAR data acquired between 31 October 2017 and 3 March 2022 are utilized to extract the spatial-temporal distribution of the post-event deformation of the Guang'an Village Landslide. Since the landslide area in this study is located in mountainous regions covered by relatively heavy vegetation, the presence of vegetation results in a severe temporal correlation phenomenon. In addition, as the water vapor issues are very serious, strong atmospheric artifacts are contained in the interferometric signals. Moreover, the elevation difference of the study area is over 700 m between the toe and the head of the landslide, which leads to severe geometric distortion in radar coordinates. To overcome the above problems, a new TS-InSAR approach has been developed in this study by integrating LiDAR-derived DEM into the conventional processing chain. The main features of the approach can be summarized as follows: (1) a multi-looking process is not conducted, which minimizes the effect caused by the loss of resolution; (2) the latter parts of the TS-InSAR analysis are conducted over the map coordinates. This allows for the minimizing of the influence of geometric distortion at the slant-range during the spatial phase unwrapping stage, since the spatial distribution of measurements is more in line with prior geological prior knowledge in map coordinates.

This paper is arranged as follows. Section 2 introduces the background of the study area and the datasets used in this work (i.e., Sentinel-1A SAR imagery and LiDAR-derived DEM). Section 3 describes the processing strategy used in this study. Section 4 provides the results of deformation measurements and the analysis of the potential mobility and failure mechanism of the landslide revealed by the results. Section 5 presents the main concluding remarks of this study.

2. Study Area and Datasets

The Guang'an village is located in Wuxi County in the northeast of Chongqing, China with an approximate location at 31.54° N, 109.61° E. Wuxi County is the most abundant forest resources region of Chongqing. In 2021, the forest coverage over the county was larger than 70% where the study area is mainly covered by frutex. The geological structure of this area is complex and diverse, and the terrain is quite steep (Figure 1a). The Guang'an Village Landslide is located on the slope at the left bank of the Xixi River, with steep terrain at the middle part and relatively terrain gentle at the upper and bottom parts. This whole slope where the landslide is located has an elevation difference of approximately 925 m, with an elevation at the top of the slope of about 1200 m and an elevation at the bottom of about 275 m, at the riverbed of Xixi River (Figure 1b). The slope angle at the upper part (elevation above 1100 m), the middle part (elevation between 650 m and 1100 m), and the bottom part (elevation below 650 m) of the slope is about 35°~50°, 45°~60°, and 25°~35°, respectively. The lithology of the landslide region is complex and is mainly composed of the Quaternary, Triassic, Permian, and Silurian formations, from top to bottom. The exposed bedrock is mainly composed of medium-thick- to thick-layered limestone and thin-layered shale.

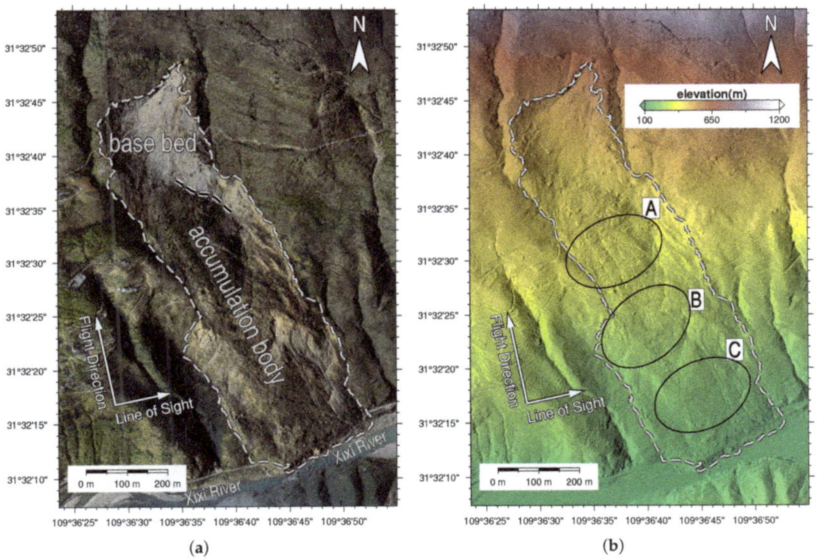

Figure 1. (**a**) The airborne true color orthoimage covering the Guang'an Village Landslide. (**b**) The elevation map at the study area. The white double broken line indicates the boundary between the base bed and the accumulation body to be used as the defined branch cuts in the phase unwrapping process. The white arrows indicate the flight direction and radar line-of-sight direction of the Sentinel-1 data. The three accumulation bodies 1, 2, and 3 are indicated by black ovals marked as A, B, and C, respectively.

The Guang'an Village Landslide occurred on 21 October 2017. The initial landslide deposits have caused the riverbed to uplift, which blocked the river channel for about 160 m in distance over the surface and formed a barrier lake. As can be seen in Figure 1a, the landslide region can be initially classified into two areas based on the morphological features observed from the recent optical images: base bed and accumulation body.

According to the field investigation and longitudinal profile data provided in [1], the landslide region can also be divided into two areas based on their failure characteristics: base bed area and accumulated area. The upper part of the study area is the base bed area and the lower part is the accumulation area. The initial base bed area has a length of about 550 m and a width of 150~280 m. The average thickness and the volume of the initial base bed area are about 30 m and 360×10^4 m^3, respectively. The sliding accumulation area is scraped and thrusted by the based bed, forming an accumulation body of about 1100 m in length and 300 m in width. The average thickness and volume of the accumulated area are about 25 m and 825×10^4 m^3, respectively. In addition to the analysis above, this study roughly divided the study area into four areas with the field survey information provided in [1]: (1) base bed, (2) accumulation body 1, (3) accumulation body 2, and (4) accumulation body 3 (the three accumulation bodies are indicated by black ovals marked as A, B, and C in Figure 1b).

In this study, the analysis is conducted mainly based on three datasets, i.e., the Sentinel-1 SAR data, LiDAR DEM, and the rainfall data. The average daily rainfall data, recorded between 1 January 2018 and 1 April 2022, was collected from the locally installed rainfall gauge near the study area. The ascending track of the C-band Sentinel-1 SAR VV polarized data acquired in the Terrain Observation by Progressive Scans (TOPS) mode were collected for time-series InSAR analysis. A total of 128 Sentinel-1 data acquired between 31 October 2017 and 3 March 2022 (Path 84 Frame 100) were selected for analysis. The flight direction and radar line-of-sight direction (LOS) of the Sentinel-1 data are shown in Figure 1. As can be seen from this figure, the slope direction of the landslide is almost parallel to the flight direction of the Sentinel-1 ascending track. The LOS displacement obtained from TS-InSAR analysis is mainly composed of the landslide displacement in the horizontal and vertical directions. The airborne LiDAR data used in this work were acquired in 2021 to provide high-resolution and up-to-date elevation data for this analysis.

As pointed out in [38], rainfall is one of the most important factors triggering landslide failures. The average daily precipitation data between 1 January and 11 April 2022 were therefore collected from the nearest locally installed gauge (approximately 1.5 km away from the study area). In general, flood season lasts from early April to early October in the study area. According to the rainfall data, the precipitation intensities of 2018, 2019, 2020, and 2021 are 1496.8 mm, 1057.6 mm, 1666.3 mm, and 1712.2 mm, respectively. In total, 6089.2 mm rainfall was recorded during the observation period.

3. Methodology

3.1. Preparation of the Interferogram Stack

In order to generate an image stack for time-series analysis, all Sentinel-1 images have to be co-registered to the same image grid. A primary image acquired on 19 March 2019 was chosen for the generation of a Sentinel-1 image stack. To generate the SAR image stacks, all secondary images are coregistered to the image grid of the primary image. Because of the variations in Doppler centroid frequency in Sentinel-1 TOPS mode, a coregistration accuracy of 1/1000 pixel is required for the interferometric phase error to be negligible between bursts. The enhanced spectral diversity (ESD) method, which is able to achieve such accuracy, was utilized. However, since ESD derives mis-coregistration from the cross-interferometric phase, a fair amount of high coherence pixels are needed at burst overlaps area for the ESD process to achieve the required coregistration accuracy [39]. The high ratio of decorrelated pixels over the landslide region due to the temporal decorrelation could degrade the achieved accuracy from ESD. To avoid this issue, the whole Sentinel-1 scene is processed with the one arc-second Shuttle Radar Topography Mission (SRTM) Digital Elevation Model (DEM) and the Sentinel-1 precise orbits. The purpose of processing the full scene instead of the sub-scene covering only the landslide and nearby areas is to ensure that there are enough high coherence pixels for the ESD process.

Once all SAR images are coregistered to the same grid, a differential interferogram can be formed. In order to ensure an accurate inversion of surface deformation from the time-series InSAR analysis, it is necessary to ensure that the minimum possible decorrelated InSAR pairs are selected. An investigation was conducted prior to the selection of interferometric pairs to obtain the optimal combination of interferometric pairs for time-series analysis. It was found that temporal decorrelation is the dominating decorrelation source after analyzing all possible combinations of interferometric pairs over the study area. There are noticeable outliers that exhibit poor coherence even with a temporal baseline of 24 days. Therefore, the interferometric pairs are selected with the shortest temporal baseline available to guarantee that coherence is preserved in all the image pairs to some extent. One hundred and twenty-seven differential interferograms were generated with respect to the primary image using the conventional two-pass DInSAR approach [40]. A one arc-second SRTM DEM was used to remove the topography phase from the interferograms.

Once the differential interferogram stack is formed, a filtering operation can be conducted to enhance the coherence. Since the location of the deformation zone to be monitored is expected to be small in spatial extent, full-resolution processing is preferred instead of processing with the multi-looked interferogram. The two-sample Kolmogorov–Smirnov (KS) test was carried out to determine the adaptive filtering window, i.e., the statistically homogeneous areas (SHP), for each pixel. The initial searching window of 15 × 45 pixels (approximately 0.07 km^2) is used here.

3.2. Atmospheric Correction with Masking of the Deforming Area

The differential phase in interferogram k, in the pixel of azimuth and range coordinates (x, y) from acquired times t_B and t_A is given by:

$$\Delta\phi_{x,y}^k \approx -\frac{4\pi}{\lambda}\left(d_{x,y}^{t_A} - d_{x,y}^{t_B}\right) - \frac{4\pi}{\lambda}\left(\frac{B_{\perp x,y}^k \Delta h_{x,y}}{R_{x,y}\sin\theta_{x,y}}\right) + \left(\phi_{atm}t_{x,y}^{t_A} - \phi_{atm}t_{x,y}\right) + \Delta n_{x,y}^k \quad (1)$$

where λ is the wavelength of the radar signal and $\left(d_{x,y}^{t_A} - d_{x,y}^{t_B}\right)$ is the LOS displacement between the acquisition times t_B and t_A. The second term in Equation (1) is the phase induced by inaccuracy in the reference DEM. $\Delta h_{x,w}$, $B_{\perp_{x,y}}$, $R_{x,y}$, and $\theta_{x,y}$ are the DEM error, mean perpendicular baseline, slant-range distance, and local incidence angle, respectively, between the pixel (x, y) and the reference point. The third term is induced by the atmospheric inhomogeneities in times t_B and t_A. The last term, $\Delta n_{x,y}^k$, is induced by many factors, including thermal noise, mis-coregistration, inaccurate orbit data, decorrelation, etc.

It is worth noting that the post-failure surface deformation of the Guang'an Village Landslide is highly non-linear, especial within the short time frame. Therefore, the deformation cannot be accurately estimated using the common linear or the seasonal models. For this reason, atmospheric correction must be removed before the spatial phase unwrapping operation to minimize the phase ambiguity problem over the sliding area. In the time-series InSAR analysis, the atmospheric component in Equation (1) is often assumed to be spatially correlated and temporally uncorrelated. Since all InSAR pairs are selected with the shortest possible temporal baseline, the atmospheric phase of each pair can be expressed as:

$$\begin{bmatrix} \Delta\phi_{LP_{x,y}^1} \\ \vdots \\ \Delta\phi_{LP_{x,y}^K} \end{bmatrix} = \begin{bmatrix} 1 & -1 & & \\ & \ddots & \ddots & \\ & & 1 & -1 \end{bmatrix} \begin{bmatrix} \phi_{atm_{x,y}^{t_0}} \\ \vdots \\ \phi_{atm_{x,y}^{t_K}} \end{bmatrix} \quad (2)$$

where $\Delta\phi_{LP_{x,y}^K}$ is the spatially correlated phase in interferogram k and $\phi_{atm_{x,y}^{t_j}}$ is the atmospheric phase at acquisition time j for the pixel of azimuth and range coordinates (x, y). In this study, $\Delta\phi_{LP}$ is calculated based on the spatial low pass filtering on the wrapped interferogram. It is clear that the design matrix in Equation (2) is rank-deficient. The approach proposed in [41] is adopted here for the atmospheric correction. The interferogram with minimal variance in the interferogram stacks is first obtained and selected and then the selected interferogram is used to calibrate the others based on Equation (2).

Since the post-failure surface deformation is correlated in the spatial domain to some extent, some of the deformation signals can be incorrectly treated as the atmospheric phase using the procedure above. Therefore, refinement of the preliminary atmospheric signal is necessary. The refinement procedure used in this work is as follows: (1) spatial phase unwrapping operation is applied to each interferogram (after the initial atmospheric correction); (2) the accumulated deformation can be obtained by adding up all of the unwrapped interferograms; (3) the deformation area is identified and a mask is created to mask out the sliding area; (4) the atmospheric correction procedure is reapplied with the masked interferograms to estimate the atmospheric phase component; (5) the spatial interpolation is applied to estimate the atmospheric phase at the masked area; (6) the refined atmospheric phases are subtracted from the original differential interferograms.

3.3. Terrain Correction and Spatial Phase Unwrapping

After the atmospheric correction process, the remaining phase in the interferogram is expected to be contributed to mainly by the deformation and DEM error. Since the InSAR pairs are selected with the minimal temporal baseline possible, most pairs are separated by 12 days. Given that the temporal baseline between each pair is relatively short and the landslide deformation is non-linear in time, DEM error can be estimated using the stacking approach [42] without considering the influence of deformation. Once the DEM error is estimated and corrected, spatial phase unwrapping can be applied to the refined interferogram stack for deformation analysis.

Unlike many other TS-InSAR studies, the variation of elevation in the study area is over 700 m within a relatively small spatial extent, i.e., 1.5 km $\times$ 0.85 km. The steepness of the slope is up to $\sim$60°. The geometric distortion caused by high elevation and extremely undulating terrain, such as strong layover and foreshortening effects, can induce strong phase distortion. This makes the spatial phase unwrapping process challenging without correcting the terrain effects. Figure 2 shows the amount of geometric distortion observed using the 1 arc-second STRM DEM. The look-up table for geographic-to-radar coordinate projection based on the SRTM DEM and the LiDAR have been generated independently. The amount of geometric distortion is derived by the projected range coordinate difference between the SRTM and the LIDAR look-up tables. These differences can hence be considered to be the projection bias of the SRTM look-up table. It can be clearly observed that largest difference is larger than 20 pixels. Since this study requires processing the data at a full-resolution scale, such a bias will lead to a relatively large distortion in the geographic pattern of the interferometric signals. Although the elevation information obtained from the TS-InSAR analysis can be used to compensate for the DEM error, the variation in the perpendicular baseline of the Sentinel-1 InSAR stack is too small to obtain the required accuracy. In order to deal with this issue, an airborne LIDAR DEM with an expected accuracy of 5 cm is used to assist the phase unwrapping process. The LiDAR DEM with a much higher resolution (0.5 m $\times$ 0.5 m) is able to derive the land feature in more detail, which allows the defining of the correct branch-cut to correct the phase unwrapping error, and at the same time provides accurate geolocation information for each pixel. The InSAR interferograms are resampled to a grid with a resolution of 8 m $\times$ 8 m based on the LiDAR DEM. The reason for choosing such a resolution is because the number of elements from the map grid and the slant-range grid is approximately the same. Therefore, this allows the phase information to be retained as much as possible without degrading the coherence.

Once all interferograms are resampled to the map coordinate, the Goldstein filter [43] is applied to ensure the continuity of the phase data for spatial phase unwrapping. The pixels with temporal coherence, i.e., the sum of complex coherence throughout the stack, larger than 0.75 are considered as coherent pixels for phase unwrapping. The minimum cost network flow phase unwrapping algorithm (MCF) is performed on all the interferograms to derive the unwrapped phase stack [44]. The costs for each flow are derived from the coherence obtained from the space adaptive filter (SHP) in the early stage. The phase unwrapping results are then investigated and the results that contained serious tiling effects [45,46] are identified. Correction is conducted by adding the defined branch cuts to the minimum cost flow network to avoid such an issue. With the data resampled into the map coordinates, branch cuts can be easily defined based on the morphological features from the LiDAR data and the optical images. Figure 3 shows an example of the unwrapped phase in map coordinate with and without the correction using the proposed method. As can be seen in Figure 3a, there is an area with an obvious phase unwrapping error observed in the phase unwrapping result that is obtained without the defined branch cuts between the base bed and the accumulation body (highlighted by the black circle in Figure 3). The phases have integrated from the base bed to the accumulation body at one end and vice versa at the other end. Figure 3b shows the unwrapped result obtained with the defined branch cut between the base bed and the accumulation body (see Figure 1a for the defined branch cut at the boundary). This clearly shows that the phase unwrapping error has been corrected and the phases no longer pass the boundary between the base bed and the accumulation body. This suggests that the proposed method can improve the correctness of the phase unwrapping results.

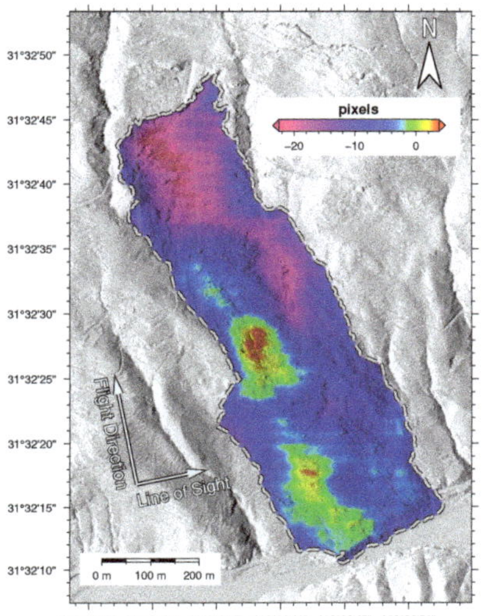

Figure 2. The projected range coordinate difference between the 1 arc-second SRTM and the LiDAR look-up tables.

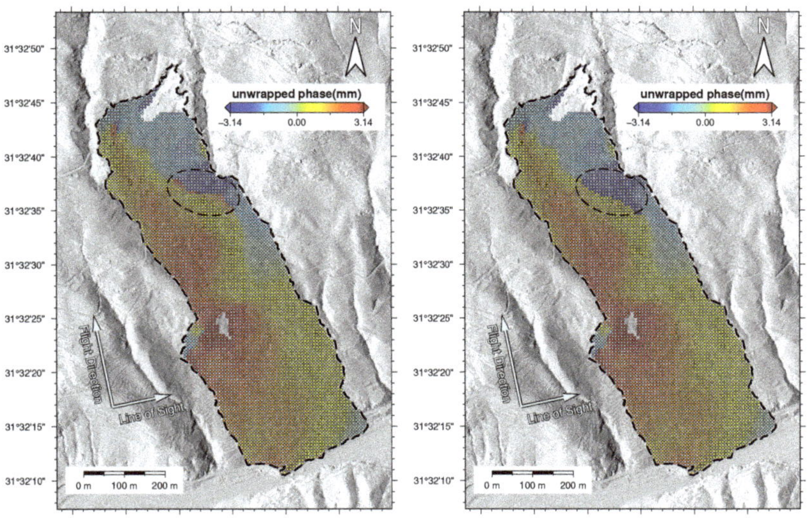

(**a**) (**b**)

Figure 3. Unwrapped phase in map coordinates (**a**) without and (**b**) with the correction using the proposed method. The area with the obvious phase unwrapping error observed is highlighted by the black circle.

3.4. Time-Series Deformation Analysis

Once the refined interferograms are unwrapped and the DEM error is corrected, network integration in the time domain can be conducted to obtain the time-series deformation information by solving the linear system:

$$
\begin{bmatrix} 0 \\ \phi_{defo^{1}_{x,y}} \\ \vdots \\ \phi_{defo^{K}_{x,y}} \end{bmatrix} = \begin{bmatrix} 1 & & & \\ 1 & -1 & & \\ & \ddots & \ddots & \\ & & 1 & -1 \end{bmatrix} \begin{bmatrix} -\frac{4\pi}{\lambda} D_{LOS^{t_0}_{x,y}} \\ \vdots \\ \vdots \\ -\frac{4\pi}{\lambda} D_{LOS^{t_K}_{x,y}} \end{bmatrix} \quad (3)
$$

where $\phi_{defo^{K}_{x,y}}$ is the unwrapped deformation phase in refined interferogram k, and $D_{LOS^{t_K}_{x,y}}$ is the cumulative deformation along the LOS direction at times corresponding to the acquisition of the secondary images. Finally, a temporal low pass filter with kernel windows of 96 days is applied to the time-series deformation data to remove the decorrelation noise.

4. Results: Deformation Velocity Measured by Time-Series InSAR Analysis

The LOS displacement time-series obtained from the TS-InSAR processing of the 128 Sentinel-1 data are shown in Figures 4 and 5. Figure 6 represents the accumulated displacement. The measurement points (MPs) with red color (positive values) indicate that targets are moving away from the satellite along the LOS direction. On the other hand, the blue color counterparts (negative values) indicate that targets are moving towards the satellite. Figures 4 and 5 show that the study area experienced continuous post-event deformation between 2017 and 2021, after the initial Guang'an Village Landslide in October 2017. Since noticeable deformation is observed in this region, a cumulative deformation map was generated in order to provide a more comprehensive understanding of the spatial distribution of the deformation (Figure 6). A total of around 4700 MPs, covering more than 90% of the landslide area, were obtained. This suggests that single-pixel processing (without multi-looking processing) can effectively avoid the loss of interference signal and maximize the number of MPs in this region. In addition, because the TS-InSAR processing is conducted at the ground distance with the assistance of high-resolution LiDAR DEM, the MPs are more uniformly distributed in the displacement map. This effectively minimized the problem of uneven distribution of MPs due to large topographic relief. According to the TS-InSAR results, it was found that there are two areas, i.e., near the base bed and point 4 (located in accumulation body 2), that experienced strong decorrelation and from which limited MPs can be obtained. Comparing the deformation map with the SAR intensity images and the LiDAR data suggests that the decorrelation near the base bed is mainly due to the strong shortening and layer effect in the region. The main reason for the decorrelation near point 4 in the accumulation body 2 is expected to be the strong SAR shadowing effect.

As shown in Figure 6, there are five active deformation zones. The maximum cumulative LOS deformations observed at the areas near the five points are approximately 80 mm, 80 mm, −200 mm, 200 mm, and 150 mm, respectively, of which point 2 has the maximum accumulated LOS deformation. As mentioned in the previous section, the direction of the flight path is parallel to the main sliding direction. Therefore, all points except for point 2 are experiencing the ground surface moving away from the satellite, where InSAR-derived deformation in these regions is expected mainly in the vertical direction. On the other hand, point 2 is experiencing the ground surface moving towards the satellite, suggesting that the InSAR-derived deformation is expected mainly in horizontal displacement perpendicular to the flight direction.

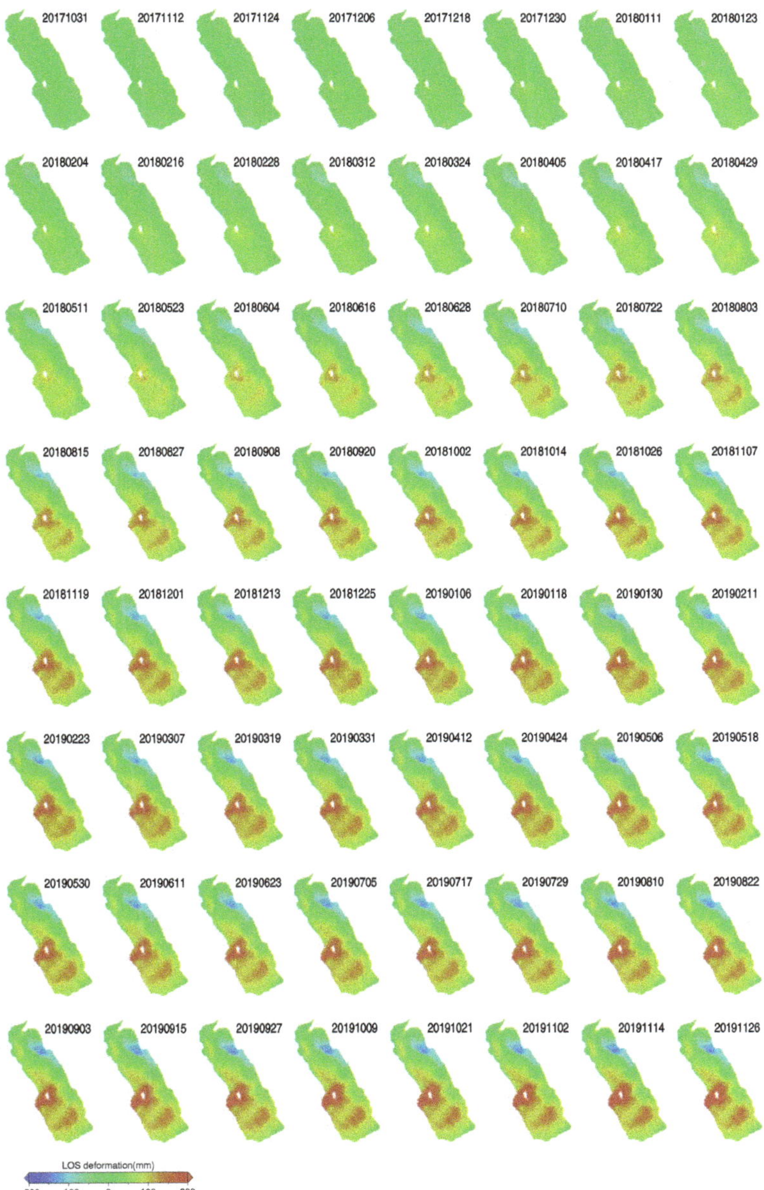

Figure 4. LOS displacement time-series maps for the Guang'an Village Landslide at the 128 Sentinel-1 acquisition date, referring to the first SAR image acquired on 31 October 2017 (from 31 October 2017 to 26 November 2019). The number at the upper right of each subplot represents the image acquisition date (in yyyymmdd format).

Figure 5. LOS displacement time-series maps for the Guang'an Village Landslide at the 128 Sentinel-1 acquisition date, referring to the first SAR image acquired on 31 October 2017 (from 8 December 2019 to 3 March 2022). The number at the upper right of each subplot represents the image acquisition date (in yyyymmdd format).

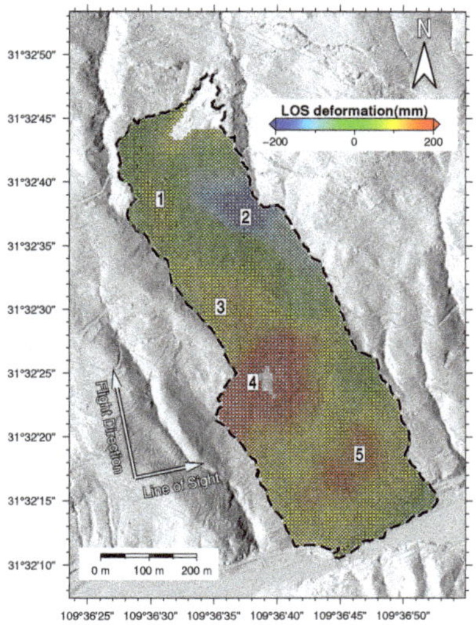

Figure 6. Cumulative LOS displacement between 31 October 2017 and 3 March 2022 for the Guang'an Village Landslide. The numbers 1 to 5 indicate the five active deformation zones for Figure 7.

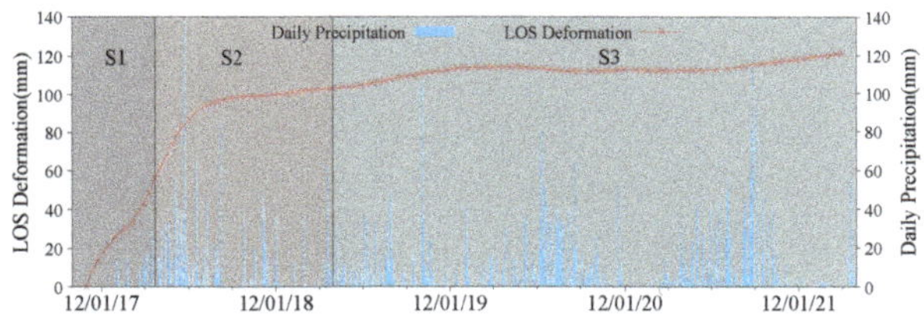

(a)

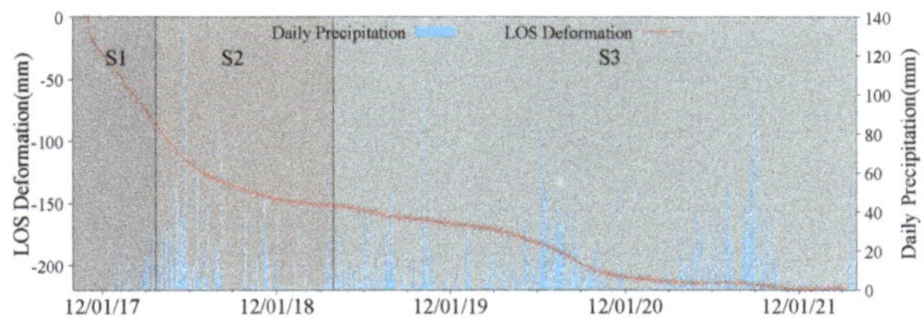

(b)

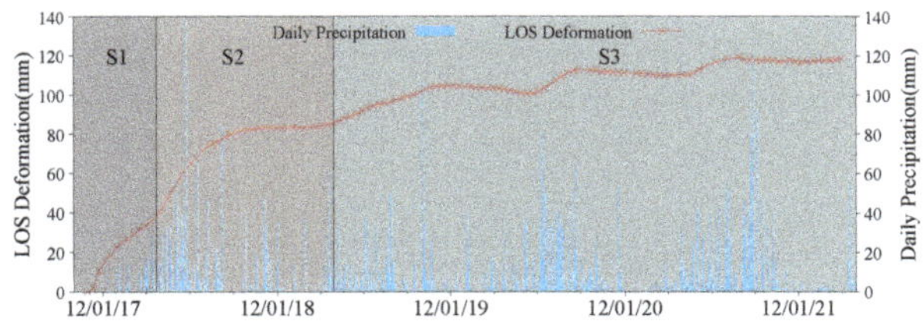

(c)

Figure 7. *Cont.*

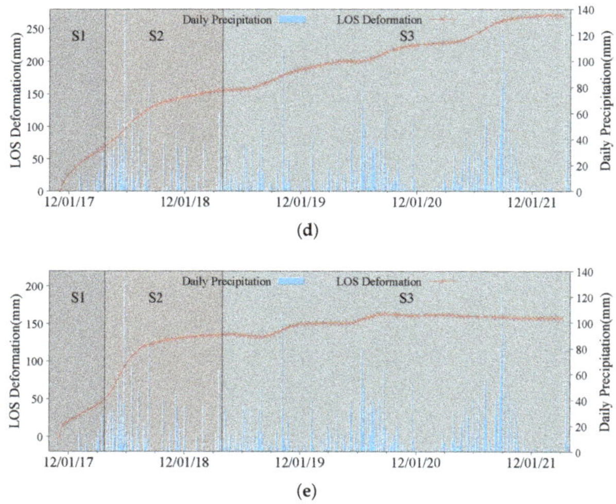

Figure 7. Time-series of the post-event displacement and rainfall for selected MPs from the five active deformation zones: (**a**) point 1, (**b**) point 2, (**c**) point 3, (**d**) point 4, (**e**), and point 5, from the cumulative displacement map (Figure 6). The three landslide movement stages S1, S2, and S3 in Figure 8 are highlighted by three-color backgrounds in the time-series plot.

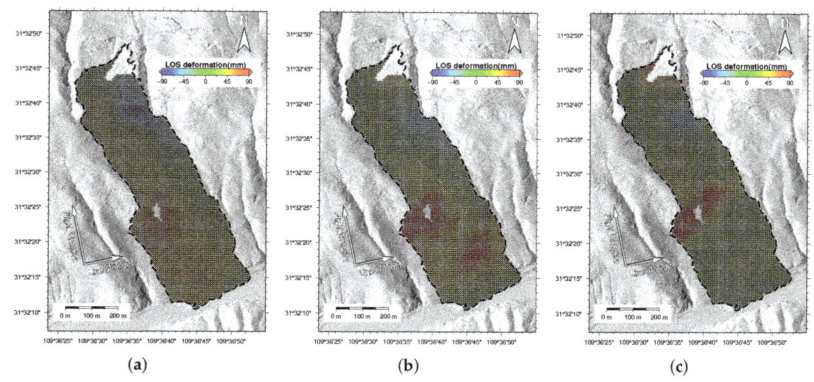

Figure 8. The cumulative LOS displacement maps of the three landslide movement stages: (**a**) S1: post-failure stage, (**b**) S2: post-failure + reactivation stage, and (**c**) S3: reactivation stage.

Figure 7 shows the deformation trend of MPs with the highest displacement from each of the five active deformation zones. It can be seen that all the five MPs have ladder-like deformation characteristics, but have a different degree of correlation with rainfall.

5. Discussion

5.1. Temporal–Spatial Evolution of the Post-Event Deformation of the Guang'an Village Landslide

In this section, the spatial-temporal evolution of the post-event slope stability at the active deformation zones are analyzed. As can be seen from the displacement time-series in Figure 7, during the observation period, point 3 and 4 show an obvious periodic "acceleration-creep" deformation pattern before and after the flood season every year, indicating that the deformation trends at these active deformation zones are strongly correlated with the rainfall in the flood season. Apart from the two points mentioned, the acceleration of deformation is strongly correlated to the rainfall intensity at point 5 from 2018 to 2020. However, once the accumulated deformation reached its peak after the flood season in 2020, the accumulated deformation remained stable even under the action of the rainfall during the flood season in 2021. Continuous deformation has been observed at point 1 since the initial landslide incident, but the deformation rate significantly decreased after the flood season in 2018. Even though the point continues to deform, the acceleration in deformation observed during flood season rainfall is relatively low and the correlation between the deformation and rainfall during the flood season is decreased significantly compared with that in 2018. Compared to the rest of the MPs, point 2 shows a unique continuous deformation trend with an opposite sign (moving towards the satellite). As discussed previously, the deformation at point 2 is mainly in horizontal displacement in the direction perpendicular to the flight direction. Rapid deformation was observed after the initial landslide incident. However, the rate of deformation decreased significantly along with the weakening of rainfall after the flood season in 2018. Because the amount of rainfall in the flood season of 2019 was lower compared to the other years, the deformation rate at point 2 was relatively stable after the flood season of 2018 until the beginning of the flood season in 2020. An obvious acceleration in deformation was observed under the action of rainfall in the flood season of 2020. However, although the peak daily rainfall intensity was higher during the flood season in 2021 than in 2020, only a small "acceleration-creep" deformation trend was observed.

According to [47], the landslide movement can be classified into four stages: (1) a pre-failure stage including a deformation process leading to failure; (2) the onset of failure characterized by the formation of a continuous shear surface through the entire soil mass; (3) a post-failure stage starting from failure until the mass stops; (4) a reactivation stage when sliding occurs on a pre-existing shear surface. By taking the four landslide movement stages proposed by [47] into account, the temporal evolution of the post-event deformation at the five selected MPs is classified into three stages here based on correlation analysis between the change in deformation rate and rainfall: (1) post-failure stage (S1: until 24 March 2018), (2) post-failure and reactivation stage (S2: between 25 March 2018 and 31 March 2019), and (3) reactivation stage (S3: after 31 March 2019). The criteria used for such a classification can be summarized as follows. (1) The flood season over the Wuxi County starts from the beginning of April. (2) According to Figure 7, the displacement observed on each point during 2018 is clearly larger than that observed during 2019, 2020, and 2021. (3) For the cases of point 1 and point 5, the deformation evaluations generally remain stable after 2018. The three landslide movement stages S1, S2, and S3 at the five selected MPs are highlighted in Figure 7.

In order to analyze the spatial distribution of deformations at the three stages, the cumulative deformation corresponding to the three stages is computed (Figure 8). As can be seen in Figure 8, the spatial extent of the deformation areas has gradually reduced and most of the affected area was stabilized from S1 to S3. During stage S1, the deformation was mainly concentrated in the active deformation zones point 2 and point 4. The non-active deformation areas were mainly affected by horizontal movement at the active deformation zones point 2 and by surface movement in slip directions at the active deformation zone point 4. During stage S2, the deformation was mainly concentrated in the active deformation zones point 4 and point 5. The deformation in the landslide area tended to be stable. During stage S3, the front edge of the landslide area tended to be stable. However, noticeable deformation was observed in the active deformation zones point 2 and point 4, and the spatial extent of the deformation zones gradually extended from the west boundary of the landslide to the east boundary.

5.2. Deformation Mechanism Analysis and Risk Evaluation of the Guang'an Village Landslide Area

According to the cumulative displacement at the reactivation stage (Figure 8c), there are two zones (points 2 and 4) in the study area where intensive deformation was observed, which may become the potential trigger source for the secondary landslide. Based on the above analysis, it was found that there is a clear difference between the deformation phenomenon in active deformation zone point 2 (source area) and the rest of the deformation zones. To investigate the reason why the deformation trend was opposite in point 2, the gravity trend direction over the region was computed based on the least square solution of the eight-neighbor elevation difference from the high-precision LiDAR DEM (Figure 9). It can be seen that there is a certain angle between the gravity trend direction of the active deformation zone point 2 and the main sliding direction of the slope. This may lead to the abnormal deformation trend in the zone. In addition, by analyzing the study area and its surroundings using the LiDAR data, it was found that there is a long crack appearing at the north-east of the study area (Figure 10), which suggests that it is highly likely there is deformation occurring in that area. This finding has been verified with the field investigation, and the slope instability in that area is named the Yaodunzi Landslide.

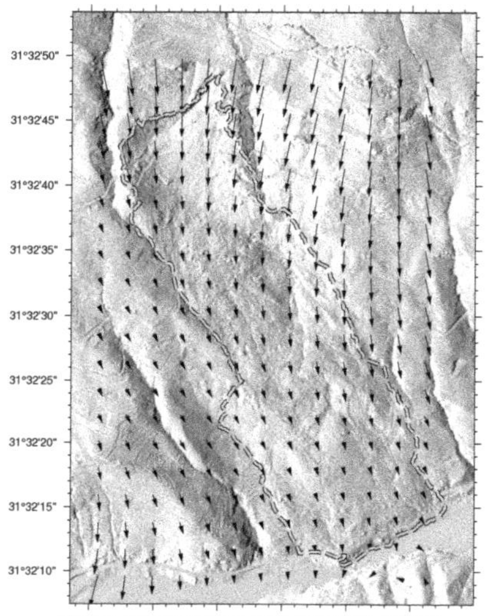

Figure 9. The quiver plot of the gravity trend direction for the Guang'an Village Landslide and surroundings.

As the Yaodunzi Landslide area moves in the slope direction, it can induce the pressure on the accumulation body in the Guang'an Village Landslide, leading to horizontal displacement perpendicular to the main sliding direction (moving towards the satellite) in the active deformation zone (Figure 10). In fact, the active deformation zone where point 2 is located becomes the main anti-sliding section of the Yaodunzi Landslide. The anti-sliding force in this area is mainly composed of two parts: (1) the friction between the accumulation body and the sliding bed in this area; (2) the lower area acting as the anti-sliding force to this area. According to the profile (see [1]), since the initial Guang'an Village Landslide, the thickness of the accumulation layer in this area is relatively shallow and the normal stress of the sliding bed is relatively limited. Therefore, the contribution of the anti-sliding force from the second part is likely to be larger than the first part.

This can also be observed in the deformation time-series of the active deformation zones point 3, point 4, and point 5 (Figure 7) that the deformation trends of these zones are relatively synchronous. However, the InSAR results in Figure 8c show that the cumulative surface deformation in S3 stage is highest at the accumulation body 2 (where point 4 is located). Since the sliding direction of the landslide is almost perpendicular to the Sentinel-1 LOS direction, the actual deformation in the accumulation 2 region may be much larger than the InSAR LOS displacement values. Based on the above analysis, the Guang'an Village Landslide may have two chain failure modes under extreme rainfall conditions: (1) rainfall infiltrates from the cracks at the rear edge of the Yaodunzi Landslide under heavy rainfall, resulting in the reduction of the anti-sliding force from the sliding soil of the Yaodunzi Landslide. The anti-sliding force from active deformation zone point 2 then can no longer maintain the stability of the Yaodunzi landslide, causing a failure of the slope; (2) the accumulation 2 region continues to deform under the heavy rainfall and the continuous deformation in this area leads to the slope instability of accumulation 1 region, and eventually to the slope failure of the Yaodunzi Landslide.

Figure 10. The effect of the Yaodunzi Landslide on the active deformation zone point 2. The arrow indicates the force induced by the Yaodunzi landslide. The red and blue arrows represent the indirect force and direct force, respectively, to the active deformation zone point 2. The white arrows show the location of the cracks at the rear edge of Yaodunzi landslide.

6. Concluding Remarks

This study developed an approach, based on the conventional TS-InSAR method, to monitor the post-event deformation of the Guang'an Village Landslide, which integrates the LiDAR data into the TS-InSAR processing chain. This approach has two main features: (1) a multi-looking process is not conducted to minimize the loss of interferometric signal due to the down-sampling issue; (2) TS-InSAR analysis is conducted on the ground distance, which minimized the influence of geometric distortion in slant-range coordinates on the geocoded deformation results. This allows the spatial distribution of measurement points to be in line with prior geological knowledge, and hence prior knowledge can be applied to assist the spatial phase unwrapping process. A total of 128 Sentinel-1 SAR images, acquired from October 2017 to March 2022, were analyzed to obtain the long-term evolution of the post-event deformation of the landslide area. The InSAR-derived deformation information, LiDAR DEM, and the rainfall data were jointly used to analyze and explore the deformation characteristics of the study area. It was found that: (1) the study area tends to be stable in general, but there are two areas that continue to suffer from deformation due to rainfall in the flood season; (2) the post-event deformation between October 2017 and March 2022 can be divided into three main stages: the post-failure stage (October 2017 to March 2018), the post-failure and reactivation stage (March 2018 to March 2019), and the reactivation stage (March 2019 to March 2022); (3) the results show that the study area can be further divided into four main areas: base bed, accumulation body 1, accumulation body 2, and accumulation body 3. Among the four areas, it is observed that there are signs of abnormal accumulation of lateral deformation at the trailing edge of accumulation 1 area. A possible reason for such deformation is that the upper part of accumulation 1 is suffering from downward movement and compression from another landslide (Yaodunzi Landslide), resulting in lateral deformation occurring at the upper part of the accumulation 1 area; (4) although the study area is currently in the reactivation stage, active deformation zones have been found in both accumulation body 1 and accumulation body 2 areas. These active deformation zones are strongly correlated to rainfall intensity in the flood season. These two deformation zones may become the origin of a secondary landslide triggered by heavy rainfall in the future. On-going monitoring of these areas is therefore essential for the detection and early warning of further failure and for planning an emergency response.

Author Contributions: Conceptualization, K.Z. and F.G.; methodology, K.Z. and L.L.; software, K.Z. and L.L.; validation, F.G. and A.H.-M.N.; formal analysis, K.Z. and P.L.; investigation, L.L., A.H.-M.N., and P.L.; resources, K.Z. and A.H.-M.N.; data curation, K.Z., F.G., and L.L.; writing—original draft preparation, F.G.; writing—review and editing, K.Z.; visualization, A.H.-M.N. and P.L.; supervision, K.Z.; project administration, K.Z.; funding acquisition, K.Z. All authors have read and agreed to the published version of the manuscript.

Funding: This research was funded by the National Natural Science Foundation of China (Grant No. 42074034) and the Program for Guangdong Introducing Innovative and Entrepreneurial Teams (2019ZT08L213).

Institutional Review Board Statement: Not applicable.

Informed Consent Statement: Not applicable.

Data Availability Statement: Sentinel-1 data were analyzed in this study. This data can be found here: https://asf.alaska.edu/ (accessed on 29 March 2022).

Acknowledgments: The authors are very grateful to the European Space Agency for providing Sentinel-1 data.

Conflicts of Interest: The authors declare no conflict of interest.

References

1. Wang, L.; Yin, Y.; Huang, B.; Zhang, Z.; Wei, Y. Formation and characteristics of Guang'an Village landslide in Wuxi, Chongqing, China. *Landslides* **2019**, *16*, 127–138. [CrossRef]
2. Bamler, R.; Hartl, P. Synthetic aperture radar interferometry. *Inverse Probl.* **1998**, *14*, R1. [CrossRef]
3. Gabriel, A.K.; Goldstein, R.M.; Zebker, H.A. Mapping small elevation changes over large areas: Differential radar interferometry. *J. Geophys. Res. Solid Earth* **1989**, *94*, 9183–9191. [CrossRef]
4. Massonnet, D.; Feigl, K.L. Radar interferometry and its application to changes in the Earth's surface. *Rev. Geophys.* **1998**, *36*, 441–500. [CrossRef]
5. Rosen, P.A.; Hensley, S.; Joughin, I.R.; Li, F.K.; Madsen, S.N.; Rodriguez, E.; Goldstein, R.M. Synthetic aperture radar interferometry. *Proc. IEEE* **2000**, *88*, 333–382. [CrossRef]
6. Simons, M.; Rosen, P. Interferometric synthetic aperture radar geodesy. In *Treatise on Geophysics-Geodesy*; Elsevier: Amsterdam, The Netherlands, 2007.
7. Ferretti, A.; Prati, C.; Rocca, F. Permanent scatterers in SAR interferometry. *IEEE Trans. Geosci. Remote Sens.* **2001**, *39*, 8–20. [CrossRef]
8. Berardino, P.; Fornaro, G.; Lanari, R.; Sansosti, E. A new algorithm for surface deformation monitoring based on small baseline differential SAR interferograms. *IEEE Trans. Geosci. Remote Sens.* **2002**, *40*, 2375–2383. [CrossRef]
9. Mora, O.; Lanari, R.; Mallorqui, J.J.; Berardino, P.; Sansosti, E. A new algorithm for monitoring localized deformation phenomena based on small baseline differential SAR interferograms. *IEEE Int. Geosci. Remote Sens. Symp.* **2002**, *2*, 1237–1239.
10. Ferretti, A.; Fumagalli, A.; Novali, F.; Prati, C.; Rocca, F.; Rucci, A. A new algorithm for processing interferometric data-stacks: SqueeSAR. *IEEE Trans. Geosci. Remote Sens.* **2011**, *49*, 3460–3470. [CrossRef]
11. Dong, J.; Zhang, L.; Tang, M.; Liao, M.; Xu, Q.; Gong, J.; Ao, M. Mapping landslide surface displacements with time series SAR interferometry by combining persistent and distributed scatterers: A case study of Jiaju landslide in Danba, China. *Remote Sens. Environ.* **2018**, *205*, 180–198. [CrossRef]
12. Blanco-Sanchez, P.; Mallorquí, J.J.; Duque, S.; Monells, D. The coherent pixels technique (CPT): An advanced DInSAR technique for nonlinear deformation monitoring. In *Earth Sciences and Mathematics*; Birkhäuser Verlag AG: Basel, Switzerland, 2008 pp. 1167–1193.
13. Blanco, P.; Mallorqui, J.; Duque, S.; Navarrete, D. Advances on DInSAR with ERS and ENVISAT data using the coherent pixels technique (CPT). In Proceedings of the 2006 IEEE International Symposium on Geoscience and Remote Sensing, Denver, CO, USA, 31 July–4 August 2006; pp. 1898–1901.

14. Ng, A.H.M.; Ge, L.; Li, X.; Zhang, K. Monitoring ground deformation in Beijing, China with persistent scatterer SAR interferometry. *J. Geod.* **2012**, *86*, 375–392. [CrossRef]
15. Ge, L.; Ng, A.H.M.; Li, X.; Abidin, H.Z.; Gumilar, I. Land subsidence characteristics of Bandung Basin as revealed by ENVISAT ASAR and ALOS PALSAR interferometry. *Remote Sens. Environ.* **2014**, *154*, 46–60. [CrossRef]
16. Ng, A.H.M.; Ge, L.; Li, X. Assessments of land subsidence in the Gippsland Basin of Australia using ALOS PALSAR data. *Remote Sens. Environ.* **2015**, *159*, 86–101. [CrossRef]
17. Werner, C.; Wegmuller, U.; Strozzi, T.; Wiesmann, A. Interferometric point target analysis for deformation mapping. In Proceedings of the IGARSS 2003. 2003 IEEE International Geoscience and Remote Sensing Symposium. Proceedings (IEEE Cat. No. 03CH37477), Toulouse, France, 21–25 July 2003; Volume 7, pp. 4362–4364.
18. Biggs, J.; Wright, T.; Lu, Z.; Parsons, B. Multi-interferogram method for measuring interseismic deformation: Denali Fault, Alaska. *Geophys. J. Int.* **2007**, *170*, 1165–1179. [CrossRef]
19. Wang, H.; Wright, T.J.; Yu, Y.; Lin, H.; Jiang, L.; Li, C.; Qiu, G. InSAR reveals coastal subsidence in the Pearl River Delta, China. *Geophys. J. Int.* **2012**, *191*, 1119–1128. [CrossRef]
20. Costantini, M.; Falco, S.; Malvarosa, F.; Minati, F.; Trillo, F. Method of persistent scatterer pairs (PSP) and high resolution SAR interferometry. In Proceedings of the 2009 IEEE International Geoscience and Remote Sensing Symposium, Cape Town, South Africa, 12–17 July 2009; Volume 3, pp. III-904–III-907.
21. Hooper, A. A multi-temporal InSAR method incorporating both persistent scatterer and small baseline approaches. *Geophys. Res. Lett.* **2008**, *35*. [CrossRef]
22. Hooper, A.; Segall, P.; Zebker, H. Persistent scatterer interferometric synthetic aperture radar for crustal deformation analysis, with application to Volcán Alcedo, Galápagos. *J. Geophys. Res. Solid Earth* **2007**, *112*. [CrossRef]
23. Kampes, B.M. *Radar Interferometry*; Springer: Dordrecht, The Netherlands, 2006; Volume 12.
24. Zhang, L.; Ding, X.; Lu, Z. Ground settlement monitoring based on temporarily coherent points between two SAR acquisitions. *ISPRS J. Photogramm. Remote Sens.* **2011**, *66*, 146–152. [CrossRef]
25. Zhang, L.; Lu, Z.; Ding, X.; Jung, H.s.; Feng, G.; Lee, C.W. Mapping ground surface deformation using temporarily coherent point SAR interferometry: Application to Los Angeles Basin. *Remote Sens. Environ.* **2012**, *117*, 429–439. [CrossRef]
26. Dong, J.; Liao, M.; Xu, Q.; Zhang, L.; Tang, M.; Gong, J. Detection and displacement characterization of landslides using multi-temporal satellite SAR interferometry: A case study of Danba County in the Dadu River Basin. *Eng. Geol.* **2018**, *240*, 95–109. [CrossRef]
27. He, K.; Li, J.; Li, B.; Zhao, Z.; Zhao, C.; Gao, Y.; Liu, Z. The Pingdi landslide in Shuicheng, Guizhou, China: Instability process and initiation mechanism. *Bull. Eng. Geol. Environ.* **2022**, *81*, 1–17. [CrossRef]
28. Shi, X.; Xu, Q.; Zhang, L.; Zhao, K.; Dong, J.; Jiang, H.; Liao, M. Surface displacements of the Heifangtai terrace in Northwest China measured by X and C-band InSAR observations. *Eng. Geol.* **2019**, *259*, 105181. [CrossRef]
29. Zhang, J.; Zhu, W.; Cheng, Y.; Li, Z. Landslide Detection in the Linzhi–Ya'an Section along the Sichuan–Tibet Railway Based on InSAR and Hot Spot Analysis Methods. *Remote Sens.* **2021**, *13*, 3566. [CrossRef]
30. Zhou, C.; Cao, Y.; Yin, K.; Wang, Y.; Shi, X.; Catani, F.; Ahmed, B. Landslide characterization applying sentinel-1 images and InSAR technique: The muyubao landslide in the three Gorges Reservoir Area, China. *Remote Sens.* **2020**, *12*, 3385. [CrossRef]
31. Dong, J.; Zhang, L.; Li, M.; Yu, Y.; Liao, M.; Gong, J.; Luo, H. Measuring precursory movements of the recent Xinmo landslide in Mao County, China with Sentinel-1 and ALOS-2 PALSAR-2 datasets. *Landslides* **2018**, *15*, 135–144. [CrossRef]
32. Intrieri, E.; Raspini, F.; Fumagalli, A.; Lu, P.; Del Conte, S.; Farina, P.; Allievi, J.; Ferretti, A.; Casagli, N. The Maoxian landslide as seen from space: Detecting precursors of failure with Sentinel-1 data. *Landslides* **2018**, *15*, 123–133. [CrossRef]
33. Kuang, J.; Ng, A.H.M.; Ge, L. Displacement Characterization and Spatial-Temporal Evolution of the 2020 Aniangzhai Landslide in Danba County Using Time-Series InSAR and Multi-Temporal Optical Dataset. *Remote Sens.* **2022**, *14*, 68. [CrossRef]
34. Li, M.; Zhang, L.; Dong, J.; Tang, M.; Shi, X.; Liao, M.; Xu, Q. Characterization of pre-and post-failure displacements of the Huangnibazi landslide in Li County with multi-source satellite observations. *Eng. Geol.* **2019**, *257*, 105140. [CrossRef]
35. Lu, Z.; Kim, J. A Framework for Studying Hydrology-Driven Landslide Hazards in Northwestern US Using Satellite InSAR, Precipitation and Soil Moisture Observations: Early Results and Future Directions. *GeoHazards* **2021**, *2*, 17–40. [CrossRef]
36. Ciampalini, A.; Raspini, F.; Lagomarsino, D.; Catani, F.; Casagli, N. Landslide susceptibility map refinement using PSInSAR data. *Remote Sens. Environ.* **2016**, *184*, 302–315. [CrossRef]
37. Scaioni, M.; Longoni, L.; Melillo, V.; Papini, M. Remote sensing for landslide investigations: An overview of recent achievements and perspectives. *Remote Sens.* **2014**, *6*, 9600–9652. [CrossRef]
38. Sun, D.; Gu, Q.; Wen, H.; Shi, S.; Mi, C.; Zhang, F. A Hybrid Landslide Warning Model Coupling Susceptibility Zoning and Precipitation. *Forests* **2022**, *13*, 827. [CrossRef]
39. Iglesias, R.; Mallorqui, J.J.; López-Dekker, P. DInSAR pixel selection based on sublook spectral correlation along time. *IEEE Trans. Geosci. Remote Sens.* **2013**, *52*, 3788–3799. [CrossRef]
40. Massonnet, D.; Rossi, M.; Carmona, C.; Adragna, F.; Peltzer, G.; Feigl, K.; Rabaute, T. The displacement field of the Landers earthquake mapped by radar interferometry. *Nature* **1993**, *364*, 138–142. [CrossRef]
41. Ng, A.H.M.; Ge, L.; Yan, Y.; Li, X.; Chang, H.C.; Zhang, K.; Rizos, C. Mapping accumulated mine subsidence using small stack of SAR differential interferograms in the Southern coalfield of New South Wales, Australia. *Eng. Geol.* **2010**, *115*, 1–15. [CrossRef]

42. Zhang, K.; Li, Z.; Meng, G.; Dai, Y. A very fast phase inversion approach for small baseline style interferogram stacks. *ISPRS J. Photogramm. Remote Sens.* **2014**, *97*, 1–8. [CrossRef]
43. Goldstein, R.M.; Werner, C.L. Radar interferogram filtering for geophysical applications. *Geophys. Res. Lett.* **1998**, *25*, 4035–4038. [CrossRef]
44. Costantini, M. A novel phase unwrapping method based on network programming. *IEEE Trans. Geosci. Remote Sens.* **1998**, *36*, 813–821. [CrossRef]
45. Chen, C.W.; Zebker, H.A. Two-dimensional phase unwrapping with use of statistical models for cost functions in nonlinear optimization. *JOSA A* **2001**, *18*, 338–351. [CrossRef] [PubMed]
46. Zhang, K.; Ge, L.; Hu, Z.; Ng, A.H.M.; Li, X.; Rizos, C. Phase unwrapping for very large interferometric data sets. *IEEE Trans. Geosci. Remote Sens.* **2011**, *49*, 4048–4061. [CrossRef]
47. Leroueil, S. Geotechnics of slopes before failure. *Landslides Eval. Stab. Lacerda Ehrlich Fontoura Sayao (Eds)* **2004**, *1*, 863–884.

forests

MDPI

Communication

Nucleation Process of the 2017 Nuugaatsiaq, Greenland Landslide

Zhenwei Guo [1,2,3], Xinrong Hou [1,2,3], Dawei Gao [1,2,3,*] and Jianxin Liu [1,2,3]

1 Key Laboratory of Metallogenic Prediction of Nonferrous Metals and Geological Environment Monitoring, Ministry of Education, Changsha 410083, China
2 Key Laboratory of Non-Ferrous Resources and Geological Hazard Detection, Changsha 410083, China
3 School of Geosciences and Info-Physics, Central South University, Changsha 410083, China
* Correspondence: daweigao@csu.edu.cn

Abstract: Seismic precursors prior to the failure of rocks are essential for probing the nucleation process and mitigating hazards. However, such precursory events before large landslides are rarely reported possibly due to the lack of near-source observations. The 2017 Nuugaatsiaq, Greenland landslide that was preceded by an abundance of small earthquakes and captured by a local seismic station is a notable exception and offers us a valuable opportunity to investigate how a large landslide initiated. Prior work suggests that accelerated creeping plays an important role during the landslide nucleation process. However, by analyzing the temporal evolution of the waveform similarities, waveform amplitudes, and inter-event times of the seismic precursors, we find that the Nuugaatsiaq landslide was very likely triggered by a series of accelerated and migratory small earthquakes approaching the nucleation area of the upcoming landslide, thus providing important insights into the failure initiation of massive landslides.

Keywords: nucleation process; landslide; waveform similarity; repeating earthquakes; neighboring earthquakes

1. Introduction

Precursory signals preceding catastrophic failure of brittle rocks such as earthquakes and landslides are of great importance in providing critical insights into the nucleation process and, hence, are essential for hazard prediction and mitigation [1–5]. Although foreshocks before significant earthquakes are widely reported and well-recognized [1,4,6,7], seismic precursors prior to a large landslide are rarely documented [8], in part because these signals are small, and unfortunately near-source observation is typically rare. One notable exception is the widely studied 2017 Nuugaatsiaq, Greenland landslide (Figure 1) with abundant seismic precursors (Figure 2) captured by a local seismic station NUUG (Figure 1), at a ~30 km distance [8,9]. This landslide occurred on the evening of 17 June and generated the largest documented tsunami wave (runup height ~90 m) in Greenland to date and caused serious casualties [10].

Although the seismic precursors of the Nuugaatsiaq landslide are of weak amplitudes (Figure 2), their waveforms are highly similar as revealed by both the conventional match filtering (MF) method [8,9] and unsupervised deep learning [11]. Based on the similar waveforms, the seismic precursors are believed to be stick-slip repeating events [8,9,11] and, hence, are interpreted as the manifestation of aseismic slip on the failure surface responsible for the occurrence of the massive Nuugaatsiaq landslide, according to an earlier study [9]. However, a growing body of literature suggests that high waveform similarity may only imply close source areas [1,12–14] and/or similar focal mechanisms [15], but not necessarily repeated ruptures.

Citation: Guo, Z.; Hou, X.; Gao, D.; Liu, J. Nucleation Process of the 2017 Nuugaatsiaq, Greenland Landslide. *Forests* **2023**, *14*, 2. https://doi.org/10.3390/f14010002

Academic Editor: Junwei Ma

Received: 15 November 2022
Revised: 13 December 2022
Accepted: 17 December 2022
Published: 20 December 2022

Figure 1. Map of our study area. Red star and lime triangle mark the Nuugaatsiaq landslide and seismic station NUUG, respectively.

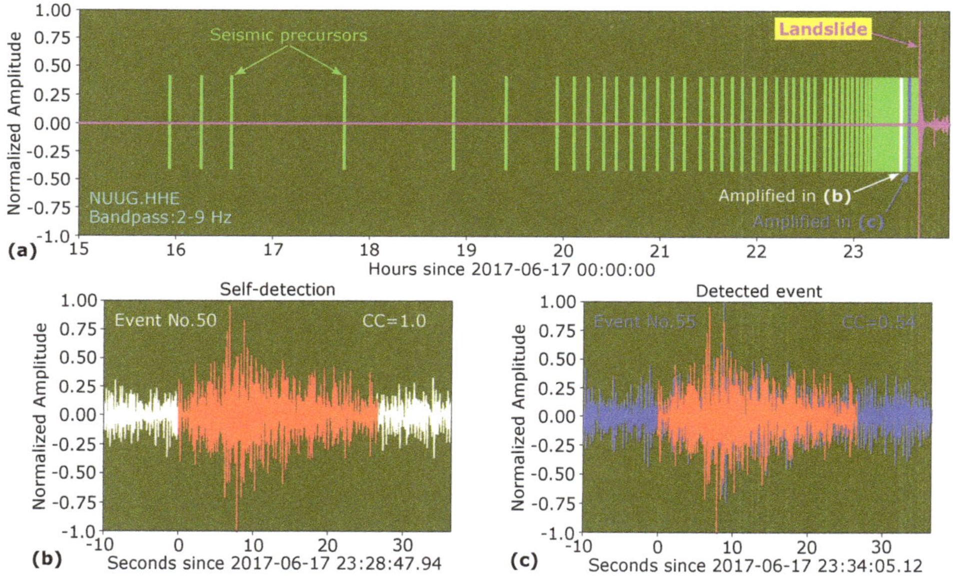

Figure 2. Seismic waveform. (**a**) East component (fuchsia line) of station NUUG. The waveform is normalized and band-pass filtered between 2 and 9 Hz [9]. Colored (lime, white, and blue) vertical bars denote the seismic precursors. (**b**) Self-detection (event No. 50). (**c**) An example of detected event (No. 55). For both (**b**,**c**), the template waveform (event No. 50, red line) is superposed at the location of the best match according to the MFMC method.

Therefore, we revisit the 2017 Nuugaatsiaq landslide with a recently developed match-filtering with multi-segment cross-correlation (MFMC) technique [16]. This technique quantifies the waveform similarity through an averaged cross-correlation coefficient (CC) by equally incorporating the contributions from various segments of the waveforms (see Section 2) and, hence, is very sensitive to the spatial difference between closely spaced seismic sources [16]. In this study, we first employ the MFMC technique to scan through the 24 h data of station NUUG (Figure 1) before the landslide to construct a reliable dataset of seismic precursors. Then we analyze the temporal evolution of the CC values, waveform amplitudes, and inter-event times of the seismic precursors. Our observations indicate that the 2017 Nuugaatsiaq landslide was very likely triggered by a cascade process with a series of small seismic events approaching the nucleation area of the upcoming landslide, providing important insights into the failure initiation of large landslides.

2. Methods

To identify seismic precursors, we utilize the MFMC technique [16] instead of the conventional MF method which can be severely biased by the presence of large-amplitude phases (e.g., S wave and surface waves) [16–19]. Compared with the conventional MF method with only one segment, the MFMC technique divides the template into a series of consecutive segments during the cross-correlation process. Such a procedure aims to mitigate the influence of the large-amplitude phases and essentially assigns more weights to important low-amplitude phases such as depth phases [20,21] and coda waves [22,23] which contain additional source location information. Hence, the MFMC technique is more reliable in differentiating the source location difference between earthquakes with similar waveforms [16].

In this study, we choose the same template (i.e., event No. 50 in Figure 2b) as a previous study [9] because this event has relatively high signal-to-noise ratio (SNR) (namely, can be visually identified) and more importantly it is isolated from other seismic precursors in the waveform time series [9]. Although the S-P times of the seismic precursors are only about 4.6 s [9], the precursory signals have long surface wave trains (Figure 2b,c) [9,11] due to the shallow source depths. Here we set the template window length (T_{win}) to be 26.5 s covering much of the signal, similar to the choice of prior work [9]. In our MFMC CC calculation, we first band-pass filter the seismic data between 2 and 9 Hz to mitigate the impact of noise [9]. Then we split the template waveform into N_{seg} segments of equal length where N_{seg} is determined by the cycles of the longest period wave ($1/f_{min}$) in the band-pass filtered waveform (i.e., $N_{seg} = T_{win} \times f_{min}$) [16]. Finally, we shift all the segments together one sample point at a time along the continuous waveform. The cross-correlation calculation is performed individually for each segment, and the CC value at each sample point is determined by the average of all segments. Once the computed CC exceeds a certain threshold, an event is declared.

3. Results

In total, we have identified 72 seismic precursors (Figure 2) with the classical detection threshold of 8 times the median absolute deviation (MAD) [24–27] with the MFMC technique. Note the number of identified precursors is slightly fewer than that in prior work [9] as our method only detects events which are located closely to the template event. Overall, the CC values of the detected precursors are low mainly because the precursor signals are of small amplitudes (Figure 2). If we take the template event No. 50 (see Section 2) as the reference, it is quite interesting to see that both the CC values (i.e., waveform similarity) and waveform amplitudes of the seismic precursors initially increase towards the reference event (Figure 3a,b). Previous studies [8,9] hypothesize that the seismic events are repeating earthquakes repeatedly initiated from the same asperity, yet with a growing rupture dimension (Figure 4a). Therefore, the increase of waveform similarity may be due to the rise of signal-to-noise ratio (SNR) [11]. However, this hypothesis does not hold as the CC values and the waveform amplitudes exhibit obviously contrasting trends after the reference event (Figure 3a,b).

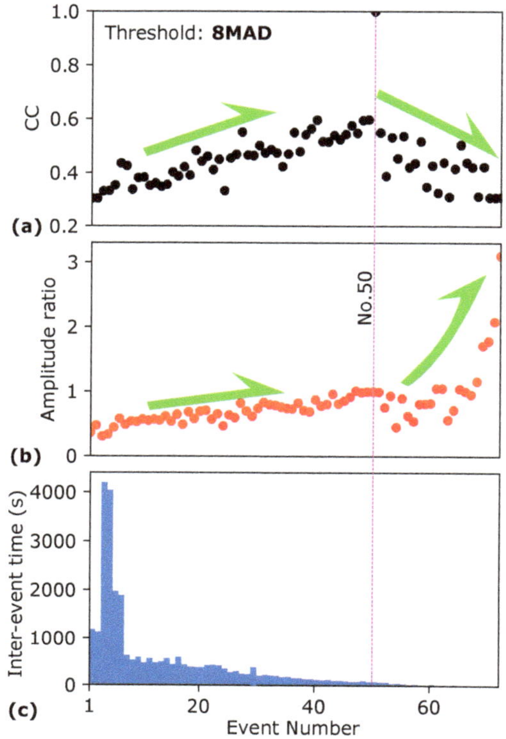

Figure 3. Characteristics of the seismic precursors. (**a**) CC values derived from MFMC. (**b**) Amplitude ratio of the seismic precursors with respect to the reference event No. 50. (**c**) Inter-event times of the seismic precursors. The dashed vertical fuchsia line marks the reference event (No. 50).

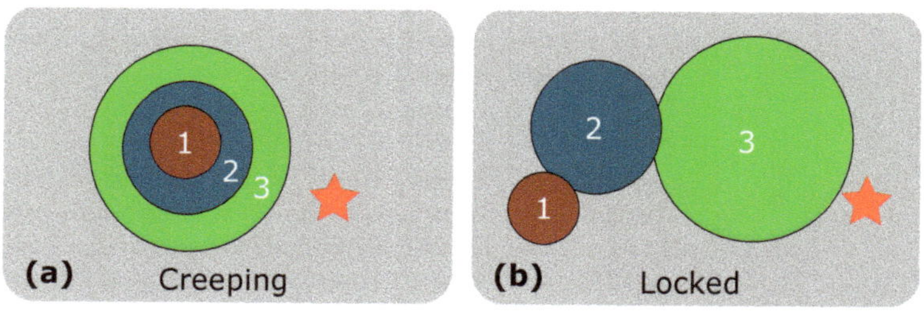

Figure 4. Schematic illustration of the hypothetical nucleation models. (**a**) Aseismic triggering hypothesis [9]. (**b**) Accelerated and migratory triggering hypothesis. In both (**a**,**b**), colored circles represent the seismic precursors; red star marks the nucleation point of the landslide.

Notice that the recurrence times of well-studied repeating earthquakes are widely considered to be on the order of months/years [28–30]. Hence, it is extremally difficult to imagine that the slip surface can repeatedly fail and heal tens of times within only a few hours (Figure 2a). Moreover, recurrence times of true repeating events are nearly constant [28,29] yet the inter-event times of the seismic precursors observed here drop dramatically towards the occurrence of the large landslide (Figure 3c).

Given the similar seismic waveforms, increasing waveform amplitudes, and extremely short and obviously decreasing inter-event times (Figure 3), our observations suggest that the seismic precursors are very likely to be neighboring events progressively triggering larger weak asperities with barely overlapped source areas (Figure 4b). Thus, the overall trend of CC changes can be fully explained by the migration of the seismic precursors, namely that the CC gets higher when the neighboring events migrate towards the reference event, and vice versa [12,13]. Moreover, both the rising of waveform amplitudes and the shortening of inter-event times can be explained by the accelerating of the inter-event triggering of larger asperities before the landslide, as the sliding surface overall is becoming more and more unstable. Finally, we note that using a detection threshold slightly lower (7.5 MAD) or higher (8.5 MAD) yields similar results and does not change our conclusion.

4. Discussion and Conclusions

Both repeating and neighboring earthquakes are characterized by highly similar waveforms, yet they have totally opposite implications for the nucleation process of catastrophic failure [1,3,12,13]. To unambiguously differentiate these two kinds of events would require sufficient near-source observations to precisely locate the seismic source and calculate the rupture dimension [12,13,31]. With limited data, prior works [8,9,11] have interpretated the seismic precursors as repeating events simply based on waveform similarity. However, recent studies have shown that waveform similarity alone is insufficient to identify true repeating earthquakes [1,12,13] and similar seismic signals with very short inter-event times are commonly taken as neighboring events [12,32].

Although we cannot resolve the source properties of the seismic precursors (e.g., source location, rupture dimension, and stress perturbation) given the single-station waveform data with poor SNR, the observed temporal evolution of the CC values, waveform amplitudes, and inter-event times provides compelling evidence that a series of accelerated and migratory seismic precursors occurred immediately before the 2017 Nuugaatsiaq landslide inherently suggesting a causal relationship likely through a cascade of stress perturbation between neighboring asperities. Notice that our inferred cascade of stress transfer triggering process (Figure 4b) has also been documented to be responsible for some large earthquakes such as the 1999 Mw 7.6 Izmit (Turkey) [1] and 1999 Mw 7.1 Hector Mine (USA) [6]. Our findings bring important insights into the nucleation process of landslides, suggesting that massive landslides may initiate in a similar way as many large earthquakes. However, we note that whether our hypothesis can be generalized to other landslides remains to be tested. Finally, our study highlights the significance of near-source observations in capturing weak precursor signals. Monitoring these tiny signals may not only improve our understanding of the catastrophic failure initiation process but also contribute to hazard preparedness.

Author Contributions: Conceptualization: D.G.; investigation: Z.G., X.H. and D.G.; writing—original draft: Z.G. and X.H.; writing—review & editing: Z.G., X.H., D.G. and J.L. All authors have read and agreed to the published version of the manuscript.

Funding: This research is jointly funded by Open Research Fund Program of Key Laboratory of Metallogenic Prediction of Nonferrous Metals and Geological Environment Monitoring (Central South University), Ministry of Education (Grant No. 2022YSJS16) and National Natural Science Foundation of China (Grant Nos. 42130810 and 42204067).

Data Availability Statement: Waveform data used in this study were downloaded from the Incorporated Research Institutions for Seismology (http://ds.iris.edu/ds/nodes/dmc/, last accessed on 12 July 2022). Seismic data are processed with Obspy [33]. Figures are made with GeoMapApp [34] and Matplotlib [35].

Acknowledgments: We thank two anonymous reviewers for their constructive comments.

Conflicts of Interest: The authors declare no conflict of interest.

References

1. Ellsworth, W.L.; Bulut, F. Nucleation of the 1999 Izmit earthquake by a triggered cascade of foreshocks. *Nat. Geosci.* **2018**, *11*, 531–535. [CrossRef]
2. Tape, C.; Holtkamp, S.; Silwal, V.; Hawthorne, J.; Kaneko, Y.; Ampuero, J.P.; Ji, C.; Ruppert, N.; Smith, K.; West, M.E. Earthquake nucleation and fault slip complexity in the lower crust of central Alaska. *Nat. Geosci.* **2018**, *11*, 536–541. [CrossRef]

3. Gomberg, J. Unsettled earthquake nucleation. *Nat. Geosci.* **2018**, *11*, 463–464. [CrossRef]
4. Zhu, G.; Yang, H.; Tan, Y.J.; Jin, M.; Li, X.; Yang, W. The cascading foreshock sequence of the Ms 6.4 Yangbi Earthquake in Yunnan, China. *Earth Planet. Sci. Lett.* **2022**, *591*, 117594. [CrossRef]
5. Wang, K.; Chen, Q.-F.; Sun, S.; Wang, A. Predicting the 1975 Haicheng earthquake. *Bull. Seismol. Soc. Am.* **2006**, *96*, 757–795. [CrossRef]
6. Yoon, C.E.; Yoshimitsu, N.; Ellsworth, W.L.; Beroza, G.C. Foreshocks and mainshock nucleation of the 1999 M w 7.1 Hector Mine, California, Earthquake. *J. Geophys. Res. Solid Earth* **2019**, *124*, 1569–1582. [CrossRef]
7. Meng, H.; Fan, W. Immediate foreshocks indicating cascading rupture developments for 527 M 0.9 to 5.4 Ridgecrest earthquakes. *Geophys. Res. Lett.* **2021**, *48*, e2021GL095704. [CrossRef]
8. Bell, A.F. Predictability of landslide timing from quasi-periodic precursory earthquakes. *Geophys. Res. Lett.* **2018**, *45*, 1860–1869. [CrossRef]
9. Poli, P. Creep and slip: Seismic precursors to the Nuugaatsiaq landslide (Greenland). *Geophys. Res. Lett.* **2017**, *44*, 8832–8836. [CrossRef]
10. Strzelecki, M.C.; Jaskólski, M.W. Arctic tsunamis threaten coastal landscapes and communities–survey of Karrat Isfjord 2017 tsunami effects in Nuugaatsiaq, western Greenland. *Nat. Hazards Earth Syst. Sci.* **2020**, *20*, 2521–2534. [CrossRef]
11. Seydoux, L.; Balestriero, R.; Poli, P.; de Hoop, M.; Campillo, M.; Baraniuk, R. Clustering earthquake signals and background noises in continuous seismic data with unsupervised deep learning. *Nat. Commun.* **2020**, *11*, 3972. [CrossRef] [PubMed]
12. Gao, D.; Kao, H.; Wang, B. Misconception of waveform similarity in the identification of repeating earthquakes. *Geophys. Res. Lett.* **2021**, *48*, e2021GL092815. [CrossRef]
13. Gao, D. Comprehensive Study of Seismic Waveform Similarity: Applications to Reliable Identification of Repeating Earthquakes and Investigations of Detailed Source Process of Induced Seismicity. Ph.D. Dissertation, University of Victoria, Victoria, BC, Canada, 2021.
14. Cheng, X.; Niu, F.; Silver, P.G.; Horiuchi, S.; Takai, K.; Iio, Y.; Ito, H. Similar microearthquakes observed in western Nagano, Japan, and implications for rupture mechanics. *J. Geophys. Res. Solid Earth* **2007**, *112*, B04306. [CrossRef]
15. Kilb, D.; Rubin, A. Implications of diverse fault orientations imaged in relocated aftershocks of the Mount Lewis, ML 5.7, California, earthquake. *J. Geophys. Res. Solid Earth* **2002**, *107*, ESE 5-1–ESE 5-17. [CrossRef]
16. Gao, D.; Kao, H. Optimization of the match-filtering method for robust repeating earthquake detection: The multisegment cross-correlation approach. *J. Geophys. Res. Solid Earth* **2020**, *125*, e2020JB019714. [CrossRef]
17. Kraft, T.; Deichmann, N. High-precision relocation and focal mechanism of the injection-induced seismicity at the Basel EGS. *Geothermics* **2014**, *52*, 59–73. [CrossRef]
18. Myhill, R.; McKenzie, D.; Priestley, K. The distribution of earthquake multiplets beneath the southwest Pacific. *Earth Planet. Sci. Lett.* **2011**, *301*, 87–97. [CrossRef]
19. Li, L.; Niu, F.; Chen, Q.-F.; Su, J.; He, J. Post-seismic velocity changes along the 2008 M 7.9 Wenchuan earthquake rupture zone revealed by S coda of repeating events. *Geophys. Suppl. Mon. Not. R. Astron. Soc.* **2016**, *208*, 1237–1249. [CrossRef]
20. Ma, S.; Atkinson, G.M. Focal depths for small to moderate earthquakes (m N$\geq$ 2.8) in Western Quebec, Southern Ontario, and Northern New York. *Bull. Seismol. Soc. Am.* **2006**, *96*, 609–623. [CrossRef]
21. Ma, S. Focal depth determination for moderate and small earthquakes by modeling regional depth phases sPg, sPmP, and sPn. Bull. *Seismol. Soc. Am.* **2010**, *100*, 1073–1088. [CrossRef]
22. Snieder, R.; Vrijlandt, M. Constraining the source separation with coda wave interferometry: Theory and application to earthquake doublets in the Hayward fault, California. *J. Geophys. Res. Solid Earth* **2005**, *110*, B04301. [CrossRef]
23. Robinson, D.; Sambridge, M.; Snieder, R. A probabilistic approach for estimating the separation between a pair of earthquakes directly from their coda waves. *J. Geophys. Res. Solid Earth* **2011**, *116*, B04309. [CrossRef]
24. Shelly, D.R.; Beroza, G.C.; Ide, S. Non-volcanic tremor and low-frequency earthquake swarms. *Nature* **2007**, *446*, 305–307. [CrossRef] [PubMed]
25. Shelly, D.R.; Skoumal, R.J.; Hardebeck, J.L. S/P Amplitude Ratios Derived from Single-Component Seismograms and Their Potential Use in Constraining Focal Mechanisms for Microearthquake Sequences. *Seism. Rec.* **2022**, *2*, 118–126. [CrossRef]
26. Thomas, A.M.; Bostock, M.G. Identifying low-frequency earthquakes in central Cascadia using cross-station correlation. *Tectonophysics* **2015**, *658*, 111–116. [CrossRef]
27. Kurihara, R.; Obara, K.; Takeo, A.; Maeda, T. Migration of deep low-frequency tremor triggered by teleseismic earthquakes in the southwest Japan Subduction Zone. *Geophys. Res. Lett.* **2018**, *45*, 3413–3419. [CrossRef]
28. Uchida, N.; Matsuzawa, T.; Ellsworth, W.L.; Imanishi, K.; Shimamura, K.; Hasegawa, A. Source parameters of microearthquakes on an interplate asperity off Kamaishi, NE Japan over two earthquake cycles. *Geophys. J. Int.* **2012**, *189*, 999–1014. [CrossRef]
29. Abercrombie, R.E. Stress drops of repeating earthquakes on the San Andreas fault at Parkfield. *Geophys. Res. Lett.* **201**, *41*, 8784–8791. [CrossRef]
30. Bohnhoff, M.; Wollin, C.; Domigall, D.; Küperkoch, L.; Martínez-Garzón, P.; Kwiatek, G.; Dresen, G.; Malin, P.E. Repeatin Marmara Sea earthquakes: Indication for fault creep. *Geophys. J. Int.* **2017**, *210*, 332–339. [CrossRef]
31. Waldhauser, F.; Schaff, D.P. A comprehensive search for repeating earthquakes in northern California: Implications for fault cree, slip rates, slip partitioning, and transient stress. *J. Geophys. Res. Solid Earth* **2021**, *126*, e2021JB022495. [CrossRef]

32. Li, L.; Chen, Q.f.; Niu, F.; Su, J. Deep slip rates along the Longmen Shan fault zone estimated from repeating microearthquakes. *J. Geophys. Res. Solid Earth* **2011**, *116*, B09310. [CrossRef]
33. Beyreuther, M.; Barsch, R.; Krischer, L.; Megies, T.; Behr, Y.; Wassermann, J. ObsPy: A Python toolbox for seismology. *Seismol. Res. Lett.* **2010**, *81*, 530–533. [CrossRef]
34. Ryan, W.B.; Carbotte, S.M.; Coplan, J.O.; O'Hara, S.; Melkonian, A.; Arko, R.; Weissel, R.A.; Ferrini, V.; Goodwillie, A.; Nitsche, F. Global multi-resolution topography synthesis. *Geochem. Geophys. Geosyst.* **2009**, *10*, Q03014. [CrossRef]
35. Hunter, J.D. Matplotlib: A 2D graphics environment. *Comput. Sci. Eng.* **2007**, *9*, 90–95. [CrossRef]

forests

MDPI

Article

Effect of Particle Form and Surface Friction on Macroscopic Shear Flow Friction in Particle Flow System

Yu Huang [1,2,*], Yi'an Wang [1] and Suran Wang [1]

1 Department of Geotechnical Engineering, College of Civil Engineering, Tongji University, Shanghai 200092, China; mxxc007@163.com (Y.W.); wangsuran@tongji.edu.cn (S.W.)
2 Key Laboratory of Geotechnical and Underground Engineering of the Ministry of Education, Tongji University, Shanghai 200092, China
* Correspondence: yhuang@tongji.edu.cn; Tel.: +86-21-6598-2384; Fax: +86-21-6598-5210

Abstract: The damage caused by landslide disasters is very significant. Among them, landslides after forest fires have been widely concerned by scholars in recent years due to their particular physical and chemical properties. This large-scale shear flow of particulate matter has similarities to fluid systems. However, due to the discontinuity of the particle system, its flow process has significant random characteristics. To investigate the random properties of particle systems, this study conducted a series of ring shear tests on four particle systems. The effects of the particle shape, normal stress, and shear velocity on the systems' shear rheological features were investigated using experimental data. The particle form has an important effect on the macroscopic properties of the system. In a spherical particle system, the macroscopic friction fluctuation is determined by the friction of the particle surface and the system's normal stress. The shear velocity has a minor effect on this characteristic. Three elements simultaneously influence the macroscopic friction fluctuation of a breccia particle system: the particle surface friction, system normal stress, and shear velocity. The origins of macroscopic frictional fluctuations in particle systems with various shapes are fundamentally distinct. This study contributes to a better understanding of the causes of particle system fluctuations, and establishes the theoretical foundation for the future development of disaster prevention technology.

Keywords: granular flow; ring-shear test; fluctuation characteristics

Citation: Huang, Y.; Wang, Y.; Wang, S. Effect of Particle Form and Surface Friction on Macroscopic Shear Flow Friction in Particle Flow System. *Forests* **2022**, *13*, 1107. https://doi.org/10.3390/f13071107

Academic Editors: Haijia Wen, Weile Li, Chong Xu and Hiromu Daimaru

Received: 12 May 2022
Accepted: 13 July 2022
Published: 14 July 2022

1. Introduction

Avalanches and landslides are examples of the aggregate movement of an enormous number of solid materials exhibiting considerable fluid-like qualities in nature. These ubiquitous particle flow systems can potentially be disastrous. A series of rain-induced landslides occurred in the Mocoa region on 31 March 2017, and resulted in 333 deaths, 398 injuries, and 76 missing people [1]. On 1 July 2017, a landslide occurred in Ningxiang County, Hunan, China, and nine people were killed in the process of searching for survivors [2]. A statistical study considering the Chittagong area in Bangladesh reported that 730 landslides occurred over 17 years, causing enormous economic losses [3].

Abundant vegetation conditions have been shown in several studies to be of great significance for slowing/suppressing landslides [4,5]. However, forest areas face a very specific and dangerous risk: post-fire mudslides. Researchers used alluvial stratigraphy to reconstruct 32 fire-related alluvial events, indirectly demonstrating the prevalence and hazard of debris flow disasters after major fires [6]. A study on the debris flow after the fire in Australia pointed out that for the same area, the regional peak flow before and after burning can differ by more than 30 times, and the sediment concentration increases by three orders of magnitude [7]. After the fire, the physical and chemical properties of the soil will change greatly [8,9]. How this change in particle properties affects landslides is unclear.

Particle flow occurs naturally in various situations, including landslides and other natural calamities [10,11]. Studies have discovered that particle flow exhibits complicated random properties in practical engineering [12], and is a classic example of an unstable system in a flow process. In practical engineering, the unstable properties exacerbate the difficulty of predicting landslides. Ceccato et al. calculated the impact force of particle flow on stiff retaining walls and discovered that the impact force fluctuation of a discrete system is substantially larger than that of a continuous system. Zhang et al. [13] investigated the flow impact characteristics of several particle systems by conducting an indoor model experiment and discrete element simulation. They discovered that the temporal history of the impact force is significantly unstable. The peak difference in the impact force between two particle systems operating under the same boundary condition may be more than three-fold. Owing to the uneven nature of the granular system's flow mechanism, more precise landslide control is required.

Most recent studies on the flow characteristics of granular systems have focused on the average macroscopic parameters [14,15]. Numerous particle flow constitutive models can accurately describe the velocity distribution [16], friction coefficients [17], blocking circumstances [18], and flow state transitions [19,20] occurring during the system's flow process. With the advance of research on particle flow systems in recent years, the unstable properties of the particle flow systems' macroscopic motion have been increasingly receiving attention. Artoni et al. [21] investigated the velocity fluctuation and self-diffusion impact of particles in dense particle flow through the discrete element simulation of the ring shear model. They concluded that the fluctuation of a single particle is strongly connected to the system's overall velocity. Similar investigations have been conducted by Meng et al. [22], who established that the boundary effect influences the fluctuation of local particles. Lu et al. [23] conducted a discrete element simulation of overland flow and demonstrated that there exists a considerable correlation between the dynamic characteristic fluctuation of the particle flow system and the system particle size. Huang et al. [24] conducted a series of ring shear experiments to investigate the relationship between particle fragmentation and the fluctuations in the macroscopic shear behavior of particle flow systems. However, the source of the variation in the flow properties of a particle system has not been elucidated to date, and experimental data are lacking. Therefore, additional experimental studies are required to fully clarify the variation mechanism.

Most previous studies have established particle flow shear and rheological models by considering simple spherical particle systems. However, research has revealed that the breccia system used in actual engineering has significantly different compressibility, shear friction, and other properties compared with a simple spherical system. Sun et al. [25] reported that, in non-spherical particle systems, the particle motion is mainly caused by sliding friction and energy dissipation resulting from the great increase of particle rotation/dislocation. Jiang et al. [26] conducted ring-shear experiments on glass beads/quartz sand and demonstrated that the volume compressibility, peak strength, shear residual strength, and other properties of particle systems with varying shapes change significantly during the shear rheological process. To better guide engineering practice, it is very important to investigate complex particle systems. A detailed explanation of the effect of the particle characteristics on the macroscopic features of the system can help in understanding the sources of the random motion characteristics of particle flows. This study investigated the effect of particle features on the shear wave behavior of macroscopic systems by conducting 24 groups of annular shear experiments on various materials' particles with different shapes.

2. Experimental Design of Ring Shear Test

This study used a GCTS ring shear instrument, which can provide significant shear distance without affecting the shear area, and is, therefore, ideal for investigating massive deformation systems. The sample had an exterior diameter of 150 mm, inner diameter of 100 mm, and height of 20 mm to satisfy the H > 10D requirement. Four types of glass beads quartz sand, spherical corundum, and brecciated corundum were used to investigate the effect of the particle form and surface friction on the system's shear characteristics. Corundum is a widely used granular industrial abrasive with consistent chemical properties, high surface roughness, and low cost. Therefore, this unique material was selected to experimentally investigate the high friction of particles. Each sample was initially screened to a particle size of 2–2.5 mm. In the ring shear experiment, the normal stress gradient was adjusted to 100–200–300 kPa, the shear speed was set to 5°/min–30°/min–90°/min, and the shear distance was set to 360°. There are few velocity groupings and normal stress groupings set in this paper. This allows subsequent data analysis to only achieve trend analysis. In future research, we will use experimental or simulated methods to further supplement the data set to give more detailed conclusions. Table 1 lists the experimental groups and their associated numbers. Experimental groups with repeated trials are highlighted in the table. The sampling rate was set to 10–100 Hz, depending on the duration of the experiment. Pre-experiments were conducted to determine the filling quality of each material and ensure that each sample had the same initial height under a positive pressure of 100 kPa. Finally, the weight of each group of glass beads and quartz sand samples was determined as 330 g, whereas the weight of each group of the corundum samples was determined as 212 g. To ensure consistent filling, the zonal filling approach was adopted. Figure 1 shows the experimental equipment, materials and zonal filling method. This figure is modified from a previously published article by the author [24]. After completing the zonal filling, normal stress was applied to achieve system pre-consolidation and retain strong particle contact. Shear stress was added after the system volume had stabilized. Only the stationary shear part of the system was evaluated during data processing.

Table 1. Experimental group summary.

Number	Material	Normal Force	Shear Velocity
G1-3 *	Glass bead	100 kPa	5°/min–30°/min–90°/min
G4-6		200 kPa	5°/min–30°/min–90°/min
G7-9		300 kPa	5°/min–30°/min–90°/min
Q1-3*	Quartz sand	100 kPa	5°/min–30°/min–90°/min
Q4-6		200 kPa	5°/min–30°/min–90°/min
Q7-9		300 kPa	5°/min–30°/min–90°/min
CS1-3	Spherical corundum	100 kPa	5°/min–30°/min–90°/min
CS4-6		200 kPa	5°/min–30°/min–90°/min
CS7-9		300 kPa	5°/min–30°/min–90°/min
CB1-3	Brecciated corundum	100 kPa	5°/min–30°/min–90°/min
CB4-6		200 kPa	5°/min–30°/min–90°/min
CB7-9		300 kPa	5°/min–30°/min–90°/min

* indicates experimental groups with repeated tests.

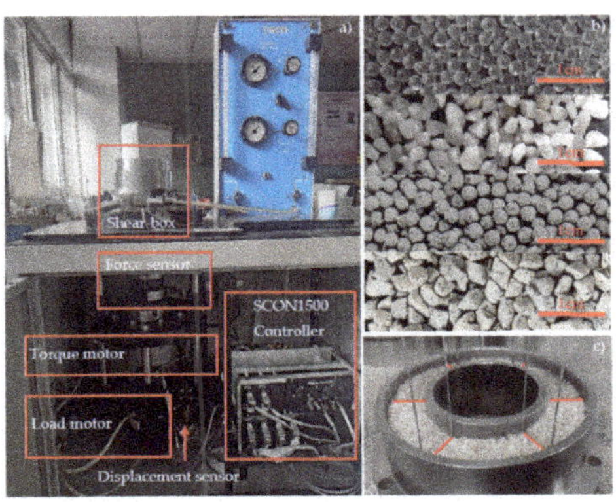

Figure 1. The basic condition of the ring-shear test. (**a**) Structure of GCTS ring-shear apparatus; (**b**) Initial photos of experimental materials; (**c**) Partition filling method. This figure was modified from Huang et al. [24].

3. Analysis of Shear Flow Characteristics of Different Particle Systems

3.1. Compression Characteristics of the Particle System

Figure 2 illustrates the typical shear compression curves of spherical particle systems with different friction coefficients. The shearing trend of the spherical particle system is not substantial, and the overall volume change rate is less than 2%. The compression of the samples mainly occurred in the pre-shear period. The change rate of the sample volume gradually decreased as the shear displacement increased. In the later shear period of the sample with low friction, the volume change rate tends to increase. The particle surface friction is proportional to the particle system's volume change rate and steady-state volume. As the particle surface friction increased, the initial volume changes became slower and the steady-state volume eventually became smaller.

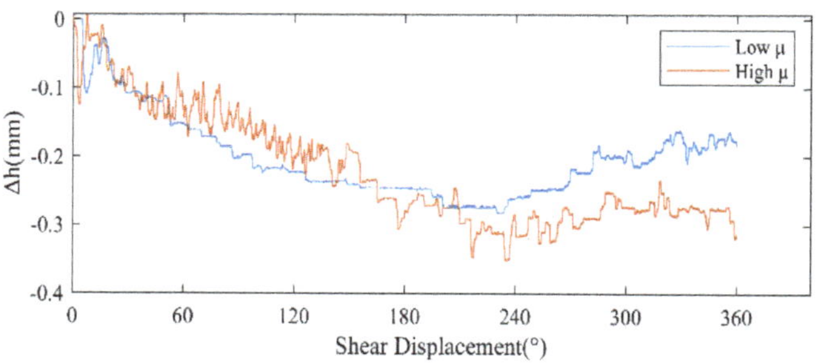

Figure 2. Vertical displacement versus shear displacement of glass beads and spherical corundum at the same shear velocity.

Figure 3 shows the typical shear compression curve of the breccia system with varying friction coefficients. The sample volume changed significantly during the shear process of the angular particle system, and the total volume change rate is approximately 4%–6%. Owing to the pre-treatment of samples, significant particle rearrangement did not occur, and therefore, "early dilatancy—late shear shrinkage" phenomena did not occur. The system's continuous shear shrinkage during shearing is linearly proportional to the shear displacement. For the breccia system, the particle surface friction is proportional to the system's volume change rate. The volume change rate of a high-friction system is much slower than that of a low-friction system.

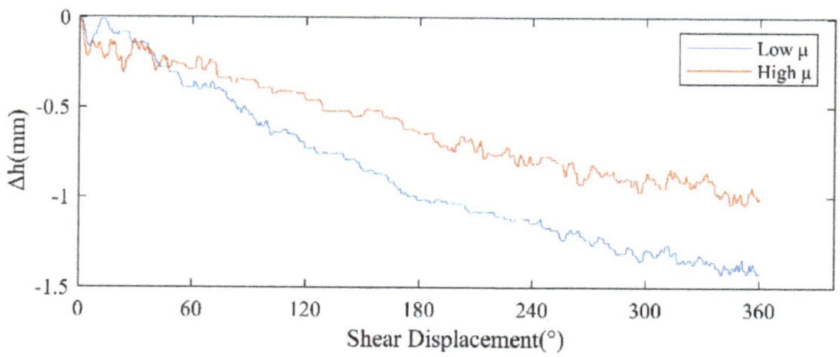

Figure 3. Vertical displacement versus shear displacement of quartz sand and brecciated corundum at the same shear velocity.

According to the particle form, the reasons of shear shrinkage vary, resulting in a variety of shear shrinkage curves for various particle systems. Owing to the highly symmetrical shape of a spherical particle, the volume change of the spherical particle system is governed solely by the particle position distribution. In a complex breccia system, the particle form is complex and the symmetry is inadequate. The volume change of the system is also affected by distribution parameters such as the long axis angle of the particles. In shear motion, the interaction of particles is more intricate. Hence, it is considered that the spherical arrangement can more easily achieve extreme compression. Notably, breccia systems are more prone to particle shear breakage, which results in system volume loss. In practice, however, it is impossible to avoid the inaccuracy induced by particle breakage, and additional research into novel granular materials is required to supplement the experimental results.

3.2. Macroscopic Friction Analysis of Particle System

Ideally, the system boundary conditions should be stable during the investigation of the shear rheological properties of granular systems. However, owing to the feedback-adjustment mechanism of the ring shear instrument's normal stress loading system, the normal stress cannot be kept constant and exhibits highly random variation. The shear stress of a system is directly proportional to the normal stress, and the normal stress fluctuation has a significant effect on the shear stress. In order to avoid such systemic inaccuracies, the subsequent analysis did not directly address shear stress, but instead statistically evaluated the system's macroscopic friction coefficient. Figure 4 shows the correlation analysis of the stress ratio and shear velocity in the spherical particle system. Figure 3a shows the experimental result for the glass bead system. The macroscopic friction coefficient is 0.33–0.39, and has little association with the shear velocity, normal tension, and other boundary conditions. Moreover, there existed substantial divergence between the macroscopic friction and the rest of the data in the low-speed shear experiment. Figure 3b shows the experimental results for the corundum particle system. The macroscopic friction is 0.385–0.41, which is much larger than that of the glass bead system. In a spherical particle system, the particle surface friction dominates the macroscopic friction coefficient because the relationship between macroscopic friction and boundary conditions is weak. The macroscopic shear friction was not affected by boundary conditions such as the shear velocity and normal tension.

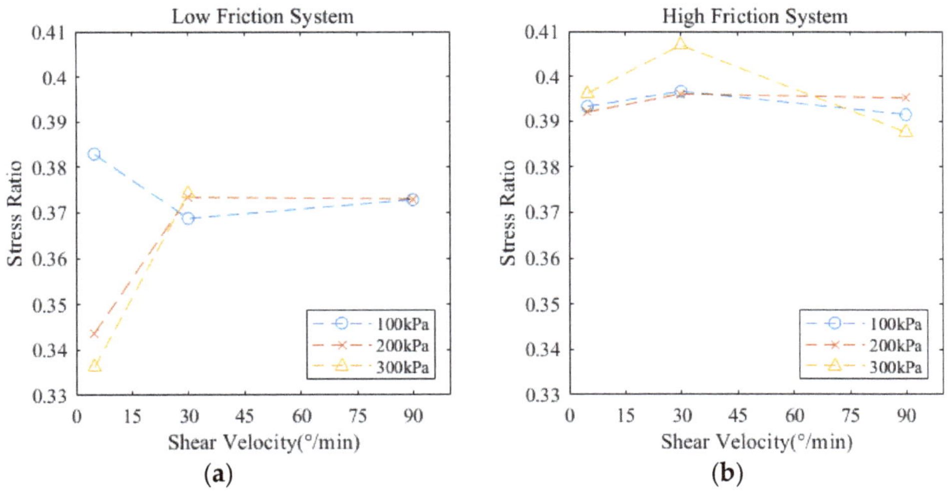

Figure 4. Relationship between shear velocity and stress ratio under different normal stresses. (**a**) Glass bead; (**b**) Spherical corundum.

Figure 5 shows the correlation analysis of the stress ratio and shear velocity in the brecciated particle system, where Figure 5a,b correspond to the quartz sand and brown corundum systems, respectively. There is a modest negative association between the macroscopic friction and shear velocity in breccia particle systems. The normal stress did not obviously influence the macroscopic friction. The macroscopic friction coefficients of the two distinct granular materials are 0.5–0.55. Therefore, the particle surface friction did not affect the macroscopic friction in the diagonal gravel system.

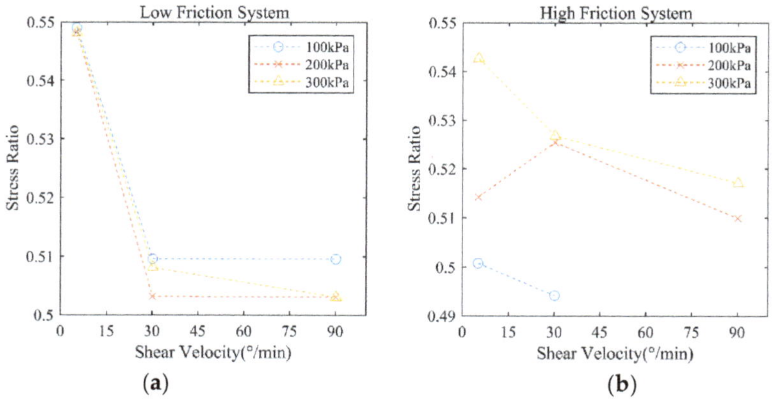

Figure 5. Relationship between shear velocity and stress ratio under different normal stresses. (**a**) Quartz sand; (**b**) Brecciated corundum.

The macroscopic friction of a particle system is mainly determined by the particle form and surface friction, but not by the boundary conditions. The particle form is complicated, and the poor sphericity of the particles increases the system's macroscopic friction. In turn, higher particle surface friction increases the macroscopic friction of a spherical particle system.

3.3. *Analysis of Macroscopic Frictional Fluctuation Characteristics of Particle System*

Figure 6 shows the investigation of the correlation of the macroscopic frictional fluctuation and shear velocity in the spherical particle system, where Figure 6a shows the glass microbead system and Figure 6b shows the brown corundum spherical system. The link between the shear velocity and the macroscopic fluctuation is low in the spherical system. The system fluctuation was significantly dampened by the normal stress. The particle surface friction significantly affects the system's macroscopic friction fluctuation, which is substantially smaller in a low-friction particle system compared with a high-friction particle system.

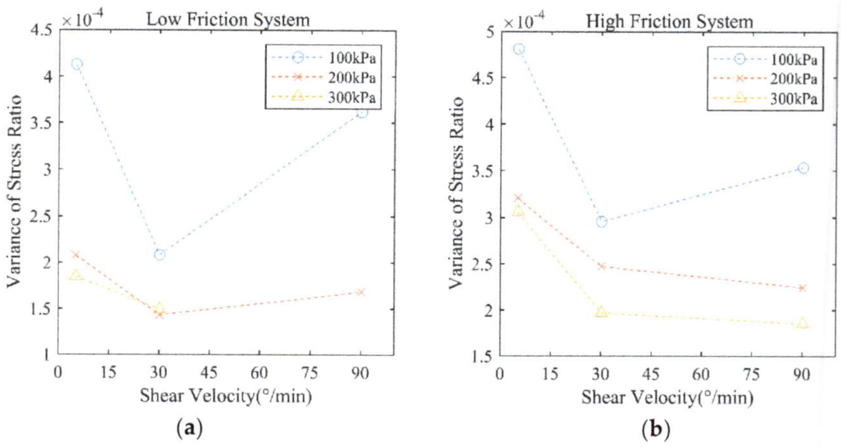

Figure 6. Relationship between shear velocity and the variance of stress ratio under different normal stresses. (**a**) Glass bead; (**b**) Spherical corundum.

Figure 7 shows the correlation analysis of the macroscopic friction fluctuation and shear velocity in the breccia particle system, where Figure 7a shows a quartz sand particle system and Figure 7b shows an angular brown corundum particle system. Higher shear velocity decreases the macroscopic friction fluctuation in breccia systems. Normal stress has a complex effect on the system's macroscopic friction. In a low-friction particle system, the normal stress is inversely proportional to the system's macroscopic friction fluctuation. Higher normal stress suppresses the system fluctuation. However, at V = 30°/min, aberrant behavior was observed, but the reason behind this phenomenon is unknown. In high-friction systems, the normal stress is positively correlated with macroscopic friction fluctuation. Specifically, the normal stress levels increase to the point where they can no longer restrain the system's volatility. Comparative investigation revealed that there exists positive correlation between particle surface friction and the system's macroscopic fluctuation. Particles with a rougher surface exhibit greater system shear fluctuation.

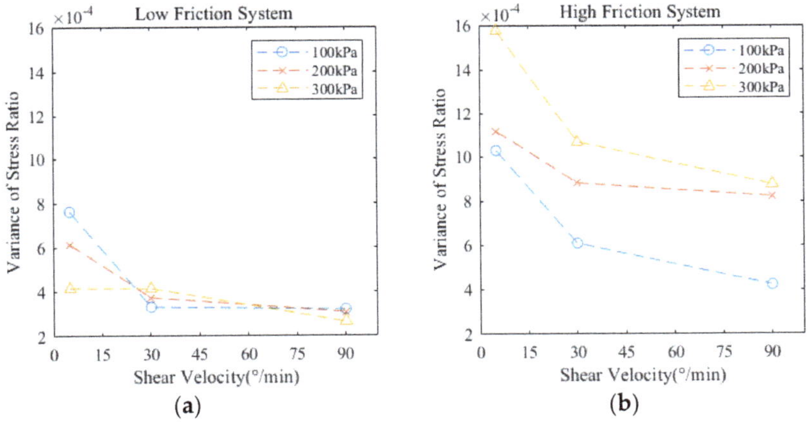

Figure 7. Relationship between shear velocity and the variance of stress ratio under different normal stresses. (**a**) Quartz sand; (**b**) Brecciated corundum.

An additional quantitative investigation revealed that the logarithmic function may be employed to fit the relationship between the macroscopic variation the and shear velocity with remarkable precision.

$$\sigma^2 = Aln(v) + B \tag{1}$$

where A and B are state parameters related to the particle form, surface friction, and stress conditions, respectively. When $A = 0$, the spherical particle system can be considered as a particular solution. The precise expression form of A/B can be further investigated using discrete element simulation and other techniques.

4. Analysis of Causes of Fluctuant Features in Particle Systems

The factors influencing the mean value of macroscopic friction and the fluctuation features of the particle systems were investigated based on the experimental data. The relevant laws are more complicated owing to the numerous factors involved. To simplify the discussion in this section, the pertinent laws obtained as described in the previous section are provided in tabular form. Table 2 summarizes the conclusions drawn for the spherical particle system.

Table 2. Conclusions in the spherical particle system.

	Particle Surface Friction	Normal Stress	Shear Velocity
Macroscopic friction	+	x	x
Macroscopic frictional fluctuation	+	-	x

+ indicates a positive correlation, - indicates a negative connection, and x indicates an irrelevant association.

According to previous studies, macroscopic variation is generated by intergranular split-layer occlusion. The primary cause is thought to be the "particle bite–slip–over–rebite" process. If this factor is significant in a spherical particle system, the macroscopic fluctuation will be positive relative to the shear velocity. However, such positive correlation does not exist in the experimental data obtained by this study. Hence, split-layer occlusion is not responsible for the macroscopic variation in a spherical particle system.

The mesoscopic mechanism of macro fluctuations may be related to the shift between rolling and sliding friction. Owing to the high degree of symmetry of spherical particles, their relative motion is mainly rolling motion. However, sliding friction between particles may exist under specific boundary conditions. This type of local dislocation increases the local stress ratio, which in turn results in the fluctuation of the system's macroscopic friction coefficient. To verify this hypothesis, discrete element modeling must be carried out to capture the system's mesoscopic motion characteristics.

Table 3 summarizes the conclusions drawn for the breccia system.

Table 3. Conclusions in the breccia system.

	Particle Surface Friction	Normal Stress	Shear Velocity
Macroscopic friction	x	x	(-)
Macroscopic frictional fluctuation	+	(+)	-

+ indicates a positive correlation, - indicates a negative connection, x indicates an irrelevant association, and ( ) indicates a weak correlation.

The biting force between particles may be the main factor controlling the macroscopic friction fluctuation in breccia systems. The system fluctuation increases when the interparticle occlusion is tighter. The roughness of the particle surface, which increases in normal stress and decreases in shear velocity, contribute to the increase in the biting force between the particles and system fluctuation. This conclusion is strongly supported by the experimental results. Hence, for the breccia system, the non-stationary "intergranular occlusal–over–reocclusal" process is the primary cause of the system's macroscopic mechanical behavior fluctuation.

This study ignored the influence of particle breakage, which is a limitation of the investigation presented herein. The experimental conditions considered in this study cannot ensure that particles are not shattered in a breccia system. Therefore, numerical tests or the development of new high-strength materials are required to further refine the experimental conclusions.

5. Conclusions

Particle systems have fluid-like properties and exhibit significant instability during the flow process. The physical and chemical properties of the soil will change significantly after forest fires, and the changes in the properties of the particles themselves have an impact on the macroscopic flow characteristics of the system, such as particle shape, particle surface friction, shear boundary conditions, and other factors. This study investigated the causes of fluctuation in four particle systems by conducting a series of ring shear experiments. The consideration of the random properties of particle flow systems in practical engineering can assist in the prevention of landslides and other common geological disasters and provides a theoretical foundation for the development of relevant engineering technology. The following main conclusions were drawn from this study:

(1) The shape of the particles has a significant effect on the trend of volume variation during the system's shear process. Compression limits exist for spherical particle systems. The sample volume of a breccia particle system diminishes as the shear distance increases.

(2) For a spherical particle system, the mean value of the macroscopic friction is independent from the boundary conditions and proportional to the particle surface friction. The macroscopic friction fluctuation of the system is determined by the particle surface friction and the system's normal stress and is completely independent of the shear velocity.

(3) In a breccia system, the mean value of the macroscopic friction is only weakly connected to the shear velocity. Additionally, three elements affect the macroscopic friction fluctuation: particle surface friction, system normal stress, and shear velocity.

(4) The fluctuating macroscopic friction of the spherical particle system is a result of the changing friction mode between particles. The "biting-over-rebiting" mechanism of the interlayer particles is responsible for the macroscopic friction fluctuation of the system in a breccia particle system.

Author Contributions: Conceptualization, Y.H. and S.W.; methodology, Y.W. and S.W.; formal analysis, Y.W.; investigation, Y.W.; resources, Y.H.; data curation, Y.W.; writing—original draft preparation, Y.W.; writing—review and editing, Y.W. and S.W; visualization, Y.W.; funding acquisition, Y.H. All authors have read and agreed to the published version of the manuscript.

Funding: This study was supported by the National Natural Science Foundation of China (Grant No. 42120104008).

Conflicts of Interest: The authors declare no conflict of interest.

References

1. García-Delgado, H.; Machuca, S.; Medina, E. Dynamic and Geomorphic Characterizations of the Mocoa Debris Flow (March 31, 2017, Putumayo Department, Southern Colombia). *Landslides* **2019**, *16*, 597–609. [CrossRef]
2. Zhang, Y.; Li, D.; Yin, K.; Chen, L.; Xu, Y.; Woldai, T.; Lian, Z. The July 1, 2017 Wangjiawan Landslide in Ningxiang County, China. *Landslides* **2018**, *15*, 1657–1662. [CrossRef]
3. Rabby, Y.W.; Li, Y. Landslide Inventory (2001–2017) of Chittagong Hilly Areas, Bangladesh. *Data* **2019**, *5*, 4. [CrossRef]
4. Rickli, C.; Bebi, P.; Graf, F.; Moos, C. Shallow Landslides: Retrospective Analysis of the Protective Effects of Forest and Conclusions for Prediction. In *Recent Advances in Geotechnical Research*; Springer Series in Geomechanics and Geoengineering; Springer: Cham, Switzerland, 2019; pp. 175–185.
5. Grima, N.; Edwards, D.; Edwards, F.; Petley, D.; Fisher, B. Landslides in the Andes: Forests Can Provide Cost-Effective Landslide Regulation Services. *Sci. Total Environ.* **2020**, *745*, 141128. [CrossRef]
6. Bigio, E.R.; Swetnam, T.W.; Pearthree, P.A. Late Holocene Fire-Climate Relationships of the Western San Juan Mountains, Colorado. *Int. J. Wildland Fire* **2017**, *26*, 944–962. [CrossRef]
7. Nyman, P.; Sheridan, G.J.; Smith, H.G.; Lane, P.N.J. Evidence of Debris Flow Occurrence after Wildfire in Upland Catchments of South-East Australia. *Geomorphology* **2011**, *125*, 383–401. [CrossRef]
8. Bowman, D.M.J.S.; Kolden, C.A.; Abatzoglou, J.T.; Johnston, F.H.; van der Werf, G.R.; Flannigan, M. Vegetation Fires in the Anthropocene. *Nat. Rev. Earth Environ.* **2020**, *1*, 500–515. [CrossRef]
9. Alcañiz, M.; Outeiro, L.; Francos, M.; Úbeda, X. Effects of Prescribed Fires on Soil Properties: A Review. *Sci. Total Environ.* **2018**, *613*, 944–957. [CrossRef]
10. Li, Y.R.; Wen, B.P.; Aydin, A.; Ju, N.P. Ring Shear Tests on Slip Zone Soils of Three Giant Landslides in the Three Gorges Project Area. *Eng. Geol.* **2013**, *154*, 106–115. [CrossRef]
11. Zhu, C.; Huang, Y.; Sun, J. Solid-like and Liquid-like Granular Flows on Inclined Surfaces under Vibration—Implications for Earthquake-Induced Landslides. *Comput. Geotech.* **2020**, *123*, 103598. [CrossRef]
12. Gray, J.M.N.T. Particle Segregation in Dense Granular Flows. *Annu. Rev. Fluid Mech.* **2018**, *50*, 407–433. [CrossRef]
13. Zhang, B.; Huang, Y. Unsteady Overflow Behavior of Polydisperse Granular Flows against Closed Type Barrier. *Eng. Geol.* **2021**, *280*, 105959. [CrossRef]
14. Radin, C.; Swinney, H.L. Phases of Granular Matter. *J. Stat. Phys.* **2019**, *175*, 542–553. [CrossRef]
15. Liao, C.C.; Hsiau, S.S.; Hu, Y.M. Density-Driven Sinking Dynamics of a Granular Ring in Sheared Granular Flows. *Adv. Powder Technol.* **2017**, *28*, 2597–2604. [CrossRef]
16. Meng, F.J.; Liu, K. Velocity Fluctuation and Self Diffusion Character in a Dense Granular Sheared Flow Studied by Discrete Element Method. *Acta Phys. Sin.* **2014**, *63*, 134502. [CrossRef]

17. Sun, X.; Kob, W.; Blumenfeld, R.; Tong, H.; Wang, Y.; Zhang, J. Friction-Controlled Entropy-Stability Competition in Granular Systems. *Phys. Rev. Lett.* **2020**, *125*, 268005. [CrossRef]
18. Sun, J.; Sundaresan, S.A. Constitutive Model with Microstructure Evolution for Flow of Rate-Independent Granular Materials. *J. Fluid Mech.* **2011**, *682*, 590–616. [CrossRef]
19. Chialvo, S.; Sun, J.; Sundaresan, S. Bridging the Rheology of Granular Flows in Three Regimes. *Phys. Rev. E* **2012**, *85*, 021305. [CrossRef]
20. Gu, Y.; Ozel, A.; Sundaresan, S. Rheology of Granular Materials with Size Distributions across Dense-Flow Regimes. *Powder Technol.* **2016**, *295*, 322–329. [CrossRef]
21. Artoni, R.; Larcher, M.; Jenkins, J.T.; Richard, P. Self-Diffusion Scalings in Dense Granular Flows. *Soft Matter* **2021**, *17*, 2596–2602. [CrossRef]
22. Meng, F.; Meng, X.; Hua, S.; Ma, S. Fluctuation and Self-Diffusion Research about Dry Granular Materials under Shearing. *J. Braz. Soc. Mech. Sci. Eng.* **2019**, *41*, 153. [CrossRef]
23. Lu, S.; Tang, H.; Zhang, Y.; Gong, W.; Wang, L. Effects of the Particle-Size Distribution on the Micro and Macro Behavior of Soils: Fractal Dimension as an Indicator of the Spatial Variability of a Slip Zone in a Landslide. *Bull. Eng. Geol. Environ.* **2018**, *77*, 665–677. [CrossRef]
24. Huang, Y.; Wang, Y.; Wang, S. Effects of Crushing Characteristics on Rheological Characteristics of Particle Systems. *Water* **2022**, *14*, 532. [CrossRef]
25. Ji, S.; Sun, S.; Chen, X. Shear Cell Test on Transition of Shear Flow States of Granular Materials. *Chin. J. Theor. Appl. Mech.* **2016**, *48*, 1061–1072. [CrossRef]
26. Jiang, Y.; Wang, G.; Kamai, T. Fast Shear Behavior of Granular Materials in Ring-Shear Tests and Implications for Rapid Landslides. *Acta Geotech.* **2017**, *12*, 645–655. [CrossRef]

forests

MDPI

Article

A Case Study on the Energy Capacity of a Flexible Rockfall Barrier in Resisting Landslide Debris

Lei Zhao [1], Lijun Zhang [1], Zhixiang Yu [1,*], Xin Qi [1], Hu Xu [1] and Yifan Zhang [1,2]

[1] School of Civil Engineering, Southwest Jiaotong University, Chengdu 610000, China
[2] Sichuan OST Slope Protection Engineering Co., Ltd., Chengdu 610000, China
* Correspondence: yzxzrq@swjtu.edu.cn; Tel.: +86-13-980-689-658

Abstract: Landslides frequently occur in forest areas with a steep hillside, especially when severely disturbed by human activities. After sustained heavy rainfall, a landslide occurred near the Tianwan tunnel entrance of the Chongqing-Huaihua railway in China. Fortunately, the landslide debris was successfully intercepted by a flexible barrier originally installed to stop rockfalls, which is, to date, the first publicly reported case of landslide debris having been successfully intercepted by a flexible barrier without any damage, in mainland of China. A field investigation was first conducted, and then a back analysis of the landslide mobility and the interaction between the landslide and the flexible barrier was carried out. The back analysis showed that the impact energy was three-times larger than the rated energy capacity of the flexible barrier. It also showed that the elongation of the brake rings and the deflection of the flexible barrier from the numerical simulation was comparable to that from the field measurements. The fact that these brake rings were not elongated to their limit indicated that the capacity of the flexible barrier still had a surplus. Finally, to investigate the maximum energy capacity of a flexible rockfall barrier in resisting landslide debris, parametric analyses of a flexible barrier impacted by landslide debris with different impact energies and velocities were carried out using a coupled ALE-FEM modeling technique. The results showed that the flexible barrier dissipated less than 40% of the total energy of the landslide debris. With an increase of impact energy, the energy dissipation ratio of the flexible barrier decreased linearly. The maximum energy capacity of a flexible rockfall barrier in resisting landslide debris is four-times that of resisting a rockfall.

Keywords: flexible rockfall barrier; energy capacity; landslide debris; field investigation; coupled numerical simulation

Citation: Zhao, L.; Zhang, L.; Yu, Z.; Qi, X.; Xu, H.; Zhang, Y. A Case Study on the Energy Capacity of a Flexible Rockfall Barrier in Resisting Landslide Debris. *Forests* **2022**, *13*, 1384. https://doi.org/10.3390/f13091384

Academic Editors: Haijia Wen, Weile Li, Chong Xu and Hiromu Daimaru

Received: 22 July 2022
Accepted: 29 August 2022
Published: 30 August 2022

1. Introduction

Forests play an important role in the ecological environment, and can effectively improve the stability of slopes and inhibit the occurrence of geological disasters, such as debris flows and landslides [1–4]. However, with the rapid development of transport infrastructure in the western mountainous regions of China, forests in these areas have been severely disturbed, resulting in the frequent occurrence of slope geological hazards, such as landslides, debris flows, and rockfalls (Figure 1). To make matters worse, destructive earthquakes occur frequently in this area, such as the M8.0 Wenchuan earthquake, M7.0 Lushan earthquake, etc. The risk of slope geological hazards after earthquakes in this regions is heightened significantly [5,6]. In addition to traditional rigid barriers [7], such as rigid walls, reinforced dams, etc., flexible barriers, which benefit from rapid construction, easy maintenance, environmental sustainability, and economic competitiveness, are increasingly being considered as a viable measure to deal with slope geological hazards [8]. It can be seen that flexible barriers are capable and help in the implementation of the main sustainable development goals (SDG) employed by the United Nations (UN) Agenda.

Figure 1. Landslide debris occurring near a high-speed railway tunnel exit [9] (Rongjiang, Guizhou, China).

According to the characteristics of flexible barriers, flexible barriers can be categorized into three types: active flexible barriers, passive flexible barriers, and attenuator systems [10]. Active flexible barriers consist of three basic elements; namely, anchors, support cables, and nets, and are used to press the soil from the moment of installation, thus preventing instabilities [11]. Passive flexible barriers are made of a cable net, structural steel posts, and special connecting components, and are used to intercept and stop blocks of rock from bouncing, rolling, and sliding along a slope [12–14]. Attenuator systems are structures made of flexible wire netting, designed to reduce the kinetic energy of a rockfall and guide its trajectory [15,16].

Over the past years, after the flexible barriers that were originally installed to intercept rockfalls were found to have successfully stopped and contained landslide debris [17,18], researchers and engineers have become interested in studying and adopting flexible barriers to mitigate debris flow.

Large-scale tests were conducted to investigate the response of flexible barriers to debris flows. Usually, a varied instrumentation was used to record flow velocities, forces on cable ropes, debris flow character, and barrier response [19–22]. A full-scale test site was also built by WSL in Illgraben, one of the most active debris flow torrents in the Swiss Alps, and V-Barrier systems without any posts were developed and installed in the channel [7]. Small-scale laboratory tests were also carried out to parametrically study the performance of flexible barriers subjected to debris impacts, such as the influence of mesh size, and the gap between the lower barrier edge and the channel's floor [23–27]. Numerical tools and methods, divided into static simulation and dynamic simulation, have been proposed and developed to model flexible barriers under debris impacts. In static models, impact pressures are applied quasi-statically. Due to the large-sliding, large-nonlinear characteristic of flexible barriers, specially developed software, namely FARO [28] and NIDA-MNN [29], are available for capturing the response of flexible barriers. In dynamic models, coupled methods are adopted to simulate the interaction between flexible barriers and debris flows. Useful coupled methods include CFD-DEM [30,31], FEM-DEM-LBM [32], ALE-FEM [33–35], DEM-MPM [36,37], etc.

Benefiting from the above studies, two design methods, namely the force approach and energy approach, have been proposed to design flexible barriers against debris flows. The force approach is the traditional method in the design of structures, which means the structure members are checked and optimized after calculating the internal forces of the structure under specific loads. The core of the force approach lies in the determination of the load, including the distribution and magnitude of the load. Thus, a load model to calculate the loads acting on a flexible barrier has been a research focus [20,22,26,28,38]. However, the value of the dynamic pressure coefficient is empirical and varies greatly, from 0.6 to 5.5 [39]. The energy approach, similar to the design of flexible rockfall barriers, means the barrier is only required to dissipate a certain amount of impact energy [40–42]. The design method was proposed by Wartmann and Salzmann [43] and described in detail by Roth et al. [44]. Compared to the force approach, the energy approach is much simpler and was also adopted in guidelines in Hong Kong [45]. Due to a lack of knowledge, a scaling factor of not exceeding 75% is adopted to reduce the energy capacity of a flexible barrier established by full-scale rockfall tests, in the case of resisting debris flows. However, Song pointed out that less than 10% of the debris impact energy was absorbed by the flexible barrier, and over 90% of the energy was dissipated through the internal and boundary shearing [46].

Due to the difference in material properties and load modes, the energy transformation and dissipation characteristics of a flexible barrier will be significantly different in resisting rockfalls and landslide debris, and it is not sufficient to evaluate the capacity of a flexible barrier in resisting landslide debris just by the ratio of energy dissipated by the flexible barrier, as mentioned above. To date, the technical and scientific knowledge of the assessment of a flexible barrier subjected to a rockfall is relatively mature, and assessment documents have been published and accepted widely [47]. Several test sites have also been built, and a large number of tests have been conducted [40,48,49]. For evaluating the capacity of a flexible barrier established using a full-scale rockfall test in the case of resisting landslide debris conveniently, it is worth revealing the relationship between the energy capacity of a flexible barrier in resisting rockfalls and debris.

In this paper, a field investigation of landslide debris successfully intercepted by a flexible barrier was first conducted. Then, a back analysis of the landslide mobility and the interaction between the landslide and the flexible barrier were carried out. The elongation of the brake rings and deflection of the flexible barrier from field measurements were used to verify the numerical simulation. Finally, parametric analyses of the flexible barrier impacted by landslide debris with different impact energies and velocities were carried out, to reveal its ultimate energy capacity.

2. Open Hillside Landslide in Chongqing, China

2.1. Field Investigation

The open hillside landslide occurred on terrain above a cut soil slope, which was formed to become a tunnel entrance of the Chongqing-Huaihua railway (Figure 2). A retaining wall was constructed to increase the stability of the slope. A flexible barrier with a total length of about 100 m was installed behind the retaining wall, to mitigate rockfalls from the hillside. The type of flexible barrier is a RXI-075, which is rated with an energy of 750 kJ. The flexible barrier is divided into ten functional modules by eleven steel posts, and the spacing of the adjacent two posts is 10 m. The height of the posts is 5 m. The posts were made of H-shape steel, with a section of 150 × 150 × 6 × 10 mm, and connected to foundations by pins. The main nets were composed of 300-mm opening rings formed by nine windings of 3-mm diameter steel wires. A twisted hexagonal wire mesh with openings of about 60 mm was attached to the ring nets, to capture small rock pieces.

The landslide occurred on the morning of 13 June 2020, due to heavy rainfall for days. After the finding of the landslide, an investigation and maintenance were carried out immediately. A realistic 3D model of the terrain was built using an DJI unmanned aerial vehicle (UAV) combined with 3D real-scene modeling software named ContextCapture. By comparing the terrain before and after the landslides, the total volume of the landslide was found to be approximately 70 m^3, and this was in accordance with the volume measured during maintenance. Figure 3 shows the plan view of the locations of the posts of the whole flexible barrier and the slide area of the landslide. The debris material was almost totally retained by the flexible barrier. No signs were found to show that the debris overflowed the barrier or passed through the net. In addition to the expected elongations of braking rings, rotations of the posts P5 and P6, and the deformation of nets, no failures of the flexible barrier were found, even for post P5, which was directly struck by the debris. In other words, the flexible barrier intercepted the landslide debris very successfully.

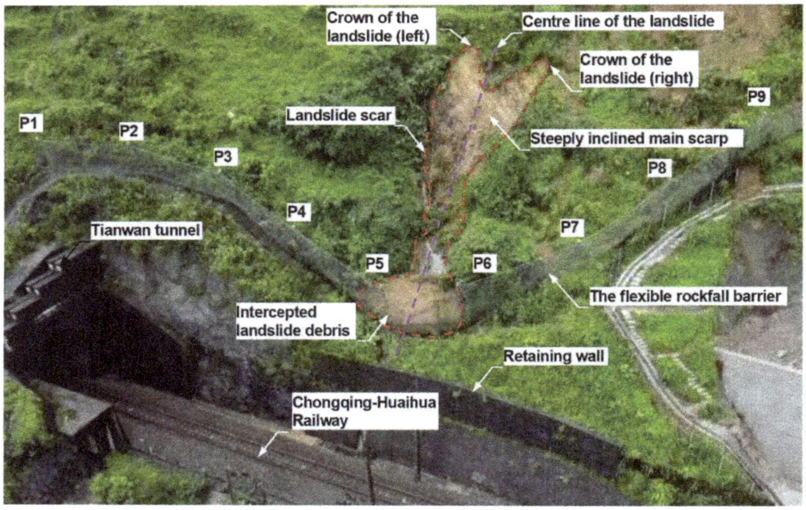

Figure 2. An overall view of the landslide site.

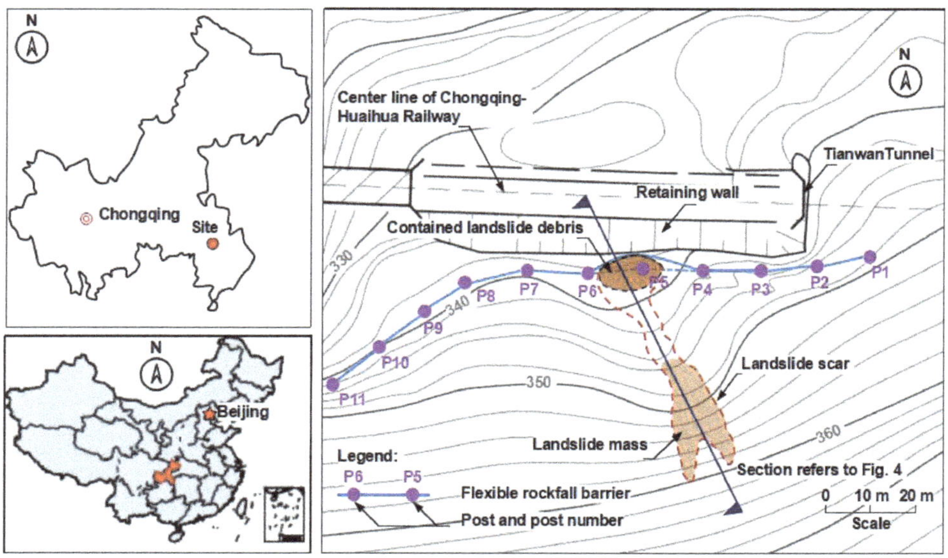

Figure 3. Location of the landslide and the posts of the flexible barrier.

The cross-section along the centerline of the landslide is shown in Figure 4. The landslide started on a steeply inclined scarp covered with unconsolidated sediments. The upper portion of the source area comprises vegetated terrain and is steeply inclined (>45°), while the lower portion comprises predominately matrix-supported debris overlying saprolite. The debris was saturated. The stacking angle of the stopped debris was about 15°. The stacking angle did not represent the internal friction angle of the debris, because it was intercepted by the flexible barrier. Obviously, the internal friction angle of the debris was less than 15°.

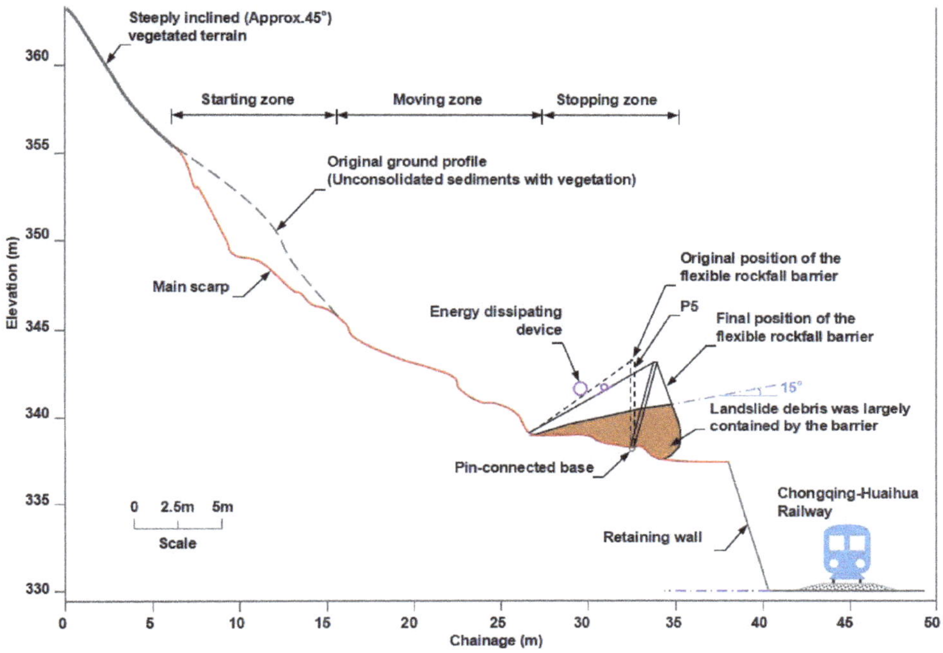

Figure 4. Cross-section along the centerline of the landslide.

After the finding of the landslide, multiple measures were adopted to maintain the flexible barrier. Most of the debris material was removed. A drainage ditch along the centerline of the landslide was excavated, to improve the drainage. Several waterproof geomembranes were layered on the top and side of the landslide-affected area, to reduce the infiltration of rainwater into the soil. The elongated braking rings and the impacted nets of the flexible barrier were replaced. The landslide site after maintenance is shown in Figure 5, and the repaired flexible barrier is shown in Figure 6.

Figure 5. An overall view of the landslide site after maintenance.

Figure 6. A front view of the impacted flexible barrier after maintenance.

2.2. Back-Analysis of the Landslide and Flexible Barrier Interaction

LS-DYNA, which has been successfully used to simulate debris mobility, the dynamic response of flexible barriers impacted by rockfalls, as well as debris and barrier interactions, was adopted to back analyze the interaction of the landslide and the flexible barrier.

2.2.1. Modeling of the Landslide Debris

The elastoplastic Drucker-Prager model [50], which has been successfully adopted to simulate the internal rheology of debris material [33,35], was also used here to simulate the landslide mass. The yield surface is shown in Equation (1):

$$f = \sqrt{J_2} - \alpha I_1 - k = 0 \quad (1)$$

where I_1 and J_2 are the first and second invariants of deviatoric stress tensor, respectively. Material constants α and k are related to the internal friction angle φ and cohesive strength c of the material, and they can be calculated using Equations (2) and (3), respectively:

$$\alpha = \frac{2\sin\varphi}{\sqrt{3}(3+\sin\varphi)} \quad (2)$$

$$k = \frac{6c\ \cos\varphi}{\sqrt{3}(3+\sin\varphi)} \quad (3)$$

As the landslide mass was saturated and moved like a flow, it was greatly deformed during the movement process. To avoid mesh distortion when simulating large deformation by the Lagrangian method, the arbitrary Lagrangian–Eulerian (ALE) formulation was adopted. In the formulations, the nodes of the computational mesh could be moved with the continuum in normal Lagrangian fashion, or be held fixed in the Eulerian manner or moved in some arbitrarily specified way, to give a continuous rezoning capability. Thus, greater distortions of the continuum could be handled than that allowed by a purely Lagrangian method, with more resolution than that afforded by a purely Eulerian approach.

The key parameters of the landslide mass are summarized in Table 1. The density was measured in the field to be roughly 1800 kg/m^3. The shear modulus and bulk modulus were assumed to be 500 kPa and 1000 kPa, respectively. As the stacking angle of the debris intercepted by the flexible barrier was about 15°, which was obviously much bigger than the internal friction angle of the debris. Therefore, the internal friction angle was assumed to be 5°. The friction coefficient between the landslide mass and the slope was assumed to be 0.4, referring to Ref. [35].

Table 1. Key parameters adopted to simulate the landslide mass.

Material Property	Adopted Value	Remarks
Density, ρ	1800 kg/m^3	Roughly measured in the field
Internal friction angle, φ	5°	
Shear modulus, G	500 kPa	By trial-and-error analysis (Evaluated according to the deformation of the barrier)
Bulk modulus, K	1000 kPa	
Cohesive strength, c	2 kPa	
Friction coefficient, μ	0.4	

2.2.2. Modeling of the Flexible Barrier

To save computational costs, only three functional modules of the flexible barrier, spanning across posts P4 to P7, were built in the model. The barrier is a proprietary product characterized by its ease of repair, as energy dissipating devices are designed as independent and replaceable units to attach to steel-wire ropes. By contacting the manufacturer of the barrier, the structural properties of the different components were confirmed.

The steel-wire ropes were modeled using discrete cable elements, which only have stiffness in axial tension. Beam elements using the plastic kinematic material model were adopted to model the posts. Energy dissipating devices were modeled using plastic tension only, as well as translational spring elements with a tri-linear load-displacement curve. Each steel-wire ring was modeled using sixteen beam elements, with a piecewise elastic-plastic stress-strain curve. The sliding characteristics between rings was explicitly modeled by the general contact algorithm. Seatbelt slip-ring elements, which work as a cable-and-pulley system, were adopted to model the sliding of support ropes on post ends. The loose connections between the edging rings and ropes with shackles were also explicitly modeled using simplified shackles, combined with a guided cable contact algorithm. The configuration and modeling method of the flexible barrier are summarized in Table 2. Generally speaking, the nonlinear and large deformation characteristics of the flexible barrier were effectively simulated in the model.

Table 2. Configuration and modeling method of the flexible barrier.

Components	Specification	Modeling Method
Post	HW 150 × 150 × 6 × 10 (Q235)	Beam elements with plastic kinematic material
Steel-wire ring net	R9/3/300	Beam elements with piecewise linear plastic material
Upper support rope	1 × 24	Discrete cable element
Lower support rope	1 × 24	
Upslope anchor rope	1 × 18	
Bypass rope	1 × 18	
Lateral support rope	1 × 18	
Lateral anchor rope	1 × 18	
Energy dissipating device attached to upper/lower support rope	2 GS-8002 brake rings in parallel	Plastic tension only, translational spring elements
Energy dissipating device attached to upslope anchor rope	1 GS-8002 brake ring	

2.2.3. Modeling of the Interaction

In the model, a special penalty-based coupling algorithm named "Constrained_lagrange_in_solid" was adopted, to recreate the interaction between the landslide and the flexible barrier, as well as the landslide and the slope. The coupling algorithm, essentially, is equivalent to placing a series of springs between the slave surface and the master surface, to limit the penetration. As this command can only represent the interaction between the Lagrangian shell and/or solid structures and the fluids modeled by ALE formulation, additional membranes modeled by shell elements with a null-type material [51] were introduced to cover the steel-wire meshes, to achieve the expected interaction. The introduced membranes could only transmit the interaction force and could not contribute any stiffness to the flexible barrier. It is worth noting that the additional membrane is impermeable, so the landslide mass could not penetrate the barrier. In fact, no signs of penetration of landslide from the flexible barrier were found in the field.

An efficient two-stage coupled modeling technique, developed by the authors [35], was adopted to build a three-dimensional model based on the rebuilt terrain, to investigate the landslide mobility and the landslide and flexible barrier interaction (Figure 7). In the first stage, only the movement of the landslide mass was simulated and the flexible barrier was totally constrained. Thus, a relatively large time step of 5×10^{-4} was sufficient to ensure the stability of the simulation. When the landslide was about to impact the barrier, the simulation of the first stage was ended and a binary file storing the model information of the last step was created. After removing the additional constraints of the barrier, the full restart technology was applied to initialize the state of the landslide with the binary file. Then, the coupled numerical simulation was launched with resetting to a much smaller time-step of 2×10^{-5}, to ensure the stability of the simulation.

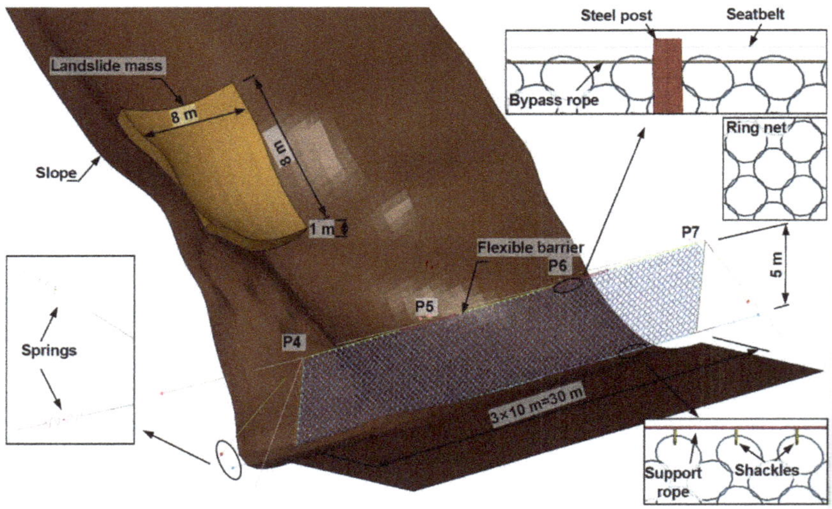

Figure 7. An isometric view of the back analyzed model.

2.3. Mobility of the Landslide

Under gravity, the landslide mass started to slide along the main inclined scarp, from being static till interception by the flexible barrier. Some typical moments of the interception process are shown in Figure 8. The velocity of the landslide at the barrier location and the kinetic energy of the landslide during the sliding are shown in Figure 9. It can be seen that when t = 2.5 s, the landslide impacted the flexible barrier with a maximum frontal velocity of around 9 m/s. After the peak value, the impact velocity dropped to zero within 2 s. The maximum kinetic energy was about 3200 kJ when t = 2.6 s, which is much bigger than the rated energy of the flexible barrier of 750 kJ, assessed using a rockfall impact. The reason for this may lie in the different processes in intercepting a rockfall or debris. The former is only a first impact and the kinetic energy is almost dissipated by the flexible barrier. The latter is successive impacts and the stopped debris may form a "dam" to dissipate the subsequent impact energy. With the development of the interaction, the kinetic energy of the landslide decreased rapidly. The interception process lasted about 2.5 s, and when t = 5.0 s, the landslide was totally stopped. The landslide mass was mainly accumulated in the middle functional span of the flexible barrier, which is generally consistent with the site observations.

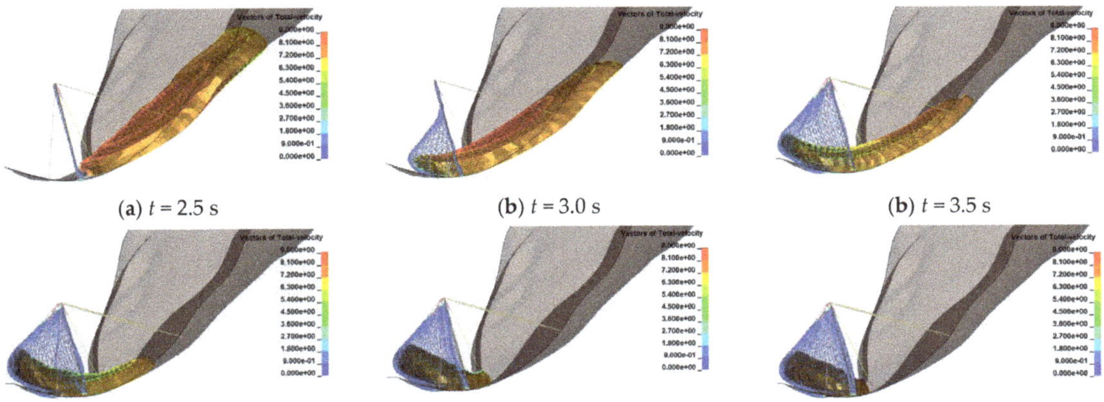

(**a**) $t = 2.5$ s (**b**) $t = 3.0$ s (**b**) $t = 3.5$ s

(**d**) $t = 4.0$ s (**e**) $t = 4.5$ s (**f**) $t = 5.0$ s

Figure 8. Typical moments of the interception process.

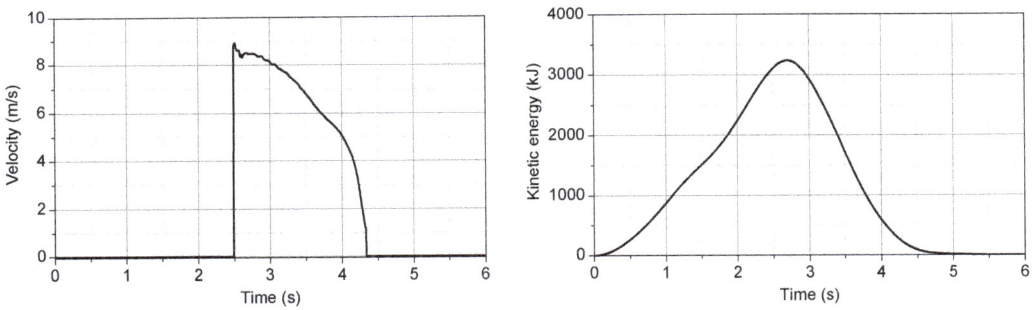

(**a**) Velocity at the barrier location (**b**) Kinetic energy

Figure 9. Moving characteristics of the landslide debris.

2.4. Response of the Flexible Barrier

2.4.1. Internal Forces of the Ropes

The ring nets were directly subjected to the impact of the landslide. Then, the impact force was transmitted to the lower support rope (LSR), upper support rope (USR), and upslope anchor ropes (UAR) in sequence. The internal forces of the ropes are shown in Figure 10. This demonstrates that the lower support rope was the first tensioned among these ropes. When $t = 2.6$ s, the internal force of the lower support rope increased rapidly, from zero to about 100 kN, which is the ideal activated force of the attached brake rings. Then, with the elongation of the attached brake rings, the internal force increased gently to the maximum value of 113 kN at $t = 3.7$ s. Then, the internal force decreased to a stable value of 53 kN, when the landslide mass was totally stopped. The internal force history of the upper support rope was almost consistent with the lower support rope. The differences mainly lie in the following two aspects: One is that the upper support rope was tensioned about 0.1 s later than the lower support rope; the other, was that the maximum value of the upper support rope was 108 kN, which was a little smaller than that of the lower support rope.

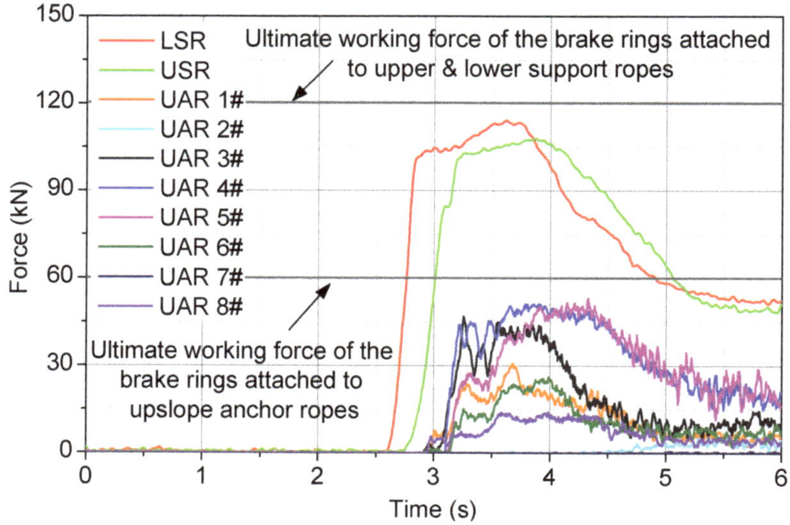

Figure 10. Internal forces of ropes of the flexible barrier (Note: UAR 1#~8# means the upslope anchor ropes from left to right).

The upslope anchor ropes were tensioned further and later than the upper support rope. The internal forces of the upslope anchor ropes connected to the middle posts P4 and P5 (Figure 7) were greater than for the other ropes, especially the upslope anchor ropes 4# and 5#, which were located at the impacted span. The maximum forces of UAR 4# and UAR 5# were close to 55 kN. As a part of the landslide mass impacted the left span of the flexible barrier, the internal force of UAS 3#, which connected the post P4 and anchored to the left span, was slightly smaller than that of UAR 4# and UAR 5#. The internal forces of other ropes were smaller than 30 kN. In particular, UAR 7# always remained loose during the interception process.

2.4.2. Elongation of Brake Rings

The elongation histories of the brake rings from the numerical simulation are shown in Figure 11. When t = 2.6 s, the brake rings attached to the left and right sides of the lower support rope began to elongate and were stable at 59 cm and 50 cm, respectively, at t = 3.6 s. When t = 2.7 s, the brake rings attached to the left and right sides of the upper support rope began to elongate and were stable at 38 cm and 15 cm, respectively, at t = 3.8 s. The elongations of brake rings attached to the support ropes were unsymmetrical, mainly due to the unsymmetrical impact of the landslide debris on the flexible barrier. In addition to the brake ring attached to the #4 upslope anchor rope, the elongations of other brake rings on the upslope anchor ropes were generally less than 10 cm. The brake ring attached to the #4 upslope anchor rope had the maximum elongation of 25 cm, due to the landslide mass directly impacting post P5, which the #4 upslope anchor rope was connected to.

The deflection of the flexible barrier and the elongation of the brake rings in the field were measured using tape before removing the landslide mass and are summarized with the simulated results in Table 3. It can be seen that the elongations derived from the numerical simulation are comparable to those from the field measurement. The deflection of the flexible barrier was 336 cm and 352 cm in the field and simulation, respectively. The difference was only 4.8%. The larger difference in the elongation of the brake rings attached to the support rope was due to the fact that the numerical model simplified the system to three functional modules, and the propagation effect of the internal force of the support rope was reduced by factors such as friction.

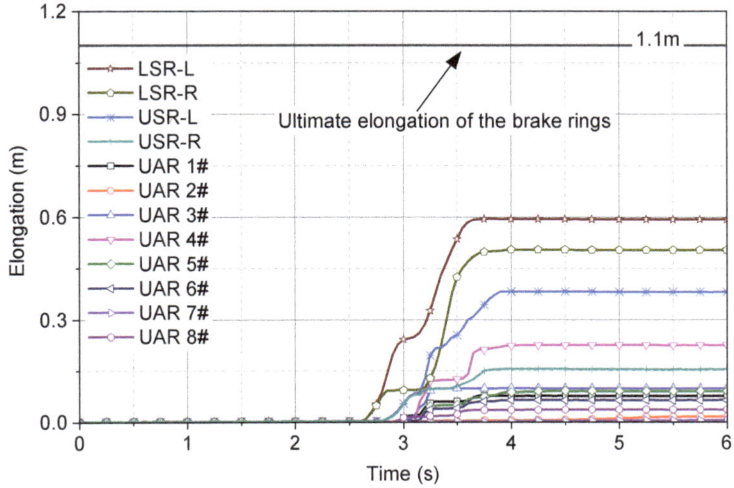

Figure 11. Elongations of brake rings in the numerical simulation.

Table 3. Deflection of the flexible barrier and elongation of the brake rings.

Components	Position		Deflection/Elongations (cm)	
			Field	Simulation
Flexible barrier	-		336	352
Brake rings	Upper support rope (Left/Right)		0/12	4/16
	Lower support rope (Left/Right)		43/38	59/50
	Upslope anchor rope	1#	7	8
		2#	2	2
		3#	9	10
		4#	25	23
		5#	8	9
		6#	7	6
		7#	0	1
		8#	3	4

The energy dissipating devices attached to the lower and upper support ropes were two GS-8002 brake rings in parallel, with an ultimate working force of 120 kN. The energy dissipating device attached to the upper slope anchor ropes was a GS-8002 brake ring, with an ultimate working force of 60 kN. The ultimate elongations of the brake rings were all 110 cm. The maximum internal forces of the lower and upper support ropes, as well as the upslope anchor ropes, were all smaller than the ultimate working force of the attached brake rings. This indicates that the brake rings did not all travel to their maximum elongation. In other words, even if the kinetic energy of the landslide of 3200 kJ was three times bigger than the rated energy of the flexible rockfall barrier of 750 kJ, the flexible barrier could intercept the landslide debris successfully, without any damage.

3. Numerical Parametric Study of the Flexible Barrier Subjected to Landslide Debris

3.1. Model Description

After investigation of the dynamic behavior of the flexible barrier impacted by rockfall and verification of the numerical simulation model, a series of simulations were carried out, to study the performance of the same flexible barrier impacted by landslide debris with different impact energies, ranging from 750 to 3000 kJ. Impact velocities ranging from 4 to 10 m/s were also investigated. A total number of 20 simulation cases are summarized in

Table 4. The coupled model of the landslide and the flexible barrier is shown in Figure 12. The angle of the slope was fixed to 15°. The width of the slope was 10 m, which is the post spacing of the flexible barrier. The thickness was also fixed to a common value, in practice of 1.0 m. The density was consistent with the measured value of 1800 kg/m^3 in the field investigation. In practice, flexible barriers are commonly installed vertically or sub-vertically on slopes to stop rockfalls or landslides. In the model, the flexible barrier was also set vertically at the end of the slope and in front of the landslide to save computational cost.

Table 4. Simulation schedule for a parametric study.

Case No.	Impact Velocity (m/s)	Landslide Volume (m^3)	Landslide Mass (kg)	Impact Energy (kJ)	Gravitation Potential Energy (kJ)	Total Energy (kJ)
1	4	52.1	9.4×10^4	750	537.1	1287.1
2		69.4	1.3×10^5	1000	852.6	1852.6
3		104.2	1.9×10^5	1500	1612.6	3112.6
4		138.9	2.5×10^5	2000	2496.0	4496.0
5		208.3	3.8×10^5	3000	6114.7	9114.7
6	6	23.1	4.2×10^4	750	156.1	906.1
7		30.9	5.6×10^4	1000	243.8	1243.8
8		46.3	8.3×10^4	1500	458.3	1958.3
9		61.7	1.1×10^5	2000	761.8	2761.8
10		92.6	1.7×10^5	3000	1560.2	4560.2
11	8	13.0	2.3×10^4	750	75.9	825.9
12		17.4	3.1×10^4	1000	110.4	1110.4
13		26.0	4.7×10^4	1500	205.6	1705.6
14		34.7	6.3×10^4	2000	305.2	2305.2
15		52.1	9.4×10^4	3000	593.6	3593.6
16	10	8.3	1.5×10^4	750	46.4	796.4
17		11.1	2.0×10^4	1000	65.6	1065.6
18		16.7	3.0×10^4	1500	112.1	1612.1
19		22.2	4.0×10^4	2000	174.1	2174.1
20		33.3	6.0×10^4	3000	330.6	3330.6

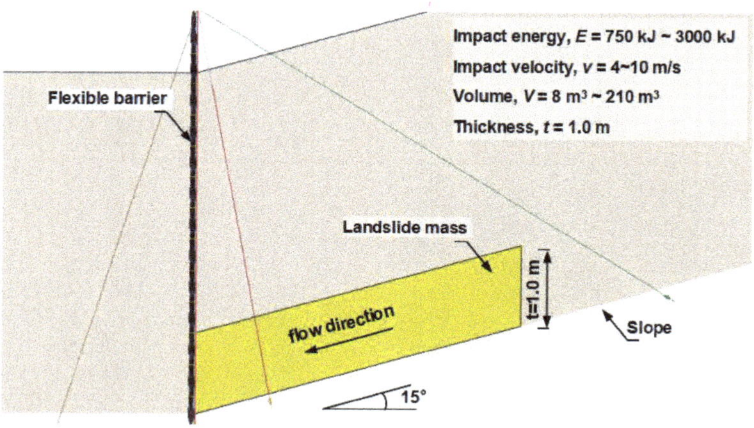

Figure 12. Side view of the simulation model for the parametric study.

3.2. *Results of the Parametric Study*

3.2.1. Elongation of the Brake Rings

The energy dissipation ratio denoted as η_s is defined as the ratio of the energy dissipated by the flexible barrier to the impact energy of the landslide mass, as follows:

$$\eta_s = \frac{E_S}{E_T} \quad (5)$$

where E_s is the energy dissipated by the flexible barrier, and E_T is the total energy of the landslide mass, including the initial kinetic energy and the gravitation potential energy.

Figure 13 shows the energy dissipation ratio of the flexible barrier subjected to landslide mass with different total energies and impact velocities. It can be seen that η_s of these cases all are less than 0.4, which means that the flexible barrier was not the main source of energy dissipation. Under the condition of a certain total energy, η_s will increase with the increase of the impact velocity. Under the condition of certain impact velocity, η_s will decrease linearly with the increase of the total energy, as shown in Equations (5)–(7):

$$\eta_s = kE_T + b \tag{5}$$

$$k = (-6.55v + 5.07) \times 10^{-6} \tag{6}$$

$$b = 0.044v - 0.043 \tag{7}$$

where a, b, and c are factors related to the impact velocity.

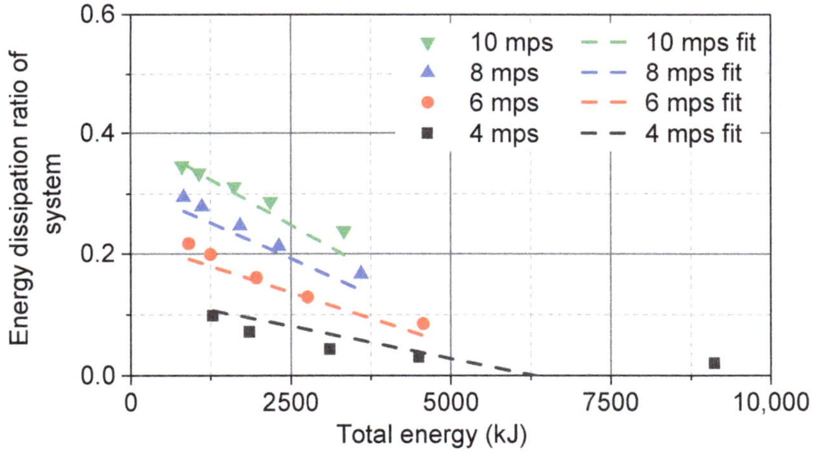

Figure 13. Energy dissipation ratio of the flexible barrier.

In particular, the η_s of the case with a total energy of 796.4 kJ and the impact velocity of 10 m/s was the biggest, and the case with a total energy of 9114.7 kJ and the impact velocity of 4 m/s was the smallest. This can be understood as follows: with the increase of total energy and the decrease of impact velocity, the volume and mass of the debris will increase, so the energy dissipated through the internal and boundary shearing will increase. It should be noted that for the cases with a total velocity of 4 m/s, the above linear decreased relationship of η_s and the total energy is not ideal. When the total energy is less than 4496.0 kJ, the η_s decrease quickly with the increase of total energy. However, the η_s of 9114.7 kJ is almost consistent with that of 4496.0 kJ. This means η_s is stable at 0.03 when the total energy is bigger than 4496.0 kJ, which is in good agreement with Song et al. [46].

3.2.2. Energy Dissipating Distribution

The energy dissipation ratio denoted as η is defined as follows:

$$\eta_s = \frac{E_s}{E_T} \tag{8}$$

where E_{dis} is the energy dissipated by each part, including the brake rings, friction energy, internal energy of landslide, and other components of the flexible barrier, in addition to the brake rings.

Figure 14 shows the distribution of the energy dissipation ratio. It can be seen that the proportion of energy dissipation, ranging from large to small, is the internal energy of landslide, friction energy, brake rings, and other components of the flexible barrier (ring net, steel posts, steel-wire ropes, etc.). The energy dissipated by other components of the flexible barrier in addition to the brake rings was less affected by the impact velocity and impact energy, and was stable at about 5%. The energy dissipated by the brake rings, friction, and internal energy ranged from 1.2% to 26.7%, 25.9 to 46.2%, and 19.2% to 74.2%, respectively. They were all significantly affected by the impact velocity and impact energy. The energy was mainly dissipated by the friction and internal energy.

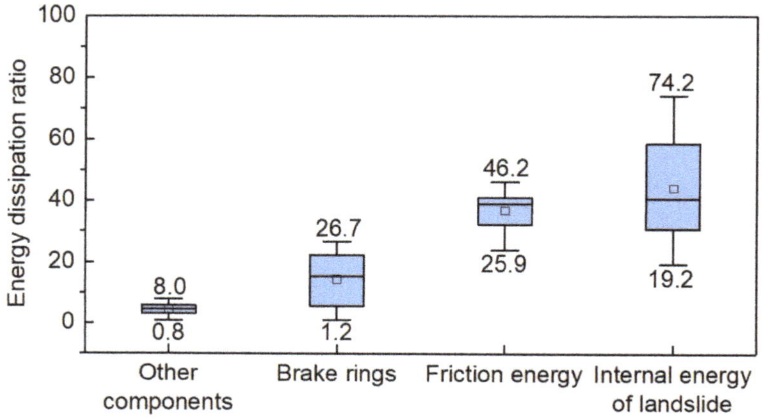

Figure 14. Energy dissipation distribution.

3.2.3. Energy Dissipation Ratio of the Brake Rings

The energy dissipation ratio of the brake rings denoted as η_b was defined, to describe the ratio of the actual energy dissipated by the brake rings to the designed energy dissipation capacity of the brake rings, as follows:

$$\eta_b = \frac{E_b}{E_T} \tag{9}$$

where E_b is the actual energy dissipation of the brake rings.

Figures 15a and 16a show the energy dissipation ratio and elongation of the brake rings attached to the upper support rope, respectively. When the impact velocity was 4 m/s, the energy dissipation was stable and less than 1%. When the impact velocity was 6~8 m/s, the energy dissipation ratio declined and the elongation increased sharply with the increase of total energy. For case 15, with an impact energy of 3000 kJ and impact velocity of 8 m/s, the elongation of the brake rings was the greatest, with a value of 1.80 m. Considering the fact that, in design practice, the elongation should be limited to 80% [41], the flexible barrier of the above cases was identified at its limit state.

Figures 15b and 16b show the energy dissipation ratio and elongation of the brake rings attached to the lower support rope, respectively. For the impact velocity of 6 m/s~10 m/s with an increase of total energy, the η_b declined quickly and the elongation increased sharply. When the total energy was larger than 2500 kJ, the elongation tended to be stable. When the impact velocity was 4 m/s, the energy dissipation ratio was less than 5%; and with the increase of total energy, the energy dissipation ratio declined and the elongation increased slightly. In particular, for the two cases of 10 m/s with the impact energies of 2000 kJ and 3000 kJ, the elongation of the brake rings was greatest, with a value of 2.2 m.

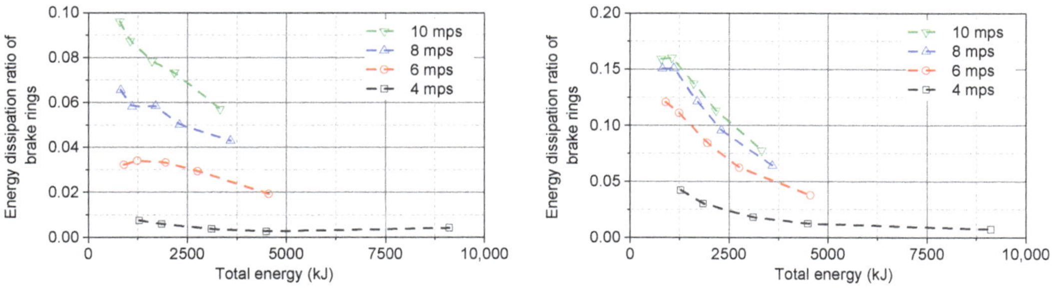

(a) Upper support rope (b) Lower support rope

Figure 15. Energy dissipation ratio of brake rings.

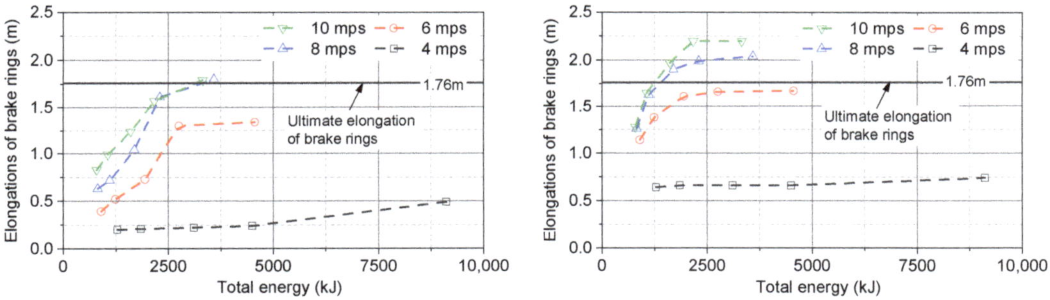

(a) Upper support rope (b) Lower support rope

Figure 16. Elongation of brake rings.

The energy dissipation ratio of the brake rings attached to the lower support rope was larger than that of the upper support rope, so the maximum intercepted capacity of the flexible barrier subjected to debris was controlled by the lower support rope, which is consistent with that when subjected to boulders [42]. From the energy dissipation ratio of the brake rings attached to the lower support rope, it can be seen that the flexible barrier can withstand a landslide debris impact with an initial impact energy of 3000 kJ, which is four-times the rated energy of a rockfall impact. In addition, with the decrease of impact velocity, the maximum capacity will increase further. Authors should discuss these results and how they can be interpreted from the perspective of previous studies and of the working hypotheses. The findings and their implications should be discussed in the broadest context possible. Future research directions may also be highlighted.

4. Conclusions

Landslide debris was successfully intercepted by a flexible rockfall barrier, without any damage at the Tianwan tunnel entrance of the Chongqing-Huaihua railway in China. Back analysis of the landslide mobility showed that the impact energy was much bigger than the rated energy capacity of the flexible rockfall barrier. To investigate the maximum energy capacity of the flexible rockfall barrier in resisting the landslide debris, parametric analyses of the flexible barrier impacted by landslide debris with different impact energies and velocities were carried out using a coupled modeling technique. The following conclusions can be drawn from this paper:

1. The total energy of landslide debris dissipated by a flexible barrier is less than 40%, most of the energy is dissipated by friction and internal energy.

2. The energy dissipation ratio of a flexible barrier decreases linearly with the increase of the impact energy.
3. The maximum energy capacity of a flexible barrier subjected to landslide debris is controlled by the lower support rope. The maximum energy capacity of a flexible rockfall barrier in resisting landslide debris is four-times that of resisting a rockfall.

In addition, with the decrease of impact velocity, the maximum energy capacity will increase further. Thus, it seems to be conservative to adopt a scaling factor not exceeding 75%, as required in the guidelines in Hong Kong [45], to reduce the energy capacity of a flexible barrier established for a rockfall, in the case of resisting debris flows.

Therefore, this research revealed the relationship of the maximum capability of a flexible barrier in intercepting a rockfall and landslide debris, in terms of energy. The conclusions will hopefully be helpful for engineers to select suitable flexible barriers rated by rockfall impact for landslide debris interception.

However, it should be noted that the findings pertain only to the particular type of flexible barrier, i.e., RXI-075, modeled and the particular impact scenarios in this study. Other impact cases and types of flexible barriers might be worth investigating in further studies. Furthermore, full-scale tests are urgently needed to investigate the behavior of flexible barriers impacted by rockfalls and landslide debris, respectively.

Author Contributions: All authors contributed to the study conception and design. Conceptualization was proposed by L.Z. (Lei Zhao) and Z.Y.; Simulations were performed by L.Z. (Lei Zhao) and L.Z. (Lijun Zhang); Investigation and data collection were performed by L.Z. (Lei Zhao), X.Q., H.X. and Y.Z. The first draft of the manuscript was written by L.Z. (Lei Zhao) and L.Z. (Lijun Zhang); and all authors commented on previous versions of the manuscript. All authors have read and agreed to the published version of the manuscript.

Funding: The work was supported by the National Key Research and Development Program of China (Grant No. 2018YFC1505405), the Department of Science and Technology of Sichuan Province (Grant No. 2020YJ0263), the Fundamental Research Funds for the Central Universities (Grant No. 2682020CX62), and Jiangxi provincial department of transportation research funding (2020H0005).

Institutional Review Board Statement: Not applicable.

Informed Consent Statement: Not applicable.

Data Availability Statement: Not applicable.

Acknowledgments: Thanks to the Research Center for Protection Structures Against Natural Hazards of Southwest Jiaotong University and China Railway 23rd Bureau Group Corporation Limited for providing us with the data, which are important to our research.

Conflicts of Interest: We declare that this manuscript entitled "A case study on the energy capacity of a flexible rockfall barrier in resisting landslide debris" is original, has not been published before, and is not currently being considered for publication elsewhere. We confirm that the manuscript has been read and approved by all named authors and that there are no other persons who satisfied the criteria for authorship but are not listed. We further confirm that the order of authors listed in the manuscript has been approved by all of us. We understand that the Corresponding Author is the sole contact for the Editorial process. He is responsible for communicating with the other authors about the progress, submissions of revisions, and final approval of proofs. Yi-fan Zhang, from Sichuan OST Slope Protection Engineering Co., Ltd., who is responsible for field investigation contribution, does not have any conflict of interest with the other authors.

References

1. Zhang, Y.; Shen, C.; Zhou, S.; Luo, X. Analysis of the influence of forests on landslides in the Bijie area of Guizhou. *Forests* **2022**, *13*, 1136. [CrossRef]
2. Grima, N.; Edwards, D.; Edwards, F.; Petley, D.; Fisher, B. Landslides in the Andes: Forests can provide cost-effective landslide regulation services. *Sci. Total Environ.* **2020**, *745*, 141128. [CrossRef] [PubMed]
3. Schmaltz, E.M.; Steger, S.; Glade, T. The influence of forest cover on landslide occurrence explored with spatio-temporal information. *Geomorphology* **2017**, *290*, 250–264. [CrossRef]

4. Griffiths, J.W.; Lukens, C.E.; May, R. Increased forest cover and limits on clear-felling could substantially reduce landslide occurrence in Tasman, New Zealand. New Zealand. *J. For. Sci.* **2020**, *50*, 1–13. [CrossRef]
5. Huang, R.Q.; Fan, X.M. The landslide story. *Nat. Geosci.* **2013**, *6*, 325–326. [CrossRef]
6. Cui, P.; Chen, X.Q.; Zhu, Y.Y.; Su, F.H.; Wei, F.Q.; Han, Y.S.; Liu, H.J.; Zhuang, J.Q. The Wenchuan Earthquake (May 12, 2008), Sichuan Province, China, and resulting geohazards. *Nat. Hazards* **2011**, *56*, 19–36. [CrossRef]
7. Grassl, H.; Bartelt, P.A.; Volkwein, A.; Wartmann, S. *Experimental and Numerical Modeling of Highly Flexible Rockfall Protection Barriers*; Soil and Rock America: Cambridge, MA, USA, 2003.
8. Boetticher, A.V.; Volkwein, A. Numerical modelling of chain-link steel wire nets with discrete elements. *Can. Geotech. J.* **2018**, *56*, 398–419. [CrossRef]
9. Available online: https://weibo.com/2615417307/Lwajt7Tto (accessed on 4 June 2022).
10. Yang, J.; Duan, S.; Li, Q.; Liu, C.Q. A review of flexible protection in rockfall protection. *Nat. Hazards* **2019**, *99*, 71–89. [CrossRef]
11. Castro-Fresno, D.; Luis, L.Q.; Bianco-Fernandez, E.; Zamora-Barraza, D. Design and evaluation of two laboratory tests for the nets of a flexible anchored slope stabilization system. *Geotech. Test. J.* **2009**, *32*, 315–324.
12. Zhu, Z.H.; Yin, J.H.; Qin, J.Q.; Tan, D.Y. A new discrete element model for simulating a flexible ring net barrier under rockfall impact comparing with large-scale physical model test data. *Comput. Geotech.* **2019**, *116*, 103208. [CrossRef]
13. Qin, J.Q.; Yin, J.H.; Zhu, Z.H.; Tan, D.Y. Development and application of new FBG mini tension link transducers for monitoring dynamic response of a flexible barrier under impact loads. *Measurement* **2020**, *153*, 107409. [CrossRef]
14. Gentilini, C.; Govoni, L.; Miranda, S.D.; Gottardi, G.; Ubertini, F. Three-dimensional numerical modelling of falling rock protection barriers. *Comput. Geotech.* **2012**, *44*, 58–72. [CrossRef]
15. Glover, J.; Denk, M.; Bourrier, F.; Volkwein, A.; Gerber, W. Measuring the kinetic energy dissipation effects of rock fall attenuating systems with video analysis. *Proc. Interpraevent Congr.* **2012**, *1*, 151–160.
16. Cui, L.M.; Shi, S.Q.; Wang, M.; Wang, W.K. Research on the impact protection performance of the rockfall attenuator system under multiposition distributed counterweights conditions. *Chin. J. Rock Mech. Eng.* **2019**, *38*, 332–342.
17. Duffy, J.D.; DeNatale, J.S. Debris flow mitigation using flexible barriers. In Proceedings of the 47th Annual Highway Geology Symposium, Cody, WY, USA, 6–9 September 1996.
18. Wendeler, C.; Volkwein, A.; Roth, A.; Denk, M.; Wartmann, S. Field measurements used for numerical modelling of flexible debris flow barriers. In Proceedings of the International Conference on Debris-Flow Hazards Mitigation: Mechanics, Prediction, and Assessment, Chengdu, China, 10–13 September 2007.
19. Bugnion, L.; McArdell, B.W.; Bartelt, P.; Wendeler, C. Measurements of hillslope debris flow impact pressure on obstacles. *Landslides* **2012**, *9*, 179–187. [CrossRef]
20. Roberto, B.; Andrea, S.; Anna, M.F. Debris flow hazard mitigation: A simplified analytical model for the design of flexible barriers. *Comput. Geotech.* **2013**, *54*, 1–15.
21. Tan, D.Y.; Yin, J.H.; Feng, W.Q.; Qin, J.Q.; Zhu, Z.H. Large-scale physical modelling study of a flexible barrier under the impact of granular flows. *Nat. Hazards Earth Syst. Sci.* **2018**, *18*, 2625–2640. [CrossRef]
22. Tan, D.Y.; Yin, J.H.; Feng, W.Q.; Zhu, Z.H.; Qin, J.Q.; Chen, W.B. New simple method for calculating impact force on flexible barrier considering partial muddy debris flow passing through. *J. Geotech. Geoenviron.* **2019**, *145*, 04019051. [CrossRef]
23. Canelli, L.; Ferrero, A.M.; Migliazza, M.; Segalini, A. Debris flow risk mitigation by the means of rigid and flexible barriers-experimental tests and impact analysis. *Nat. Hazards Earth Sys. Sci.* **2012**, *12*, 1693–1699. [CrossRef]
24. Wendeler, C.; Volkwein, A. Laboratory tests for the optimization of mesh size for flexible debris-flow barriers. *Nat. Hazards Earth Sys. Sci.* **2015**, *15*, 2597–2604. [CrossRef]
25. Ng, C.W.W.; Song, D.; Choi, C.E.; Koo, R.C.H.; Kwan, J.S.H. A novel flexible barrier for landslide impact in centrifuge. *Geotech. Lett.* **2016**, *6*, 221–225. [CrossRef]
26. Song, D. *Mechanisms of Debris Flow Impact on Rigid and Flexible Barriers*; The Hong Kong University of Science and Technology: Hong Kong, China, 2016.
27. Jiang, R.; Fei, W.P.; Zhou, H.W.; Huo, M.; Zhou, J.W.; Wang, J.M.; Wu, J.J. Experimental and numerical study on the load and deformation mechanism of a flexible net barrier under debris flow impact. *B Eng. Geol. Environ.* **2020**, *79*, 2213–2233. [CrossRef]
28 Volkwein, A. Numerical simulation of flexible rockfall protection systems. In Proceedings of the ASCE International Conference on Computing in Civil Engineering, Cancun, Mexico, 12–15 July 2005; pp. 1285–1295.
29. Chan, S.L.; Zhou, Z.H.; Liu, Y.P. Numerical analysis and design of flexible barriers allowing for sliding nodes and large deflection effects. In Proceedings of the One Day Seminar on Natural Terrain Hazard Mitigation Measures, Hong Kong, China, 16 October 2012.
30. Boetticher, A.V.; Volkwein, A.; Wüchner, R.; Bletzinger, K.; Wendeler, C. Numerical modeling of shallow landslide impacts on flexible protection systems and its validation with full scale testing. In Proceedings of the IV International Conference on Computational Methods for Coupled Problems in Science and Engineering, Kos, Greece, 20–22 June 2011.
31. Li, X.; Zhao, J. A unified CFD-DEM approach for modeling of debris flow impacts on flexible barriers. *Int. J. Numer. Anal. Methods Geomech.* **2018**, *42*, 1643–1670.
32. Leonardi, A.; Wittel, F.K.; Mendoza, M.; Vetter, R.; Herrmann, H.J. Particle-Fluid-Structure interaction for debris flow impact on flexible barriers. *Comput-Aided Civ. Inf.* **2016**, *31*, 323–333. [CrossRef]

33. Cheung, A.K.C.; Yiu, J.; Lam, H.W.K.; Sze, E.H.Y. Advanced numerical analysis of landslide debris mobility and barrier interaction. *HKIE Trans.* **2018**, *25*, 76–89. [CrossRef]
34. Kwan, J.S.H.; Chan, S.L.; Cheuk, J.C.Y.; Koo, R.C.H. A case study on an open hillside landslide impacting on a flexible rockfall barrier at Jordan Valley, Hong Kong. *Landslides* **2014**, *11*, 1037–1050. [CrossRef]
35. Zhao, L.; He, J.W.; Yu, Z.X.; Liu, Y.P.; Zhou, Z.H.; Chan, S.L. Coupled numerical simulation of a flexible barrier impacted by debris flow with boulders in front. *Landslides* **2020**, *17*, 2723–2736. [CrossRef]
36. Albaba, A. Discrete element modeling of the impact of granular debris flows on rigid and flexible structures. Ph.D. Thesis, Université Grenoble Alpes, Grenoble, France, 2015.
37. Ceccato, F.; Redaelli, I.; di Prisco, C.; Simonini, P. Impact forces of granular flows on rigid structures: Comparison between discontinuous (DEM) and continuous (MPM) numerical approaches. *Comput. Geotech.* **2018**, *103*, 201–217. [CrossRef]
38. Wendeler, C.; Volkwein, A.; McArdell, B.W.; Bartelt, P. Load model for designing flexible steel barriers for debris flow mitigation. *Can. Geotech. J.* **2018**, *39*, 1–18. [CrossRef]
39. Poudyal, S.; Choi, C.E.; Song, D.; Zhou, G.G.D.; Yune, C.Y.; Cui, Y.; Leonardif., A.; Busslingerg, M.; Wendeler, C.; Piton, G.; et al. Review of the mechanisms of debris-flow impact against barriers. In *Debris-Flow Hazards Mitigation: Mechanics, Monitoring, Modeling, and Assessment, Proceedings of the 7th International Conference on Debris-Flow Hazards Mitigation, Golden, Colorado, CO, USA, 10–13 June 2019*; Association of Environmental and Engineering Geologists: Brunswick, OH, USA, 2019.
40. Gentilini, C.; Gottardi, G.; Govoni, L.; Mentani, A.; Ubertini, F. Design of falling rock protection barriers using numerical models. *Eng. Struct.* **2013**, *50*, 96–106. [CrossRef]
41. Koo, R.C.H.; Kwan, J.S.H.; Lam, C.; Ng, C.W.W.; Yiu, J.; Choi, C.E.; Ng, A.K.L.; Ho, K.K.S.; Pun, W.K. Dynamic response of flexible rockfall barriers under different loading geometries. *Landslides* **2017**, *14*, 905–916. [CrossRef]
42. Yu, Z.X.; Zhao, L.; Liu, Y.P.; Zhao, S.C.; Xu, H.; Chan, S.L. Studies on flexible rockfall barriers for failure modes, mechanisms and design strategies: A case study of Western China. *Landslides* **2019**, *16*, 347–362. [CrossRef]
43. Wartmann, S.; Salzmann, H. Debris flow and floating tree impacts on flexible barriers. In Proceedings of the Conference on Natural Terrain—A Constraint to Development, Hong Kong, China, 14 November 2002; Institution of Mining and Metallurgy: Branch, Hong Kong, 2002.
44. Roth, A.; Kästli, A.; Frenez, T. Debris flow mitigation by means of flexible barriers. In Proceedings of the International Symposium Interpraevent 2004, Trento, Italy, 24–27 May 2004.
45. GEO. *Guidelines on Empirical Design of Flexible Barriers for Mitigating Natural Terrain Open Hillslope Landslide Hazards, (Technical Guidance Note No. 37)*; Geotechnical Engineering Office, HKSAR Government: Hong Kong, China, 2014.
46. Song, D.; Zhou, G.G.D.; Xu, M.; Choi, C.E.; Li, S.; Zheng, Y. Quantitative analysis of debris-flow flexible barrier capacity from momentum and energy perspectives. *Eng. Geol.* **2019**, *251*, 81–92. [CrossRef]
47. EOTA. Falling Rock Protection Kits. EAD 340059-00-0106. European Organization for Technical Approvals, July. 2018. Available online: https://www.nlfnorm.cz/en/ehn/6332 (accessed on 21 July 2022).
48. Xu, H.; Gentilini, C.; Yu, Z.X.; Qi, X.; Zhao, S.C. An energy allocation based design approach for flexible rockfall protection barriers. *Eng. Struct.* **2018**, *173*, 831–852. [CrossRef]
49. Volkwein, A.; Gerber, W.; Klette, J.; Spescha, G. Review of approval of flexible rockfall protection systems according to ETAG 027. *Geosciences* **2019**, *9*, 49. [CrossRef]
50. Crosta, G.B.; Imposimato, S.; Roddeman, D.G. Numerical modelling of large landslides stability and runout. *Nat. Hazards Earth Sys. Sci.* **2010**, *3*, 523–538. [CrossRef]
51. *LS-Dyna Theory Manual*; Livermore Software Technology: California, CA, USA, 2006.

forests

Article

Automatic Remote Sensing Identification of Co-Seismic Landslides Using Deep Learning Methods

Dongdong Pang [1], Gang Liu [1,2,*], Jing He [1], Weile Li [2] and Rao Fu [1]

1 College of Earth Sciences, Chengdu University of Technology; Chengdu 610059, China; pangdongdong1994@163.com (D.P.); hejing13@cdut.edu.cn (J.H.); furao@stu.cdut.edu.cn (R.F.)
2 State Key Laboratory of Geological Hazard Prevention and Geological Environment Protection, Chengdu University of Technology, Chengdu 610059, China; whylwl01@163.com
* Correspondence: liugang2014@cdut.edu.cn

Abstract: Rapid and accurate extraction of landslide areas triggered by earthquakes has far-reaching significance for geological disaster risk assessment and emergency rescue. At present, visual interpretation and field survey are still the most-commonly used methods for landslide identification, but these methods are often time-consuming and costly. For this reason, this paper tackles the problem of co-seismic landslide identification and the fact that there is little sample information in existing studies on landslide. A landslide sample dataset with 4000 tags was produced. With the YOLOv3 algorithm as the core, a convolutional neural network model with landslide characteristics was established to automatically recognize co-seismic landslides in satellite remote sensing images. By comparing it with the graphical interpretation results of remote sensing images, we found that the remote sensing for landslide recognition model constructed in this paper demonstrated high recognition accuracy and fast speed. The F1 value was 0.93, indicating that the constructed model was stable. The research results can provide reference for emergency rescue and disaster investigation of the same co-seismic landslide disaster.

Keywords: YOLOv3; deep learning; automatic landslide identification; remote sensing image

Citation: Pang, D.; Liu, G.; He, J.; Li, W.; Fu, R. Automatic Remote Sensing Identification of Co-Seismic Landslides Using Deep Learning Methods. *Forests* **2022**, *13*, 1213. https://doi.org/10.3390/f13081213

Academic Editor: Olga Viedma

Received: 29 March 2022
Accepted: 25 July 2022
Published: 1 August 2022

1. Introduction

Co-seismic landslides [1] are secondary hazards triggered by earthquakes in mountainous regions and often account for a very substantial proportion of earthquake hazards [2]. Landslides are generally catastrophic, resulting in loss of life, destroying infrastructure and houses and exerting a significant impact on the global economy [3]. For example, on 31 May 1970, Peru was struck by a 7.7 magnitude earthquake, triggering an avalanche that buried the city of Yongai and killed an estimated 23,000 people [4]. The 2008 Wenchuan earthquake triggered a landslide that directly killed nearly 30,000 people, accounting for roughly 30% of the population killed by the earthquake [5]. Therefore, the ability to obtain the information quickly and accurately about location, degree and scale of co-seismic landslides is of great significance for guiding earthquake emergency rescue and disaster risk assessment [6]. Hence, the research on remote sensing for landslides has become the main project to be addressed by various experts and scholars. In 1992, Wang [7] investigated and monitored landslides by combining the advantages of remote sensing and proposed to study and establish a landslide and debris flow information system, including the research and establishment of database and a series of application models, such as landslide and debris flow identification model, development model, impact evaluation model on engineering, disaster relief model and other concepts. Combining the interferometer radar and TM imagery, V. Singhroy et al. from Canada accurately extracted information about landslides in the lower Rocky Mountain watershed and used such information to assess the risk of the impacted area in 1995 [8]. In 2003, Wasowski [9] from Canada analyzed the opportunities brought by the development of remote sensing technology for studying landslide hazards and discussed the feasibility of using radars, ETM+, RADARASAT-1 and GPS and other multi-source data to observe landslides. In 2013, Wang [10] also proposed that if the relationship between the surface landslide wall and the landslide body with the ground slide surface and the slide bed could be determined, their spectral characteristics could be understood and a computational model could be built, and that the landslide patterns could be recognized by remote sensing technology. In 2016, Casagli et al. [11] used optical remote sensing, InSAR data and target-based means to extract landslide hazards. Gore et al. [12] proposed to extract mapping glacier features using hyperspectral remote sensing images. Based on the surface characteristics of spectral reflection characteristics, glaciers are divided into accumulation zone and ablation zone to make identification much simpler. At present, remote sensing-based technology for landslide information extraction is mainly through high-resolution, multispectral images on the texture, shape, hue and other shallow features to build extraction models [13]. Although the visual interpretation accuracy is somewhat high through this means, it is very dependent on the experience of professionals, slow identification speed and it is difficult to meet the requirements of emergency rescue, disaster timely assessment. In recent years, with the development of artificial intelligence, deep learning has been introduced into various disciplines [14], and it has been widely used in the field of remote sensing, providing a reliable solution to the identification of co-seismic landslides.

Deep learning has become a mainstream research tool in the fields of speech recognition, image recognition, image classification, target detection, etc. [15,16]. Song et al. proposed a method to transform spectral information into images and extract texture features for recognition and classification using CNN [17]. The reference [18] recognized and classified spectral features by calculating the value of the cost function through a 5-layer neural network. The reference [13] used mask region-based convolutional networks (Mask R-CNN) target detection module for automatic landslide identification. Liu et al. [19] classified frozen landslides on the Tibetan Plateau by applying migration learning. Arabameri et al. [20] created a landslide inventory map using GPS points, high-resolution satellite images, topographic maps, and historical records obtained from various machine learning methods and field analysis. Then, they compared the applicability of FLDA, RF, and ADTree models for establishing LSM in the Gallicash River basin in northern Iran near the Caspian Sea and succeeded in further applying deep learning in the field of landslide identification. Chen et al. [21] obtained a landslide sensitivity map based on the evidence confidence function model by using the slope element of hydrologic analysis and the slope element based on curvature watershed. After that, they calculated the success rate and prediction rate of the landslide susceptibility map. More and more advanced algorithms are used in landslide research, such as C5.0 decision tree (DT) model and K-means clustering algorithm, which can generate regional landslide susceptibility maps well [22]. With the development of computer vision technology [23], there are numerous algorithms for image recognition using deep learning methods, whose main objective is to pinpoint the class and location of various targets in an image or sequence of images [24]. Girshick et al. [25] proposed R-CNN convolutional neural network in 2013 for target detection using a deep convolutional neural network. Joseph Redmon, Ali Farhadi et al. [26] proposed the YOLO algorithm in 2015, a target detection system based on a single neural network. With modeling detection as a regression problem, it directly obtains the coordinates of the bounding box, the confidence of the objects contained in the bounding box and the class probability from all pixels of the whole picture, which effectively improves detection and recognition efficiency. At present, many experts and scholars have applied deep learning to the study of co-seismic landslide recognition. However, according to the most cutting-edge research results, few scholars can automatically recognize co-seismic landslide by deep learning, and many of the existing studies are about co-seismic automatic landslide extraction based on traditional machine learning or deep learning.

For the purpose of landslide recognition, this paper adopted the popular object detection algorithm YOLOv3 based on deep learning to build a convolutional neural network for co-seismic landslide recognition. TensorFlow [27] platform was used to construct the convolutional neural network, and YOLOv3 convolutional neural network to train the sample set. The convolutional neural network model with landslide characteristics was obtained, and then the landslide detection and recognition of satellite remote sensing images of co-seismic landslides were achieved. The recognition accuracy was close to that of visual interpretation in the experimental area with much faster recognition speed.

2. Study Area and Data

2.1. Study Area

The study area of this paper is in the Hokkaido region of Japan, as shown in Figure 1. Hokkaido, the largest island in Japan other than Honshu, lies between latitude 40 degrees 33′ and 45 degrees 33′ north, and longitude 139 degrees 20′ and 148 degrees 53′ east, and is bounded by Honshu Island in the south by the Tsugaru Strait and by the Soya Strait in the north by the Kuril Islands. It covers a large area, accounting for about 22% of the total area of Japan. At least 70% of the island is covered by forests. The vegetation consists mainly of bright coniferous forests and grasslands. The terrain is high in the center and low in the surroundings with volcanoes. There are mountains and ranges in the central part of Hokkaido, so the terrain is generally undulating, while the surrounding area is a vast plain. The co-seismic landslide is the major threat to people's lives and development. A 6.7 magnitude earthquake hit Hokkaido, Japan, at 3:08 am local time on 6 September 2018, with an epicenter at 42.671 N, 141.933 E, and a source depth of 35 km. The earthquake, with a maximum observed intensity of magnitude 7, was the highest magnitude earthquake observed in Hokkaido, Japan, by the Japan Meteorological Agency since 1923 [28]. During the earthquake, 44 people were killed, 32 houses were completely damaged and 18 were partially damaged. The casualties and house damage were mainly caused by lare landslides and liquefied ground settlement and seismic subsidence [29]. The disaster damage caused by some of the co-seismic landslides is shown in Figure 2.

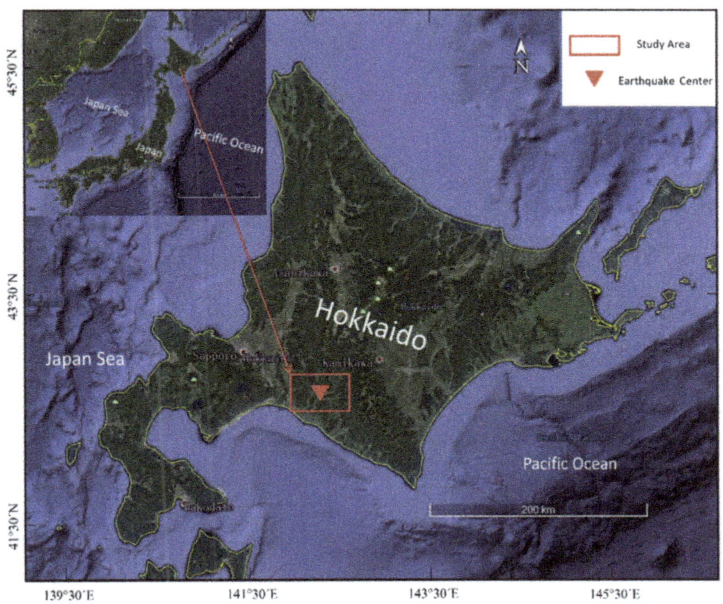

Figure 1. Study area (the red square is the study area and the inverted triangle is the earthquake center).

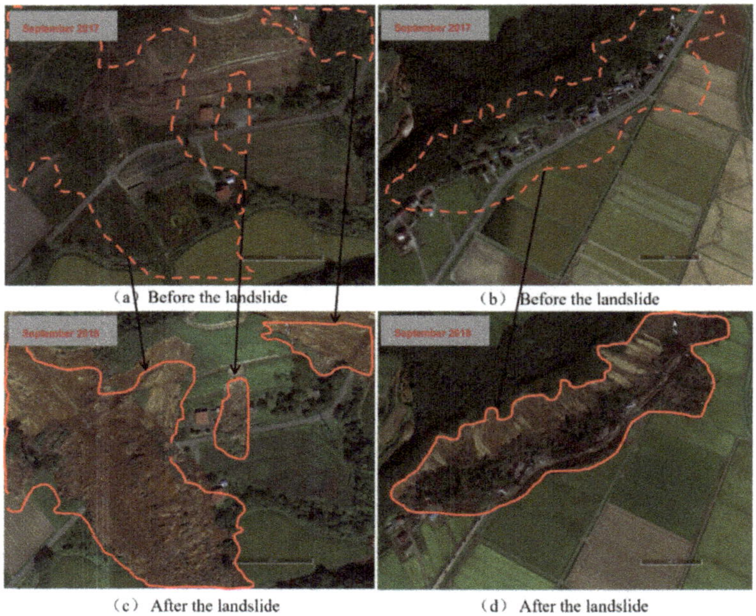

Figure 2. Comparison of the landslide before and after earthquake. (**a**,**b**) are the images before earthquake; (**c**,**d**) are the images after earthquake. The red dotted line shows the image of the surface before the landslide. The surface of the newly formed landslides is marked with a solid red line.

2.2. Data

The satellite remote sensing images in this paper were downloaded from Planet satellite (https://www.planet.com/explorer/ (accessed on 18 September 2018)), with a resolution of 3 m taken on 18 September 2018. Planet's SuperDove image, which went online in April 2020, features a 3-m resolution and eight spectral bands (temporarily open red, green, blue, near infrared and red-edge bands), integrating the advantages of FL small satellite clusters with a 3-m resolution, and RapidEye red-edge bands. ENVI 5.6 software was used to enhance the recognition of co-seismic landslides through image true-color synthesis. Finally, orthophotos were generated by combining the images of the co-seismic landslide affected area with red, green and blue bands. The image-covered area after splicing was the whole landslide region with many landslides of small scale. To provide better training samples, images were cut into 800 m × 800 m images in ArcGIS, and 120 images with the dense landslide area were selected to make data sets as training samples, including 100 as training sets and 20 as test sets. The satellite remote sensing images were transformed into image samples and XML format label samples using the LablImg label production platform and then interpreted by the experts in this field from the State Key Laboratory of Geological Hazard Prevention and Geological Environment Protection. In this paper, we use the open YOLOv3 algorithm to recognize the co-seismic landslides. The open YOLOv3 algorithm is a public algorithm, which is one of the most widely used deep learning-based object detection methods and uses the K-means cluster method to estimate the initial width and height of the predicted bounding boxes. According to the open YOLOv3 algorithm, the parameters of YOLOv3 algorithm were adjusted to make it more suitable for co-seismic landslides. Figure 3 shows some of the orthophotos of the study area and the co-seismic landslide area mapped by visual interpretation.

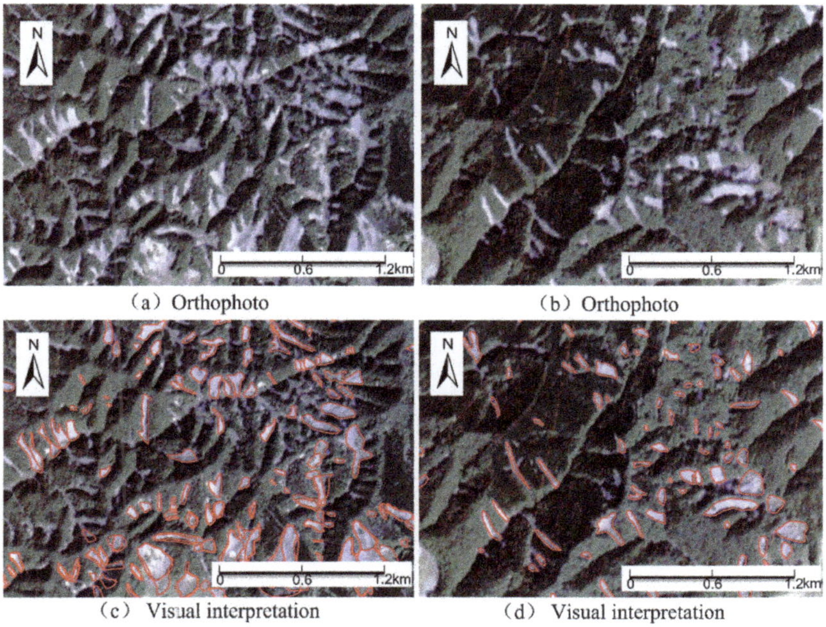

Figure 3. Orthophotos and visual interpretation maps of landslide in the study area. (**a**,**b**) are the orthophotos and (**c**,**d**) are the corresponding visual interpretation maps. The red regions in (**c**,**d**) are the landslide areas interpreted visually by experts from the State Key Laboratory of Geological Hazard Prevention and Geological Environment Protection.

3. Methodology

3.1. Research Design

The main design of the experimental study of this paper is shown in Figure 4. Firstly, the data were preprocessed as shown in Section 2.2; thus, the preprocessed data were obtained for landslide identification. With YOLOv3 algorithm as the core, a convolutional neural network was constructed using TensorFlow framework to train and validate the sample set. The convolutional neural network model with the characteristics of co-seismic landslides was obtained by constantly adjusting the parameters. Satellite remote sensing images of landslides were analyzed, tested and identified, and then the experimental results were compared with the visual interpretation results of the experts.

3.2. Procedure and Recognition Principle of YOLOv3 Convolutional Neural Network

Recently, with the rapid development of artificial intelligence, great breakthroughs have been made in target detection algorithms. According to detection ideas, common target detection algorithms based on deep learning can be divided into two categories. One is two-stage target detection algorithms based on candidate region and one-stage target detection algorithms based on regression [30]. The typical two-stage target detection algorithms include R-CNN [31] system algorithms based on region proposal, such as R-CNN, Fast R-CNN, and Faster R-CNN. The other is one-stage algorithm, such as YOLO and SSD, among which YOLO is based on an end-to-end idea. The principle of this algorithm is to model detection as a regression problem, and directly obtain the boundary box coordinates of the detection target, the confidence degree of the object contained in the boundary box and the category probability of the object from all the pixels of the image.

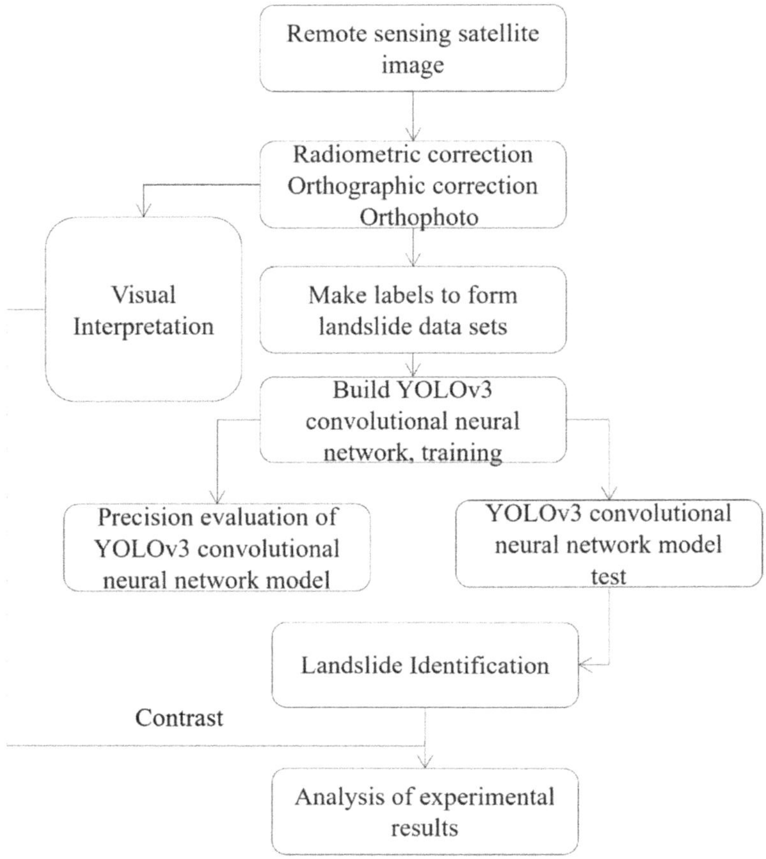

Figure 4. Design of experimental study.

Compared with other deep learning target detection algorithms, YOLO is characterized by fast detection speed. It solves object detection as a regression problem, dividing an image into S×S grid cells. If the center of an object falls in the grid, the grid will be responsible for predicting the object, and each grid for predicting B bounding boxes and C categories. In addition to returning to its own position, each bounding box should also be accompanied with a confidence value of five values, so the output tensor is S×S× (5B+C); Figure 5 shows a bounding box with scale and position prediction. YOLO detection is very fast. The standard version of YOLO can reach 45 FPS on TrainX's GPU. However, YOLO also has the disadvantages of low recall rate, low position accuracy, and poor detection effect for small objects. To overcome the above disadvantages, researchers proposed YOLOv2 and YOLOv3 [32]. YOLOv3 improves both detection accuracy and speed, demonstrating a new detection and classification network that is superior to other algorithms. According to the literature, in particular reference [31], M40 or TrainX can be detected by the same GPU (graphic processing unit), and the running speed of YOLOv3 is obviously faster than other detection methods with similar performance [33].

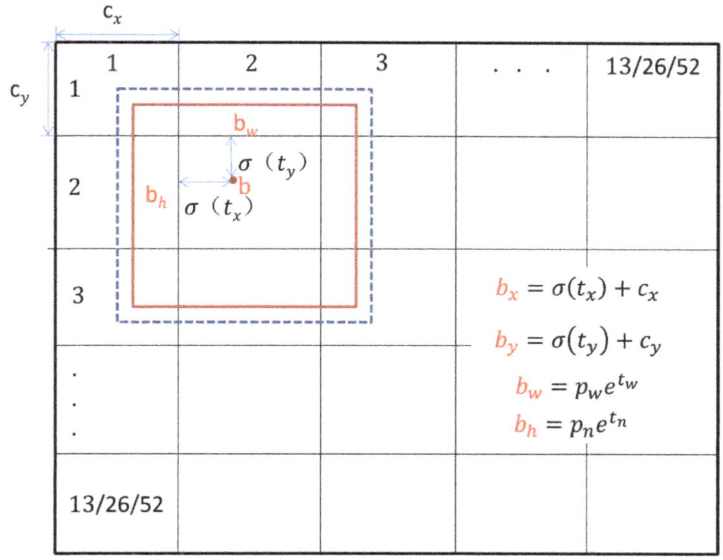

Figure 5. Bounding box with scale grid (scale 13/26/52) and location prediction.

The total square error was still used for calculation in the network training process of YOLOv3, and binary cross entropy was used for the loss function of parts other than w (the width of the bounding box) and h (the height of the bounding box). If the ground truth predicted by a coordinate is $\hat{t}_*$, the gradient is the ground truth minus our prediction, i.e., $\hat{t}_* - t_*$. This ground truth value could be easily calculated by inverting the above equation. The width and height of the bounding box were then predicted as offsets from the target center coordinates, and the position of the center coordinates of the box relative to the filter was predicted using the sigmoid function. Dimension clusters were used to fix the anchor box frame in the convolutional neural network, and four coordinates of each boundary frame were predicted, namely t_x, t_y, t_h, t_w. If the predicted bounding box at the upper left corner of the image is offset by (c_x,c_y), and the predicted width and height of the target box are p_w, p_h, then the predicted value of the target is as follows:

$$b_x = \sigma(t_x) + c_x \tag{1}$$

$$b_y = \sigma(t_y) + c_y \tag{2}$$

$$b_w = p_w e^{t_w} \tag{3}$$

$$b_h = p_h e^{t_h} \tag{4}$$

The YOLOv3 convolutional neural network uses the Darknet-53 network as the base network for feature extraction throughout YOLOv3. The Darknet-53 network achieved the highest measurement floating point operations per second, which allowed the network structure to make better use of the GPU, thus making its evaluation more efficient and faster. The base network has a total of 53 convolutional layers, with no fully connected layers. The tensor size was achieved by the step size of the convolution kernel. The whole network was a fully convolutional network, which made use of the residual module hop layer connections to reduce the negative effects of pooling gradient by removing pooling layer. In the network structure, a convolution of step size of 2 was used for down sampling. The total square error was used for the loss function of w, h and binary cross entropy for the loss function of the other parts. To enhance the detection accuracy of the YOLOv3 algorithm, the algorithm uses a complementary and fusion approach similar to

FPN, providing 3 scales of prediction frames that are divided into 3 prediction output branches, i.e., 13, 26, and 52 scales, and the structure used in the three prediction output branches is also fully convolutional, and the tensor obtained for each branch is as follows:

$$S \times S \times [3 \times (4 + 1 + N)] \tag{5}$$

where S represents different grid scales, 3 represents three predicted target boxes in each grid cell, 4 represents four coordinate values of each bounding box, 1 is the confidence value of each box, and N is the number of categories predicted by training.

3.3. Construction of YOLOv3 Convolutional Neural Network Structure for Landslide Satellite Remote Sensing Image Features

In this paper, a landslide was taken as the identification object, and the identification target was single, so there was no need to consider the category problem, thus making network construction less difficult. According to the characteristics of image data, the network model adjusted the image format of all the images involved in the training to 416 × 416 × 3 by means of down sampling before entering the training. Meanwhile, we adopted regularization to prevent overfitting by adjusting the parameters of momentum coefficient and weight attenuation regularization. Due to the difficulty in obtaining landslide data and the lack of samples, the saturation and exposure of images were adjusted to increase sample diversity. Figure 6 shows the structure of the YOLOv3 convolutional neural network for landslide satellite remote sensing image features, i.e., the target category is 1, and the number of convolutional kernels in the last convolutional layer is 18.

$$3 \times (4 + 1 + 1) = 18 \tag{6}$$

Taking the scale of 52 × 52 as an example, according to Formulas (5) and (6), the tensor of landslide data set at this scale is 52 × 52 × 18.

3.4. Accuracy Evaluation Index

The evaluation metrics of the model are generally accuracy, recall, mean average precision, IOU (intersection over union), F1 score, etc. The F1 score is a measure of the accuracy of the binary classification model, which considers the accuracy and recall of the classification model and can be regarded as a weighted average of the accuracy and recall of the model. It can be expressed as follows:

$$\text{Precision} = \frac{T_p}{T_P + F_P} \times 100\% \tag{7}$$

$$\text{Recall} = \frac{T_P}{T_P + F_N} \times 100\% \tag{8}$$

$$\text{mAP} = \frac{1}{C} \sum_{i=1}^{N} Precision(i) \Delta Recall(i) \tag{9}$$

$$\text{IOU} = \frac{\text{Detection Result} \cap \text{Ground Truth}}{\text{Detection Result} \cup \text{Ground Truth}} \tag{10}$$

$$\text{F1 score} = \frac{2 Precision \cdot Recall}{(Precision + Recall)} \tag{11}$$

In the above formula,
Precision: Accuracy
Recall: Recall rate
mAP: Average precision mean
IOU: Intersection ratio union
F1 score: The F1 score is an indicator of the accuracy of the dichotomous model
T_P: The number of samples that are correctly classified as targets

F_P: The number of samples wrongly classified as the target
F_N: The number of samples incorrectly classified as non-target objects
C: Number of target categories. The category of this paper is only landslide, so C = 1
Detection result: The predicted bounding box
Ground truth: Real bounding box

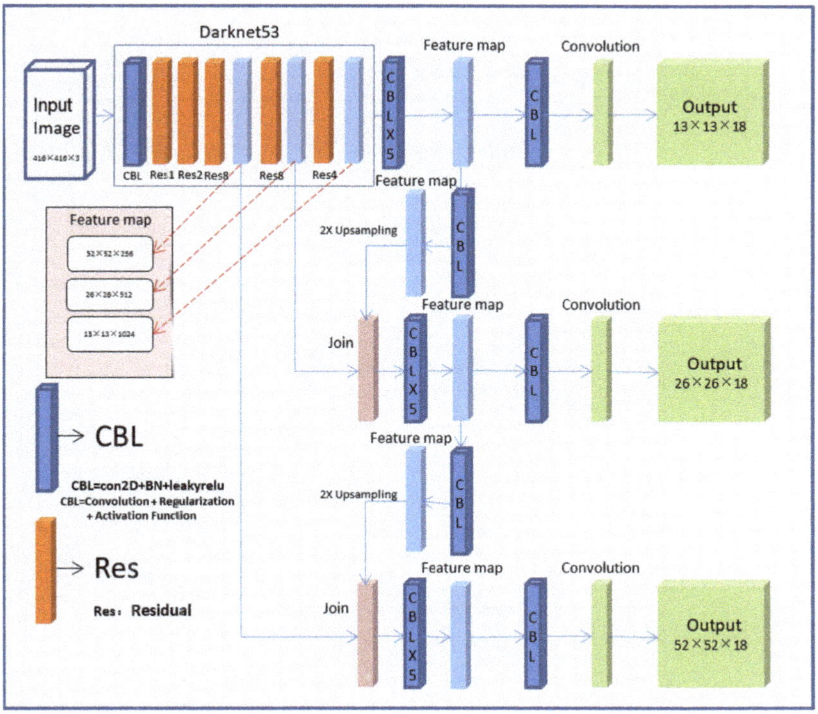

Figure 6. Structure of YOLOv3 convolutional neural network for landslide satellite remote sensing image features.

4. Experiments and Results

4.1. Network Model Training

The orthophotos after correction and stitching were segmented into 800 m × 800 m image maps in ArcGIS, and 120 images with more landslide coverage were selected to create the sample dataset for network training, and the samples were created using labeling, a tool specialized in labeling, to convert the satellite remote sensing images into image samples and label samples in XML format. The detailed experimental platform is shown in Tables 1 and 2.

Table 1. Hardware configuration for the experiment.

Type	Parameter
CPU	Intel (R) Core (TM) i9-10900K CPU @ 3.70 GHz
Memory	64 GB
GPU	NVIDIA GeForce RTX 3090

Table 2. Test all software environments.

Type	Parameter
System	Windows10 (Pro_×64. From the Redmond, WA, USA)
Operation platform	CUDA9.0 + cuDNN7
Development of language	Python3.7
Sample data enhancement	Python3.7 + OpenCV4.5.5
Sample label making	LablImg
Platforms and frameworks	Platforms: ENVI 5.6, ArcGIS10.2 Deep learning framework: TensorFlow2.0

A landslide, characterized by large area and monotonous features, was used as the research object in the experiment. Therefore, it was necessary to adjust the setting of some parameters and find out the most reasonable parameter values to train the sample set by comparing the training results for several times in the process of network training. In this training, the training times were set as 6000 times, the learning rate as 0.001, the change point of the learning rate as 500, and as 64 batches. After 64 samples were accumulated each time, the forward propagation was carried out once, and the sub batch was set as 16, that is, a batch of images were divided into 16 sub batches to complete the forward propagation of the network. The momentum coefficient was set to 0.9 and the weight decay regularization term to 0.0005 regularization in case of the change in weight. The saturation and exposure parameters were set to 1.5 and used to increase sample diversity and generate more training samples. The change in the loss value of the model training is shown in Figure 7, from which the loss value decreased sharply at the beginning of 500 training sessions, slowed down from 500 to 2500 sessions, and stabilized after 3500 training sessions, and finally decreased to 0.97 at 6000 sessions. As can be observed from the training log, 64 samples were processed in the first training, and 320,000 samples were processed in the 6000th training. This effect of reducing training loss was quite ideal, which was in line with the purpose of the experiment.

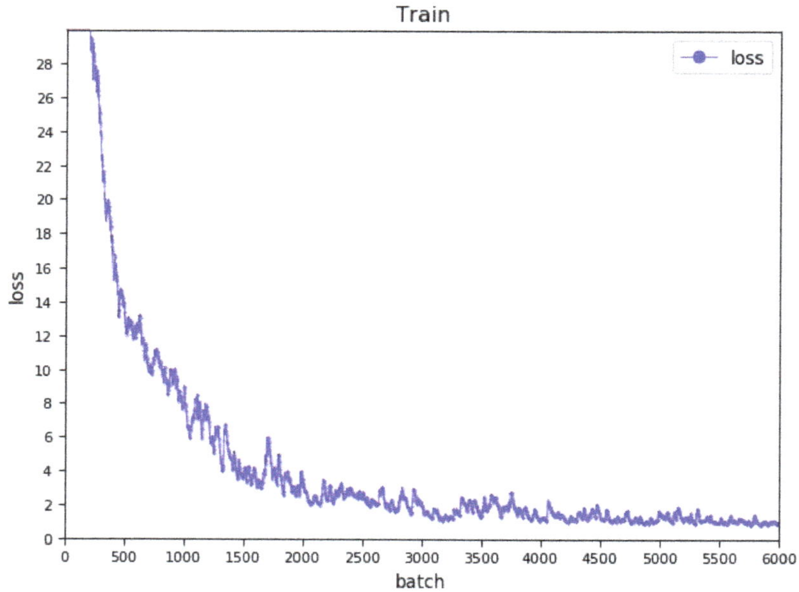

Figure 7. Plot of loss values with the number of iterations.

4.2. Accuracy Evaluation of the Network Model

According to the accuracy evaluation indexes in Section 3.3 above, in this paper, recall, IOU and F1 are selected as the accuracy evaluation indexes of the model. Recall was used as an indicator to evaluate the accuracy of the training model. The validation set data was 60 images, and the average IOU and recall of the samples were recorded when the training model was used to verify 60 images. Each image measured was recorded once, as shown in Figure 8.

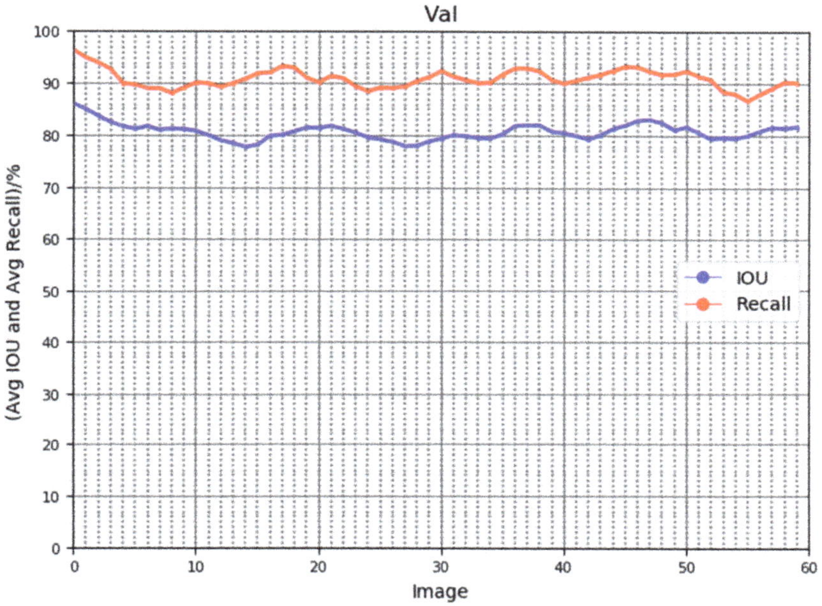

Figure 8. Average IOU and average recall.

It can be observed from Figure 8 that as the number of tested samples increased, the accuracy of recognition always fluctuated in a fixed area and was relatively stable. The IOU and recall of the first two images in the test were relatively high because the cumulative number of samples contained in the detected images was averaged and they were recorded only by the number of images. With the increase in sample size, the average crossover ratio and average recall rate tended to be stable. After testing 60 images, IOU was balanced at 80% and recall at 90%, which were ideal results.

4.3. Network Model Test

When the training part of the network was completed and the verification accuracy met the requirements, we started to test the network model for proving the usefulness of the network by randomly selecting several satellite remote sensing images in the study area that were not involved in the training for landslide recognition. The recognition results are shown in Figure 9, in which red boxes are the recognized landslides. It can be observed from the figure that good recognition results were achieved.

4.4. Analysis of Experimental Results

The experimental results show that the YOLOv3 convolutional neural network built in this paper performs well in detecting and identifying landslides. The comparison results between the network proposed in this paper and visual interpretation are shown in Figure 10.

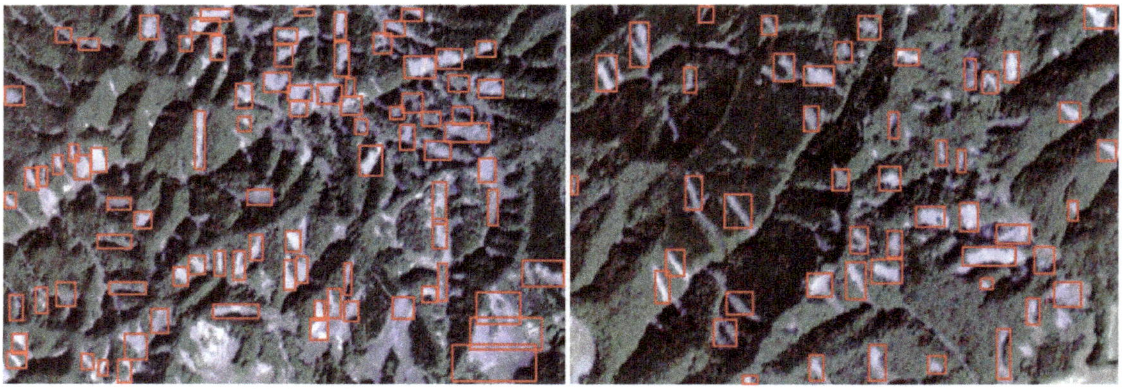

(a) Landslide identification (b) Landslide identification

Figure 9. Landslide identification results. (**a**) and (**b**) are the identification results of co-seismic landslide. The red box shows the landslide area).

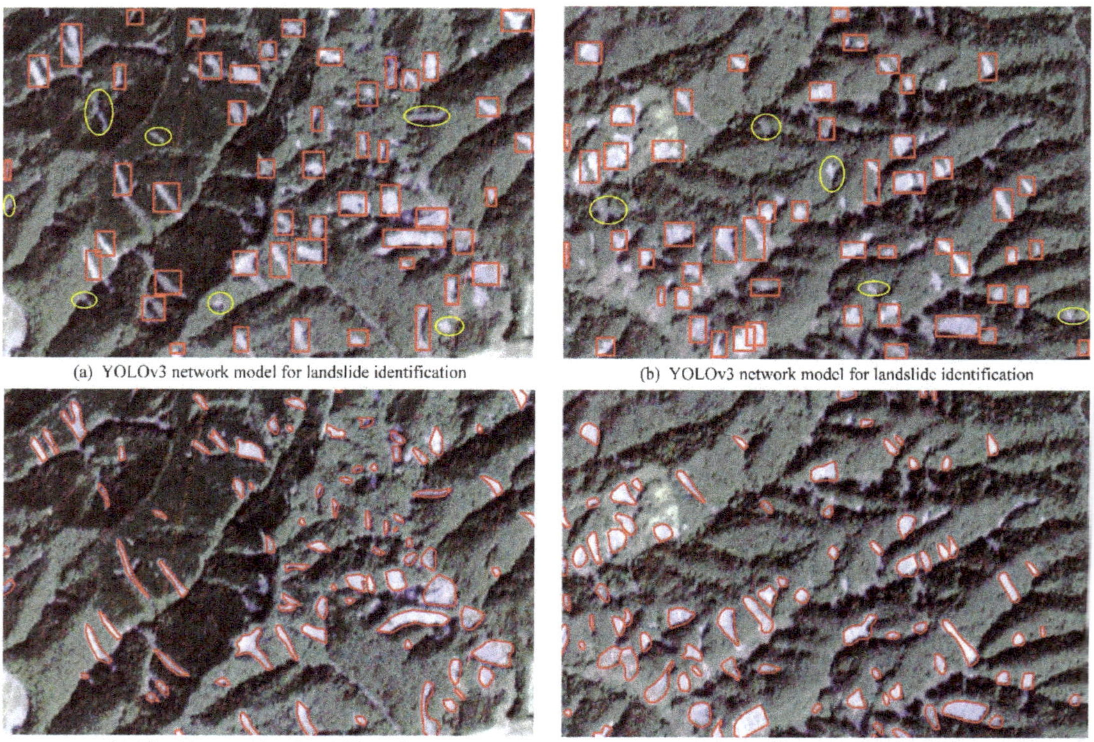

(a) YOLOv3 network model for landslide identification (b) YOLOv3 network model for landslide identification

(c) Visual interpretation of landslide results (d) Visual interpretation of landslide results

Figure 10. Comparison between landslide identification results of YOLOv3 network model and visual interpretation results. (**a**,**b**) are landslide recognition results of YOLOv3 network model, and (**c**,**d**) are the corresponding visual interpretation results. The red parts in (**c**,**d**) are the landslide area interpreted visually by experts from the State Key Laboratory of Geological Hazard Prevention and Geological Environment Protection. The yellow parts in a and b are the unrecognized co-seismic landslide area.

As can be observed from the Figure 10, the overall effect of YOLOv3 recognition on landslides was better than visual interpretation, with only 11 landslides with small areas and light colors failing to be recognized, as shown in yellow in Figure 10. The average recall was calculated as 0.88, and the average precision as 0.98 by Equations (7), (8) and (11), as shown in Figure 10a,b. The YOLOv3 network model recognized F1 was 0.93, indicating that the experiment met the research needs.

5. Discussion

The method adopted in this paper was to transform the landslide identification problem into an image processing problem. Thus, the method proposed can effectively reduce the difficulty of co-seismic landslide research, making it easier for related researchers to understand the disaster caused by co-seismic landslides. This method belongs to the category of artificial intelligence. The purpose is to solve the tedious manual interpretation problem. Therefore, it has more outstanding superiority in large-scale seismic landslide research. Furthermore, the YOLOv3 algorithm proved to have outstanding advantages in image recognition. Ref. [34] applied a neural network built by the YOLOv3 algorithm to recognize surface objects using remote sensing images and verified the efficiency of the algorithm in remote sensing image identification. In this study, the YOLOv3 algorithm was applied to satellite image recognition, and successfully identified the co-seismic landslide. The Planet image adopted in this paper is a remote sensing image with a resolution of 3 m. In addition, this method is more inclusive to different image categories, even in satellite images with different resolutions. Although the texture, color and other characteristics of the co-seismic landslide area are different, we need to build on the network model by transfer learning. By adding images of different seismic landslide features as new samples, we can identify co-seismic landslides from different satellite images.

The identification results have met the experimental requirements, and this network can quickly and accurately identify seismic landslides. There are still some small landslides that cannot be identified. The failure to identify those small landslides may be caused by the insufficient number of samples, because more images with obvious landslide characteristics in the large area were selected when making sample labels. While some parts of the YOLOv3 network were unrecognizable, YOLOv3 could identify images within seconds, while visual interpretation required professionals to manually label images based on their experience and expertise, which is a more time-consuming and costly process. Therefore, the research in this paper helps to achieve the rapid identification of landslides in the internal industry and improve the identification efficiency, which provides reference for disaster emergency rescue and investigation of co-seismic landslides. In the meanwhile, it is of great significance to the emergency rescue of landslide disasters after earthquakes.

This research realizes fast and accurate automatic recognition of co-seismic landslides. Due to the complexity of landslide remote sensing identification, there are still many problems that need to be further studied. The following are some thoughts on the research of deep learning identification of co-seismic landslides.

(1) Production of sample data sets. Training neural networks with large amounts of data can ensure that the reliability and stability of recognition results are guaranteed.
(2) Optimization of convolutional neural networks. A co-seismic landslide recognition model with less dependence on samples, higher intelligence, higher accuracy in the future and faster speed will be constructed. Although The OLOv3 algorithm has a fast speed and strong credibility of automatic recognition, the characteristics of co-seismic landslides are complex and diverse. So, it would be better to combine more network models to realize the identification of co-seismic landslides.

6. Conclusions

In this paper, the YOLOv3 algorithm was used to build a neural network model oriented to landslide identification from remote sensing satellite image features, train samples and automatically identify co-seismic landslides. In the model test, the IOU was balanced at 80% and the recall at 90%, which showed that the model had good stability. In the meantime, for the co-seismic landslide identification experiment in Hokkaido, Japan, the experimental results showed that the average recall was calculated as 0.88, and the average precision as 0.98 and the F1 as 0.93. The results also showed that the YOLOv3 network model was relatively simple to build and easy to adjust and optimize parameters with high precision and fast speed, it had excellent identification effect and could play an important role in landslide identification. Compared with expert visual interpretation results, the model adopted in this paper has a very fast speed in co-seismic landslide identification. The research in this paper effectively overcame the deficiencies of visual interpretation, such as more required time, high cost and low efficiency. In this paper, a landslide sample dataset with 4000 tags was produced. It provides data support for deep learning to identify seismic landslides.

In the future, other landslide databases caused by geological disasters can be used to train the network model to improve its accuracy. We can also extend the constructed network model for the identification of geological disasters, such as loess and rain-fall induced landslides.

Author Contributions: Conceptualization, D.P. and G.L.; methodology, D.P.; software, R.F.; validation, J.H.; data curation, W.L.; writing—original draft preparation, D.P.; writing—review and editing, D.P. and G.L. All authors have read and agreed to the published version of the manuscript.

Funding: This research was funded by the National Key National Key Research and Development Program of China (Grant No. 2021YFC3000401), State Key Laboratory of Geohazard Prevention and Geoenvironment Protection Independent Research Project (Grant No. SKLGP2018Z010), the National Natural Science Foundation of China (NSFC) (Grant No. 41871303), Sichuan Provincial Science and Technology Support Project (Grant No. 2021YFG0365), Department of Natural Resources of Sichuan Province (Grant No. kj-2021-3). Chengdu Technology Innovation R&D Project (2022-YF05-01090-SN).

Data Availability Statement: The images used in this paper are derived from Planet satellite data (https://www.planet.com/explorer/ (accessed on 18 September 2018)).

Conflicts of Interest: The authors declare that they have no known competing financial interests or personal relationships that could have appeared to influence the work reported in this paper.

References

1. Li, W.L.; Huang, R.; Xu, Q.; Tang, C.; Li, W.L.; Huang, R.; Xu, Q.; Tang, C. Rapid susceptibility mapping of co-seismic landslides triggered by the 2013 Lushan Earthquake using the regression model developed for the 2008 Wenchuan Earthquake. *J. Mt. Sci.-Engl.* **2013**, *10*, 699–715. [CrossRef]
2. Chong, X.U.; Xiwei, X.U.; Zhou, B.; Shen, L. Probability of co-seismic landslides: A new generation of earthquake-triggered landslide hazard model. *J. Eng. Geol.* **2019**, *27*, 1122–1130.
3. Ubydul, H.; Philipp, B.; Da, S.; Paula, F.; Peter, A.; Juergen, P. Fatal landslides in Europe. *Landslides* **2016**, *13*, 1545–1554.
4. Plafker, G.; Ericksen, G.E.; Concha, J.F. Geological aspects of the May 31, 1970, Perú earthquake. *Bull. Seism. Soc. Am.* **1971**, *61*, 543–578. [CrossRef]
5. Yin, Y. Research on the geo-hazards triggered by wenchuan earthquake, sichuan. *J. Eng. Geol.* **2008**, *16*, 433–444.
6. Zhang, P.; Xu, C.; Ma, S.; Shao, X.; Wen, B. Automatic Extraction of Seismic Landslides in Large Areas with Complex Environments Based on Deep Learning: An Example of the 2018 Iburi Earthquake, Japan. *Remote Sens.* **2020**, *12*, 3992. [CrossRef]
7. Wang, Z.H. Application of remote sensing technology in geological disaster investigation, monitoring and prevention. *Environ Remote Sens.* **1992**, 190–195, In Chinese.
8. Singhroy, V. Sar integrated techniques for geohazard assessment. *Adv Space Res* **1995**, *15*, 67–78. [CrossRef]
9. Singhroy, J.W.A.V. Special issue from the symposium on Remote Sensing and Monitoring of Landslides. *Eng. Geol.* **2003**, *68*, 1–2.
10. Wang, Z. A preliminary discussion on landslide pattern recognition. *J. Geo.-Inf. Sci.* **2013**, *15*, 726–733.
11. Casagli, N.; Cigna, F.; Bianchini, S.; Hölbling, D.; Füreder, P.; Righini, G.; Del Contee, S.; Friedl, B.; Schneiderbauer, S.; Iasio, C.; et al. Landslide mapping and monitoring by using radar and optical remote sensing: Examples from the EC-FP7 project SAFER. *Remote Sens. Appl. Soc. Environ.* **2016**, *4*, 92–108. [CrossRef]
12. Gore, A.; Mani, S.; Hariram, R.P.; Shekhar, C.; Ganju, A. Glacier surface characteristics derivation and monitoring using Hyperspectral datasets: A case study of Gepang Gath glacier, Western Himalaya. *Geocarto Int.* **2017**, *8*, 23–46. [CrossRef]
13. Ju, Y.; Xu, Q.; Jin, S.; Li, W.; Guo, Q. automatic object detection of loess landslide based on deep learning. *Geomat. Inf. Sci. Wuhan Univ.* **2016**, *45*, 1747–1755.

14. Hinton, G.E.; Osindero, S.; Teh, Y.W. A Fast Learning Algorithm for Deep Belief Nets. *Neural Comput.* **2014**, *18*, 1527–1554. [CrossRef] [PubMed]
15. Peng, Y.; Song, X.; Wu, H.; Zi, L. Remote sensing image retrieval combined with deep learning and relevance feedback. *J. Image Graph.* **2019**, *24*, 420–434.
16. Huang, L.W.; Jiang, B.T.; Shou-Ye, L.V.; Liu, Y.B.; De-Yi, L.I. Survey on Deep Learning Based Recommender Systems. *J. Comput. Sci.* **2018**, *41*, 1619–1647.
17. Song, X. Study on Hyperspectral Data Classification Method Based on Convolutional Neural Network. Master's Thesis, Harbin Institute of Technology, Harbin, China, 2016.
18. Zhao, M.; Ren, Z.; Gao, D.O.; Chang, W.U.; Hao, X. Convolutional neural networks for hyperspectral image classification. *J. Geomat. Sci. Technol.* **2017**, *34*, 501–507.
19. Liu, D.; Li, J.; Fan, F. Classification of landslides on the southeastern Tibet Plateau based on transfer learning and limited labelled datasets. *Remote Sens. Lett.* **2021**, *12*, 286–295.
20. Arabameri, A.; Saha, S.; Roy, J.; Chen, W.; Bui, D.T. Landslide Susceptibility Evaluation and Management Using Different Machine Learning Methods in The Gallicash River Watershed, Iran. *Remote Sens.* **2020**, *12*, 475. [CrossRef]
21. Chen, Z.; Liang, S.; Ke, Y.; Yang, Z.; Zhao, H. Landslide susceptibility assessment using different slope units based on the evidential belief function model. *Geocarto Int.* **2019**, *35*, 1641–1664. [CrossRef]
22. Guo, Z.; Shi, Y.; Huang, F.; Fan, X.; Huang, J. Landslide susceptibility zonation method based on C5.0 decision tree and K-means cluster algorithms to improve the efficiency of risk management. *Geosci. Front.* **2021**, *12*, 101249. [CrossRef]
23. Zhang, K.; SU, D.; Wang, P.B.; Chen, H.; Zhang, S.; Ye, L.J.; Zhao, N. Progress and Application of Deep Learning in Image Dense Matching. *Sci. Technol. Eng.* **2020**, *20*, 12268–12278.
24. Huang, J.; Zhang, G. Survey of Object Detection Algorithms for Deep Convolutional Neural Networks. *Comput. Eng. Appl.* **2020**, *56*, 12–23.
25. Gong, Y.; Wang, L.; Guo, R.; Lazebnik, S. *Multi-Scale Orderless Pooling of Deep Convolutional Activation Features*; Springer: Cham, Switzerland, 2014; pp. 392–407.
26. Redmon, J.; Divvala, S.; Girshick, R.; Farhadi, A. You Only Look Once: Unified, Real-Time Object Detection. *IEEE* **2016**, *5*, 1063.
27. Pang, B.; Nijkamp, E.; Wu, Y.N. Deep Learning With TensorFlow: A Review. *J. Educ. Behav. Stat.* **2020**, *45*, 227–248. [CrossRef]
28. Wang, L.; Ju, A.; Wang, L. Disaster characteristics and enlightenment of ms 6.7 Earthquake in Hokkaido, Japan. *City Disaster Reduct.* **2019**, *1*, 1–8.
29. Yamagishi, H.H.G.C.; Yamazaki, F. Landslides by the 2018 Hokkaido Iburi-Tobu Earthquake on September 6. *Landslides* **2018**, *15*, 2521–2524. [CrossRef]
30. Liu, X.; Wang, Z.; He, Y.; Liu, Q. Research on small target detection based on deep learning. *Tactical Missile Technol.* **2019**, *1*, 100–107.
31. Ren, S.; He, K.; Girshick, R.; Sun, J. Faster R-CNN: Towards Real-Time Object Detection with Region Proposal Networks. *IEEE Trans. Pattern Anal. Mach. Intell.* **2017**, *39*, 1137–1149. [CrossRef]
32. Redmon, J.; Farhadi, A. YOLOv3: An Incremental Improvement. *arXiv* **2018**, *4*, 1–3.
33. Redmon, J.; Farhadi, A. YOLO9000: Better, Faster, Stronger. *IEEE* **2017**, *6*, 6517–6525.
34. Pang, D.; Liu, G.; He, J.; Fu, R. Orange tree recognition from UAV image based on YOLOv3 network. *Geomat. Spat. Inf. Technol.* **2022**, *4*, 32–36,40.

forests

Article

Landslide Susceptibility-Oriented Suitability Evaluation of Construction Land in Mountainous Areas

Linzhi Li [1], Xingyu Chen [1], Jialan Zhang [2,*], Deliang Sun [1] and Rui Liu [1]

1 The Key Laboratory of GIS Application Research, School of Geography and Tourism, Chongqing Normal University, Chongqing 401331, China
2 Key Laboratory of New Technology for Construction of Cties in Mountainous Area, School of Civil Engineering, Chongqing University, Chongqing 400045, China
* Correspondence: lanaij@cqu.edu.cn

Abstract: The aim of the present study was to assess the suitability of mountainous areas for construction land on the basis of landslide susceptibility, to obtain the spatial distribution pattern of said suitability and to improve the existing theories and methods used to ascertain said suitability. Taking Hechuan District in Chongqing as the research area and using data relating to 754 historical landslide sites from 2000 to 2016, we selected 22 factors that influence landslides. The factors were classified into five types, namely topography and geomorphology, geological structure, meteorology and hydrology, environmental conditions and human activities. A landslide susceptibility model was constructed using the random forest algorithm, and safety factors of construction land suitability were established according to the results of landslide susceptibility, with the suitability of land for construction in mountainous areas assessed by combining the key factors (natural, social and ecological factors). The weights of the factors were determined through the use of expert approaches to classify the suitability of land for construction in the research area into five levels: prohibited, unsuitable, basically suitable, more suitable and most suitable. The results of the study show that: (1) The average accuracy of the tenfold cross-validation training set data of landslides reached 0.978; the accuracy of the test set reached 0.913; the accuracy of the confusion matrix reached 97.2%; and the area under curve (AUC) values of the training set, test set and all samples were 0.999, 0.756 and 0.989, respectively. Historical landslide events were found to be mostly concentrated in highly susceptible areas, and the landslide risk level in Hechuan District was mostly low or very low (accounting for 76.26% of the study area), although there was also a small proportion with either a high or very high risk level (9.25%). The high landslide susceptibility areas are primarily concentrated in the southern and southeastern ridge, in the valley and near water systems, with landslides occurring less frequently in the gentle hilly basin. (2) The suitability of land for construction in mountainous areas was strongly influenced by landslide susceptibility, distance from roads and distance from built-up areas; among such parameters, rainfall, elevation and lithology significantly influenced landslides in the region. (3) The land suitable for construction in the study area was highly distributed, mainly in urban areas where the three rivers meet and around small towns, with a spatial distribution pattern of high in the middle and low on both sides. Furthermore, the suitability of land for construction in Hechuan District was found to be primarily at the most suitable and more suitable levels (accounting for 84.66% of the study area), although a small proportion qualified for either the prohibited or unsuitable level (accounting for 15.72%). The present study can be extended and applied to similar mountainous areas. The landslide susceptibility map and construction land suitability map can support the spatial planning of mountainous towns, and the assessment results can assist with the development direction of mountainous towns, the layout of construction land and the siting of major infrastructure.

Keywords: landslide susceptibility; land-use suitability; random forest model; Hechuan District

Citation: Li, L.; Chen, X.; Zhang, J.; Sun, D.; Liu, R. Landslide Susceptibility-Oriented Suitability Evaluation of Construction Land in Mountainous Areas. *Forests* **2022**, *13*, 1621. https://doi.org/10.3390/f13101621

Academic Editors: Filippo Giadrossich, Haijia Wen, Weile Li, Chong Xu and Hiromu Daimaru

Received: 24 August 2022
Accepted: 28 September 2022
Published: 3 October 2022

1. Introduction

Due to the special topography, the exploitation of resources and economic development in mountainous areas have been limited by the fragile ecological environment [1]. The unreasonable exploitation of resources and the environment has led to an incompatibility between people and mountainous areas. As urban construction in China continues to expand, the state has increased efforts to protect cultivated land, resulting in difficulties in balancing development and protection [2]. There is a need to assess the suitability of mountainous areas to be used as construction land and to identify the spatial distribution of land suitable for construction and land which is not to allow for the expansion of urban development spaces and identification of high-quality cultivated land [3]. Through such means, a foundation can be laid for the spatial control of national land resources, the sustainable utilization of land resources and agricultural production. Evaluation indicators and research methods regarding the suitability of land for construction in mountainous areas are regarded as highly significant by relevant scholars. The assessment factors vary slightly depending on the research field and focus and can be divided into two main categories. The first category is assessment of the suitability of the land for construction based on key factors. For instance, Bagheri [4] combined ArcGIS and the D-AHP model to identify the risk zone of Kuala Terengganu, an eastern coastal city of Peninsular Malaysia, and constructed a land-use suitability map for disaster management. The other is the exploration of the suitability of land for construction based on selected factors, such as natural factors, ecological factors, social traffic, economic development and population density [5,6]. As an example, Ustaoglu [5] used ArcGIS and multicriteria assessment (MCA) to assess the suitability of land for urban construction in Pendik in eastern Istanbul, Turkey based on indicators such as geophysical features, accessibility, built-up areas and infrastructure, vegetation and other green and blue facilities. The focus of existing research has largely been on the evaluation of the suitability of land for construction in mountainous areas, with the aim of improving the evaluation method and enriching evaluation cases. However, the application value of suitability evaluation has not been fully explored. Therefore, continuous improvement of evaluation indicator systems and methods regarding the suitability of land for construction is worthy of additional attention, in addition to further expansion of the application field.

As one of the most common and threatening geological hazards, landslides primarily occur in mountainous areas due to the complex terrain, geological conditions and human engineering activities [7]. Landslides easily cause significant losses to towns because of the high susceptibility, frequency and speed thereof [7]. According to statistics, during the period from 2014 to 2018, landslides killed 4914 people, rendering 27,110 people homeless and resulting in asset losses totaling approximately USD 2.1 billion [8]. As a typical area in the Three Gorges reservoir area, Hechuan District in Chongqing is characterized by a large number of mountains and hills, with frequent landslides as the main geological disaster [9]. A series of explorations of landslide prediction methods have been conducted by scholars in the Three Gorges reservoir area [10]. As reported by the Hechuan District Land Resources and Housing Administration, new landslide sites develop in the area every year, none of which is within the original key monitoring areas. Thus, the exploration of machine learning methods based on landslide susceptibility in fragile ecological and geological environments will facilitate accurate identification of disaster sites and highly disaster-prone areas and is of considerable significance for the safety of local residents, development of national land and ecological protection. Landslide susceptibility based on machine learning has been extensively adopted in research on disaster prevention and mitigation in urban areas and towns. Several examples include random forest (RF) [11], logistic regression (LR) [12] and artificial neural network (ANN) models [13,14]. Such methods possess significant advantages over conventional methods in terms of assessment, verification and prediction of landslide susceptibility [9]. Among the methods, random forest is highly accurate and efficient and can process high-dimensional data while maintaining a high level of data accuracy, even if features are missing or unbalanced [8]. Depending on the geographical location, climatic conditions and the amount of available data on the researched area, an appropriate model should be selected to obtain satisfactory evaluation results. The modelling of landslide susceptibility has been widely used due to the generalization ability thereof. Analytical hierarchy process (AHP) analysis, which is used to evaluate the suitability of land for construction, is also a common weight evaluation model and was found to be applicable to the present study [15].

In summary, scholars have conducted a series of studies on landslide susceptibility and the suitability of land for construction. However, there has been a scarcity of research on the suitability of land for construction in mountainous areas with frequent disasters based on the foregoing two aspects. Thus, empirical research with in-depth and extensive discussion is needed, with a particular focus on determining how to evaluate the suitability of land for construction in mountainous areas from the perspective of disasters, as well as key problems, such as the improvement of evaluation indicator systems, evaluation criteria and technical methods concerning the suitability of land for construction from the perspective of disasters in mountainous areas.

As such, the Hechuan District of Chongqing was investigated, and random forest and AHP were adopted to explore the suitability of land for construction in mountainous areas from the perspective of landslide susceptibility. First, 754 historical landslide sites were sampled in the Hechuan District according to the theoretical method of landslide susceptibility assessment, and a landslide susceptibility model was established based on the RF model of Hechuan District; then, on the basis of the disaster safety model, an evaluation model of the suitability of land for construction in mountainous areas was established, considering social, ecological and economic factors.

Finally, the suitability for construction of different areas was rated, and the spatial distribution of land suitable for construction was revealed, providing a scientific basis for the evaluation of the suitability of land for construction in Hechuan District and constructing an accurate, operatable and generalizable evaluation model. By constructing an accurate, operatable and generalizable evaluation indicator system and research method, the present study provides a scientific basis for evaluation of the suitability of land for construction in Hechuan District, offering a reference for planning and construction of other mountainous areas with frequent disasters.

2. Area of Research and Data Source

2.1. Study Area

Located at 105°58′37″–106°40′37″ east and 29°51′02″–30°22′24″ north, Hechuan District measures 69 km in length east to west and 58 km in width north to south and is situated northwest of Chongqing [16]. The area includes 23 towns and seven subdistrict offices. In 2019, the urban population in Hechuan District was 1.43 million. The area covers a total of 2375.61 km^2, and the construction land area is 232.55 km^2, accounting for 9.92% of the total area, with approximately 7% forest land coverage.

Hechuan District is situated at a junction of hills between the hilly Sichuan Basin and the valley province of Chuandong; therefore, the terrain is roughly divided into parallel ridges and gentle hills. Situated in a transition zone between gentle hills and mountains in the basin, there are numerous slopes and deposits in the piedmont belt in the southeast (see Figure 1). The area marks the convergence of the Jialing River, the Fujiang River and the Qujiang River in the territory. The Jialing River is the largest river in the area, with well-developed water systems, abundant water resources and ample rainfall [17]. The strata in Hechuan District mainly include Paleozoic Permian (P), Mesozoic Triassic (T), Jurassic (J) and Cenozoic Quaternary (Q). At the axes of the local anticline mountains are several active Quaternary or Cenozoic faults, largely distributed in the Yunwushan area of a branch of Huaying Mountain (Line 1 of Shuangfeng-Yanjing and Sanhui-Qingping-eastern Tuchang) (Figure 1).

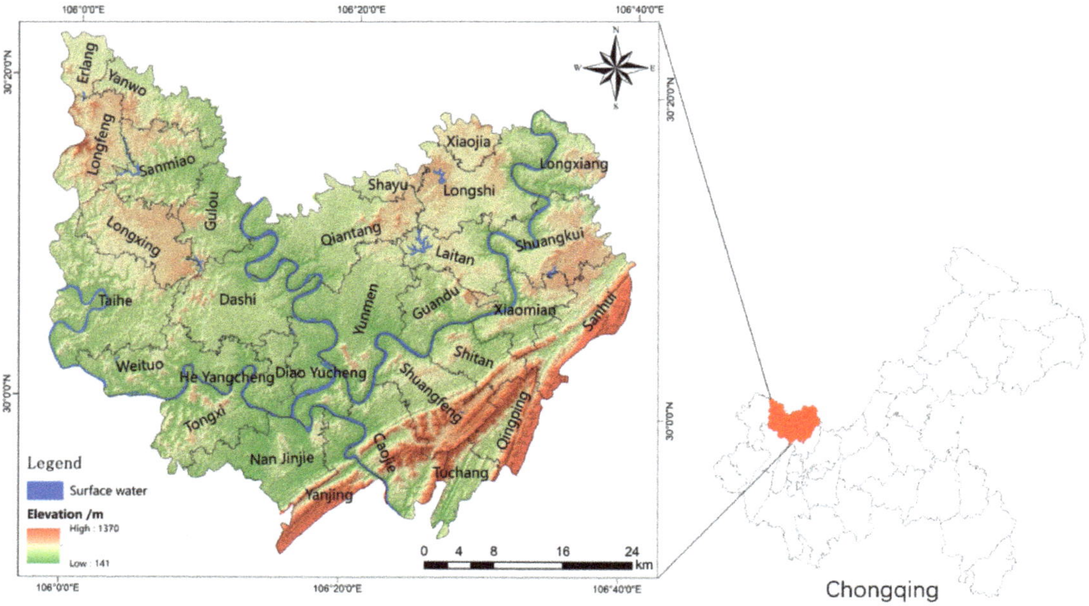

Figure 1. Topographic map of Hechuan District.

2.2. Data Sources

The data used in this research were derived from 754 historical landslides in Hechuan District (2000–2016), including: DEM raster data with a 30 m spatial resolution; geological raster data with an accuracy of 1:200,000; land-use and administrative division vector data with an accuracy of 1:100,000; satellite imagery raster data with a 30 m resolution on a geospatial data cloud platform; 1:100,000 vector data via river network 1 from the Chongqing Water Resources Bureau; a multiyear rainfall data table with an accuracy of 30 m; 1:100,000 road data from the Chongqing Municipal Transportation Commission; Chongqing POI data obtained by web crawlers; data on rural residential areas, urban built-up areas, nature reserves, ecological red lines, etc.; rural settlement data from the Land Change Investigation Database; raster data with a 30 m resolution from urban built-up areas, which were extracted from a geographical information database; and ecological red line (natural reserves and water conservation areas) statistics from Chongqing's Ministry of Natural Resources. The raster data were converted to a raster corresponding to a DEM resolution of 30 m, owing to the varied spacings and scales of the elements. Due to the restricted land conditions in Chongqing, the standard setting of urban and rural construction land areas could be adopted according to its own situation with floating coefficients. Based on the delineation standard of the resolution of districts and counties in Chongqing and in prior research [18,19], a spatial resolution of 30 m was determined in this research, which allowed for the spatial characteristics of landslides and construction land to be captured while reducing the complexity of the calculation. Additionally, continuous factors were classified for further classification and assignment of values (Table 1). Based on field surveys, expert experience and the relevant literature, the natural breakpoint method was used to determine the threshold values for each factor and subsequently adjusted to regional conditions to meet the requirements of the actual situation.

Table 1. Data and data sources for landslide susceptibility zoning.

Data	Source	Type	Accuracy
Historical landslide	Chongqing Geological Monitoring Station	Data table	
DEM	Aster satellite	Raster data	30 m
Geological	National Geological Archives Data Center	Raster data	1:200,000
Land use	Geographical Information Monitoring Cloud Platform	Vector	1:100,000
Administrative	Geographical Information Monitoring Cloud Platform	Vector	1:100,000
River network	Chongqing Water Resources Bureau	Vector	1:100,000
Yearly average rainfall	Geographical Information Monitoring Cloud Platform	Raster data	30 m
Road	Resource and Environment Science and Data Center	Vector	1:100,000
Satellite image	Geospatial Data Cloud Platform	Raster data	30 m
POI	Web crawlers	Data table	
Rural settlements	Land Change Investigation Database	Raster data	30 m
Urban built-up area	Geographical Information Database	Raster data	30 m
Ecological red line area	Chongqing Ministry of Natural Resources	Raster data	30 m

3. Development of Indicator System and Research Methodology

3.1. Indicator System for Landslide Susceptibility Assessment

Landslides are closely associated with the stratigraphy and geomorphology of mountains [20], representing one of the main geological hazards in Hechuan with respect to the safety of the area. Therefore, landslide hazards were selected in this research for the construction of a safety index system for Hechuan District. In Reichenbach's [21] study, landslide factors were classified into five major categories: geology, hydrology, land cover, landform and others. In this research, human activities were incorporated into the assessment factors through field investigations and by synthesizing the realities of the densely populated and mountainous land limitations of Hechuan District. The safety assessment factors of landslide susceptibility in Hechuan District were established as follows: topography and geomorphology (elevation, slope, degree of relief, aspect, slope position, micro-landform, synthetic curvature, profile curvature, plan curvature, terrain roughness index (TRI) and topographic wetness index (TWI)); geological conditions (slope type, distance from fault and lithology); environmental conditions (distance from rivers, rainfall, land cover and NDVI); meteorological and hydrological conditions (sediment transport index (STI) and stream power index (SPI)); and human activities (distance from roads and POI kernel density). Furthermore, to provide a basis to evaluate the suitability of land for construction in mountainous areas, a geospatial database was created and combined with historical landslide data, a random forest algorithm was used to delineate landslide susceptibility zones and a safety zone model based on RF was constructed (Table 2).

The raw data were further processed in ENVI and ArcGIS 10.6 to obtain the factor data, and the 22 processed impact factors were reclassified according to the classification thresholds presented in Table 2.

3.2. Indicator System and Data on the Suitability of Land for Construction

The assessment of the suitably of land for construction is affected by a variety of factors. In this research, we referred to relevant studies, Hechuan District in Chongqing was taken as the study area and factors were selected at four levels, namely natural, social, ecological and safety, so as to construct an index system for the assessment of the suitability of mountainous areas for construction. Hechuan District is located at the confluence of three rivers in a mountainous area. Safety has emerged as a significant influencing factor on construction land, owing to the complexity and variability of the environment. The safety factor consists of the results of the landslide susceptibility assessment. The differing climatic environments and production conditions in the region have influenced the living habits and residential choices of residents. Eight indicators (elevation, slope, degree of relief, aspect, land cover, NDVI, distance from rivers and distance to roads) were selected to build a natural factor index system that supports scientific criteria for the applicability of construction land. Location conditions affect socioeconomic development, access to information and the convenience of residents. The central city is the main supplier of major public services, whereas rural settlements are built-up areas. In this research, social factors were characterized in terms of distance from built-up areas and distance from rural settlements. The ecological red line is a spatial boundary of the national territory that is specially protected to maintain ecological Safety and ecosystem integrity [22], representing an area where development and construction are strictly prohibited. The ecological factors (nature reserves and important water-conservation areas) were restricted and designated as non-construction zones (Table 3).

Table 2. Classification of factors influencing landslide susceptibility.

Type	Impact Factor	Number of Classifications	Classification Thresholds or Criteria
Terrain topography	Elevation (m)	10	1. <241; 2. 241~279; 3. 279~316; 4. 316~355; 5. 355~398; 6. 398~461; 7. 461~545; 8. 545~679; 9. 679~902; 10. >902
	Slope (°)	9	1. <5; 2. 5~10; 3. 10~15; 4. 15~20; 5. 20~25; 6. 25~30; 7. 30~35; 8. 35~40; 9. >40
	Degree of relief (m)	7	1. <20; 2. 20~30; 3. 30~40; 4. 40~50; 5. 50~80; 6. 80~120; 7. >120
	Aspect	9	1. south; 2 southwest; southeast; 3. east; west; northwest; northeast; 4. north; 5. none
	Slope position	6	1. ridge; 2. upper slope/cliff edge; 3. mid-slope; 4. flats slope; 5. down slope/cliff base; 6. valley floor
	Micro landform	10	1. canyons, deeply incised streams; 2. mid-slope drainages, shallow valleys; 3. upland drainages, headwaters; 4. U-shaped valleys; 5. plains; 6. open slopes; 7. upper slopes, mesas; 8. local ridges, hills in valleys; 9. mid-slope ridges, small hills in plains; 10. mountain tops, high narrow ridges
	Synthetic curvature	6	1. <−1; 2. −1~−0.5; 3. −0.5~0; 4. 0~0.5; 5. 0.5~1; 6. >1
	Profile curvature	6	1. <−1; 2. −1~−0.5; 3. −0.5~0; 4. 0~0.5; 5. 0.5~1; 6. >1
	Plan curvature	6	1. <−1; 2. −1~−0.5; 3. −0.5~0; 4. 0~0.5; 5. 0.5~1; 6. >1
	TRI	5	1. <1.05; 2. 1.05~1.1; 3. 1.1~1.15; 4. 1.15~1.2; 5. >1.2
	TWI	5	1. <4; 2. 4~6; 3. 6~8; 4. 8~10; 5. >10
Geological conditions	Slope type	7	1. type I antegrade/inclined slope; 2. oblique slope; 3. oblique slopes; 4. cross slopes; 5. reverse slope; 6. type II forward/outward slopes; 7. flat slopes
	Distance from fault (m)	7	1. <500; 2. 500~1000; 3. 1000~1500; 4. 1500~2000; 5. 2000~2500; 6. 2500~3000; 7. >3000
	Lithology	8	1. T_{1f-j}; 2. P; 3. T_{21}; 4. T_{3xj}; 5. J_{1z-2x}; 6. J_{3sn}; 7. Q_p; 8. J_{2s}
Environmental conditions	Distance from rivers (m)	7	1. <100; 2. 100~200; 3. 200~300; 4. 300~400; 5. 400~500; 6. 500~600; 7. >600
	Rainfall (mm)	8	1. <1131; 2. 1131~1160; 3. 1160~1186; 4. 1186~1210; 5. 1210~1233; 6. 1233~1266; 7. 1266~1316; 8. >1316
	Land cover	9	1. woodland; 2. cultivated land; 3. water area and water conservancy facilities land; 4. woodland; 5. industrial and mining storage land; 6. garden plot; 7. other land; 8. land used for construction; 9. transportation land use
	NDVI	6	1. <0.10; 2. 0.10~0.15; 3. 0.15~0.20; 4. 0.20~0.25; 5. 0.25~0.30; 6. >0.30
Meteorological hydrology	STI	6	1. <20; 2. 20~40; 3. 40~70; 4. 70~100; 5. 100~200; 6. >200
	SPI	7	1. <15; 2. 15~30; 3. 30~45; 4. 45~60; 5. 60~100; 6. 100~1000; 7. >1000
Human activity	Distance from roads (m)	7	1. <100; 2. 100~200; 3. 200~300; 4. 300~400; 5. 400~500; 6. 500~600; 7. >600
	POI kernel density	7	1. <1; 2. 1~2; 3. 2~3; 4. 3~4; 5. 4~5; 6. 5~10; 7. >10

Table 3. Classification of contributing elements for construction land suitability evaluation.

Factor	Impact	Number of Levels	Classification Threshold or Criteria	Reasons for Classification	Remarks
Safety	Landslide susceptibility	5	1. very low; 2. low; 3. medium; 4. high; 5. very high	Expert experience (A)	The results of the landslide susceptibility assessment were combined with five levels of classification with reference to previous studies [23].
Nature	Elevation (m)	5	1. <270; 2. 270~335; 3. 335~447; 4. 447~709; 5. 709~1370	Natural breakpoints (B)	The natural breakpoint method maximises the difference between classes, and the breakpoint itself is a suitable boundary for grading [24].
	Slope (°)	5	1. <3; 2. 3~8; 3. 8~15; 4. 15~25; 5. >25	Expert experience (A)	Slope classification with reference to *Principles of Urban Planning* (3rd edition) and previous studies [5].
	Degree of relief (m)	5	1. 11; 2. 11~20; 3. 20~31; 4. 31~46; 5. 46~125	Natural breakpoints (B)	Ibid.
	Aspect (°)	5	1. south; 2. southwest, southeast; 3. east, west, northwest, northeast; 4. north; 5. none	Proprietary classification (C)	None
	Distance from rivers (m)	5	1. <300; 2. 300~500; 3. 500~1000; 4. 1000~1500; 5. >1500	Expert experience (A)	Classification based on expert experience and relevant research [25].
	Land cover	5	1. lands for construction; 2. grassland and unused lands; 3. shrub; 4. forest land and cultivated land; 5. waters, beaches, and basic farmland	Proprietary classification (C)	None
	Distance from roads (m)	5	1. <200; 2. 200~400; 3. 400~500; 4. 500~600; 5. >600	Natural breakpoints (B)	The distance of the road from the built-up area was taken into account on the basis of natural breakpoints. As aboved,
	NDVI	5	1. <0.10; 2. 0.10~0.15; 3. 0.15~0.20; 4.0.20~0.30; 5. >0.30	Natural breakpoints (B)	NDVI = 0 indicates rocky or bare soil that is not suitable for building. NDVI is graded less than 0.1 for bare land, NDVI less than 0.3 for low vegetation cover and suitable for building, and between 0.1 and 0.3 according to the percentage of area in each zone.
Society	Distance from rural settlements (m)	5	1. <500; 2. 500~1000; 3. 1000~2000; 4. 2000~3000; 5. >3000	Natural breakpoints (B)	The classification is based on the spatial distribution of the percentage of patches in rural settlements combined with natural breakpoints.
	Distance from built-up areas (m)	5	1. <500; 2. 500~1000; 3. 1000~2000; 4. 2000~3000; 5. >3000	Natural breakpoints (B)	On the basis of known patch sizes of built-up land, combined with natural breakpoints for delineation.
Ecology	Ecological red line area (nature reserve/important water-conservation sites)	1	constructive expansion prohibited zone	Proprietary classification (C)	Designated as a no-build zone.

Considering the specific characteristics of Hechuan District, an expert scoring method was adopted to comprehensively determine the grading standard value of each factor. The indicators were classified into five levels according to accepted standards established in previous studies [25]. The evaluation factors were classified as: the most suitable level, with an optimal value of "1"; medium suitability, with a value of "2"; basic suitability, with a value of "3"; unsuitable, with a value of "4"; and prohibited construction, with a value of "5". Additionally, the M_i weight values of indices at different levels were calculated; the specific grading criteria are shown in Table 3.

3.3. Research Methodology

3.3.1. Research Steps

The present research was conducted in seven steps: (1) preparation of a data inventory for the landslide and construction land suitability assessment and generation of a geospatial database in ArcGIS 10.6 software; (2) construction of an indicator system for landslide susceptibility and calculation of the weights of landslide hazard impact factors using the average Gini coefficient; (3) assessment and ranking of landslide susceptibility; (4) receiver operating characteristic (ROC) curve analysis for accuracy and model validation to verify the performance of the susceptibility assessment; (5) construction of an index system for construction site suitability assessment using the results of the landslide susceptibility assessment and the natural, social and ecological factors; (6) construction of a judgement matrix for the development of site suitability and calculation of the rankings of the variables using the AHP method; (7) and assignment of construction site suitability ratings according to the weights (Figure 2).

3.3.2. Random Forest

In 2001, Breiman [26] developed the random forest model as a modern classification and regression technique to collect data for learning and processing. Multiple samples were obtained from the original samples after resampling by bootstrapping. RF randomly samples the samples and features, thereby providing improved stability and accuracy relative to traditional landslide prediction methods [22]. The output of RF is based on multiple decision trees voting on the judgement results. The samples are trained to obtain each classification model (u1(X), u2(X), … , uk(X)1–2) and by n independent decisions (u(X,θk; K = 1,2, … N)) to form the RF model [22,27]. RF is tolerant of outliers and noise, does not overfit and achieves high prediction accuracy and stability. The categorization models are then used to build the RF model. Such an approach has been adopted in a variety of fields, such as clustering, regression, discrimination and survival analysis, in which variable evaluation is viral [22]. The RF in this research was composed of two trees (positive and negative cells), each with 22 random characteristics (22 landslide condition factors). See Equation (1) for details:

$$H(x) = arg\ max_y \sum_{i=1}^{k} I(h_i(x)) = Y \tag{1}$$

where $H(x)$ is the output classification result, h_i refers to a classifier of a single decision-making tree, Y represents the output variable and $I(h_i(x))$ denotes the indicator function.

The random forest model was constructed using the R language package "randomForest", and ROC curves were plotted using the R package "pROC". The landslide susceptibility results constructed by the RF model were assessed using ROC curve analysis. The area under the ROC curve can be used to quantify the accuracy of the model prediction; the closer the ROC curve to the top left, the higher the accuracy of the model. The area under the curve (the AUC value) refers to the area covered by the ROC curve, which can be used to quantify the accuracy of the model. The AUC value is in the range of [0, 1], with a higher value indicating higher model accuracy. See Equation (2).

$$\text{AUC} = \frac{\sum_{i=1}^{n_0} r_i - n_0 \times (n_0 + 1)/2}{n_0 \times n_1} \tag{2}$$

where the n_0 and n_1 distributions represent the number of counter and positive cases respectively, and the r_i distribution is the ranking of the i th counter case in the overall test sample.

The construction of the random forest model consists of the following main steps (Figure 3).

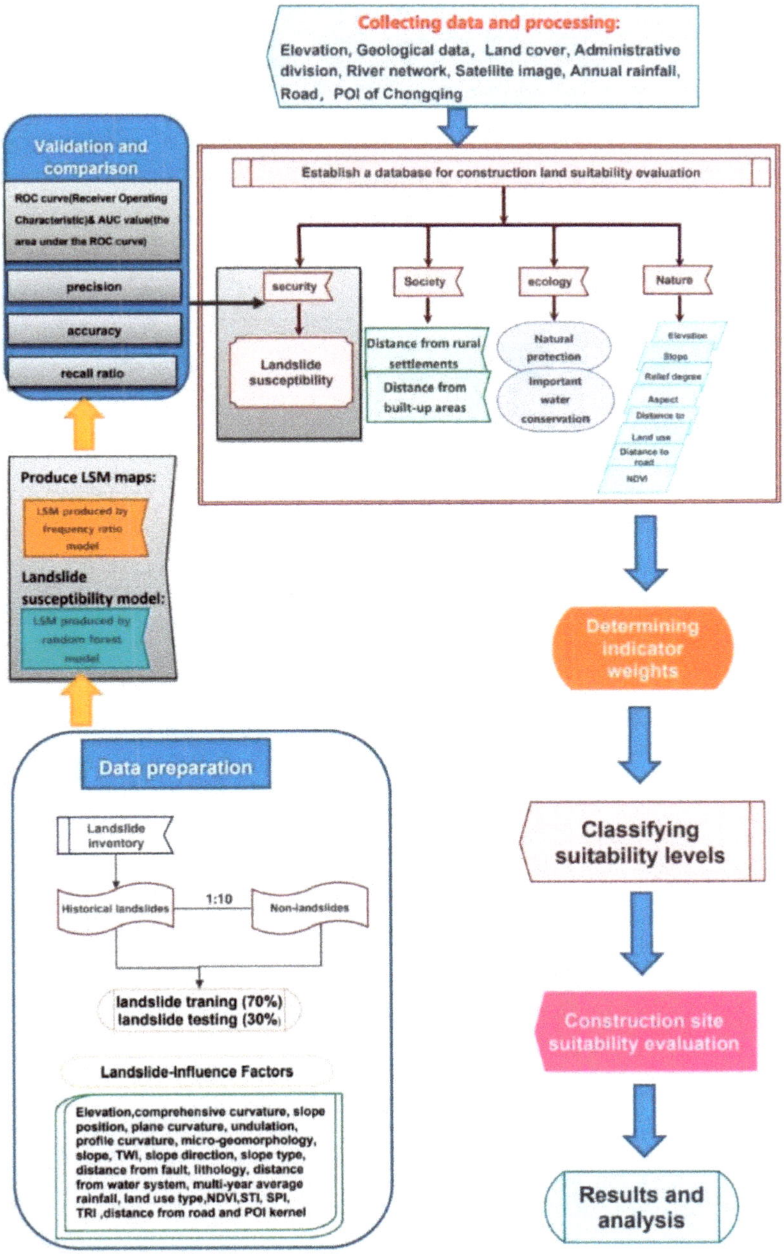

Figure 2. Flow chart of the study.

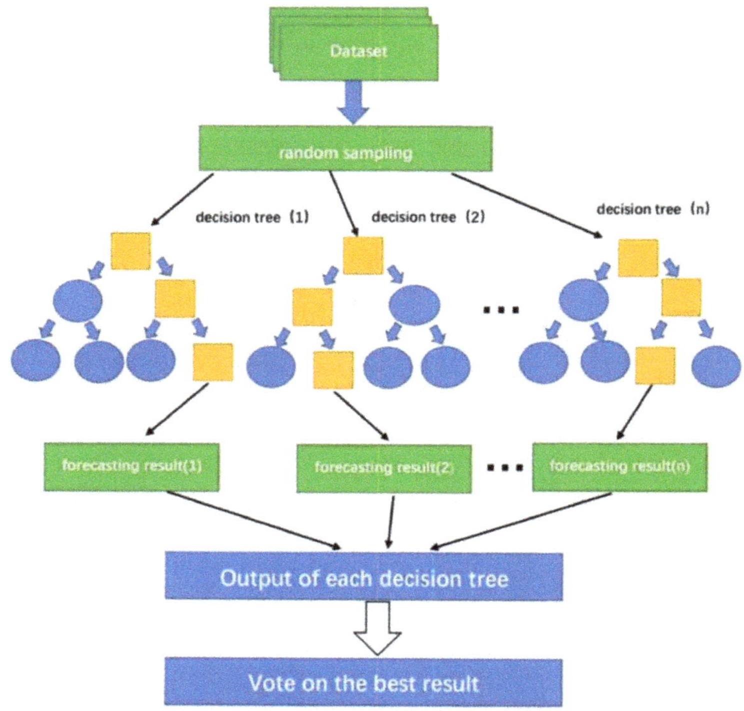

Figure 3. Schematic diagram of the RF algorithm.

A random forest model is constructed on the basis of N decision trees, and a decision tree is constructed using each subset combination. Within the constructed decision trees, node partitioning is performed using the CART algorithm. CART is based on the principle of minimizing the Gini coefficient and randomly selecting objects to be assigned to class I at node t based on the probability ($p(i \mid t)$). The estimated probability that an object actually belongs to class j is $p(j \mid t)$. See Equation (3) for details:

$$\text{Gini} = \sum_{i \neq i}^{J} (p(i|\text{t})p(j|\text{t})) \tag{3}$$

The random forest package within R was use to implement the random forest technique. The 754 historical landslide points were used as positive samples, the 500 m buffer zone of the landslide points and the river area were removed as non-landslide areas and 7540 non-landslide points were randomly selected as negative samples in a ratio of 1:10 to form the entire dataset. The training and test sets were divided into a ratio of 7:3, and for historical slippage points, there were 5806 training sets and 2488 test sets. To determine the accuracy of the ROC curve analysis, the RF model was trained and validated using the tenfold cross-validation method.

The confusion matrix is the basis for ROC curves; it is represented in a standard format for accuracy evaluation. This means of number of observations in the wrong class and the wrong class of the classification model are counted separately, and the results are presented in a table, which is shown below (Tables 4 and 5).

Table 4. Table of confusion matrices.

	Models	True Value Landslide (1)	Non-Landslide (0)	Accuracy
Predicted value	Landslide (1)	TP	FP	Accuracy
	Non-landslide (2)	TN	FN	Accuracy
		Recall rate	Recall rate	Total accuracy

Table 5. Formulae.

	Formula	Significance
Accuracy ACC	$\text{Accuracy} = \frac{TP+TN}{TP+TN+FP+FN}$	Share of all correctly judged results of the classification model among the total number of observations
Accuracy PPV	$\text{Precision} = \frac{TP}{TP+FP}$	Of all the results for which the model prediction is positive, the proportion of model predictions that are correct
Sensitivity TPR	$\text{Sensitivity} = \text{Rec} = \frac{TP}{TP+FN}$	Weight of model prediction pairs among all results for which the true value is positive
Specificity TNR	$\text{Specificity} = \frac{TN}{TN+FP}$	The proportion of model predictions that are correct among all results for which the true value is negative

3.3.3. AHP Research Method

As proposed by Professor Saaty [28] (1980), the analytical hierarchy process (AHP) method, in which multiple criteria are selected to make decisions, is a simple, adaptable and practical approach to quantitatively analyze qualitative issues. The method can split a complex problem into a number of levels and factors, thereby allowing for a comparison between two indicators to be made to determine the degree of importance and to establish a judgement matrix. In the AHP method, a hierarchical relationship is established between all the elements, and the evaluation procedure is simple and easy to operate [21]. The AHP method is used to determine the weight values of the indicators, judge the importance of each factor, conduct comparative analysis and construct a judgement matrix. The expert scoring approach determines the relative significance of the evaluation indicators, and a two-by-two judgement matrix is created to compare indicators between the layers, with the consistency of the judgement matrix then checked to ensure that there is no bias in the two-by-two comparison process. Considering that the ecological factor is incorporated into the no-construction zone, only a hierarchical structure model of safety, nature and society was constructed. In the model, each evaluation factor was normalized, the relative significance of each variable was assessed, the weight values of the three categories of factors were identified using the AHP method and further judgements on the weights of various factors within the two categories of nature and society were made, whereas the weight of the safety factor was determined through the average Gini coefficient of RF. The weight values of the indicator layer for the criterion and target layers were calculated using YAAHP software [28,29]. By overlaying the weight maps of the influencing factors obtained using ArcGIS10.6 with the AHP method, a geospatial data-based model of the suitability of land for construction in mountainous areas affected by landslides is constructed. See Equation (4).

$$T = \sum_{i=1}^{n} M_i \times R_i \tag{4}$$

where T is the comprehensive assessment value of the suitability of the assessment unit for construction land, M_i represents the weight value of factor i, derived via the hierarchical analysis method, R_i denotes the i-th single factor score corresponding to the assessment unit and n refers to the total number of factors.

4. Results

4.1. Safety Level

The q-value of the mean Gini coefficient in the random forest explains the contribution of the factor, that is, the degree of influence of the degree factor on the landslide. The results show that the three factors of average multiyear rainfall, elevation and lithology had the greatest influence on landslides (Figure 8).

Landslides are a typical dichotomous problem, and the confusion matrix can be used to analyze the accuracy of the model. Table 6 shows the confusion matrix of the entire data set of the random forest model. According to the confusion matrix, the constructed random forest model exhibited a high degree of accuracy and high predictive value (Table 6).

Table 6. Confusion matrix of RF.

RF Predicted Value	True Value		Accuracy
	Landslides	Non-Landslide	
Landslides	528	0	Accuracy: 1
Non-landslide	226	7507	Accuracy: 0.971
	Recall rate: 0.700	Recall rate: 1	Accuracy: 0.972

In addition, the landslide susceptibility results constructed by the RF model were assessed by means of ROC analysis. In this research, ROC curve analysis was performed in R Studio software using the R language. The AUC values for the training, test and all samples were 0.999, 0.756 and 0.989, respectively (Figure 4). The test AUC values were greater than 0.7, indicating that the model prediction accuracy was high and stable.

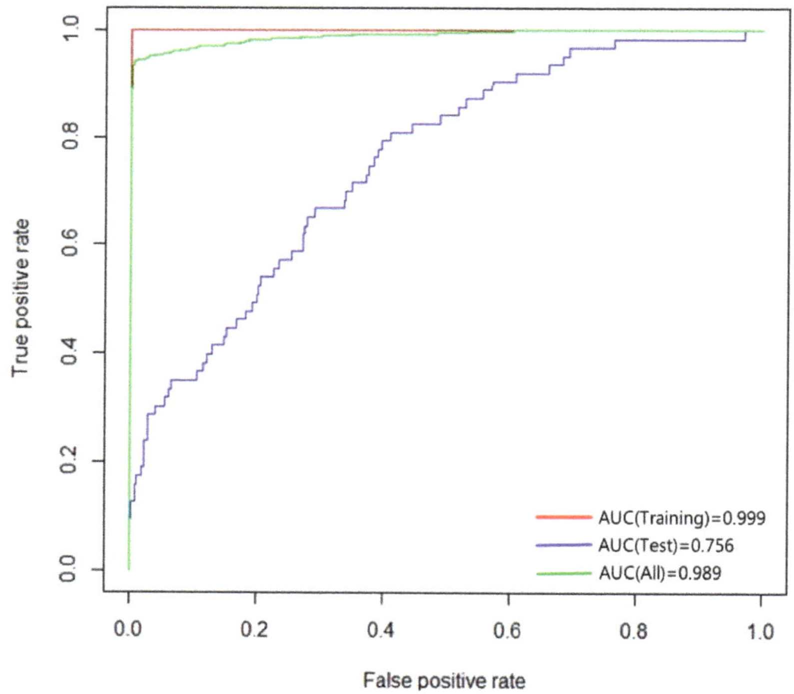

Figure 4. ROC curves and AUC values.

To evaluate the likelihood of landslides in the study area, a random forest model was applied to each grid in the area. The results of the random forest model were imported into ArcGIS 10.6, classified using the natural breakpoint method and adjusted according to the procedures described in prior research [8,22]. Landslide susceptibility was classified into the following five levels: extremely low, low, medium, high and extremely high susceptibility areas (Figure 5 and Table 7).

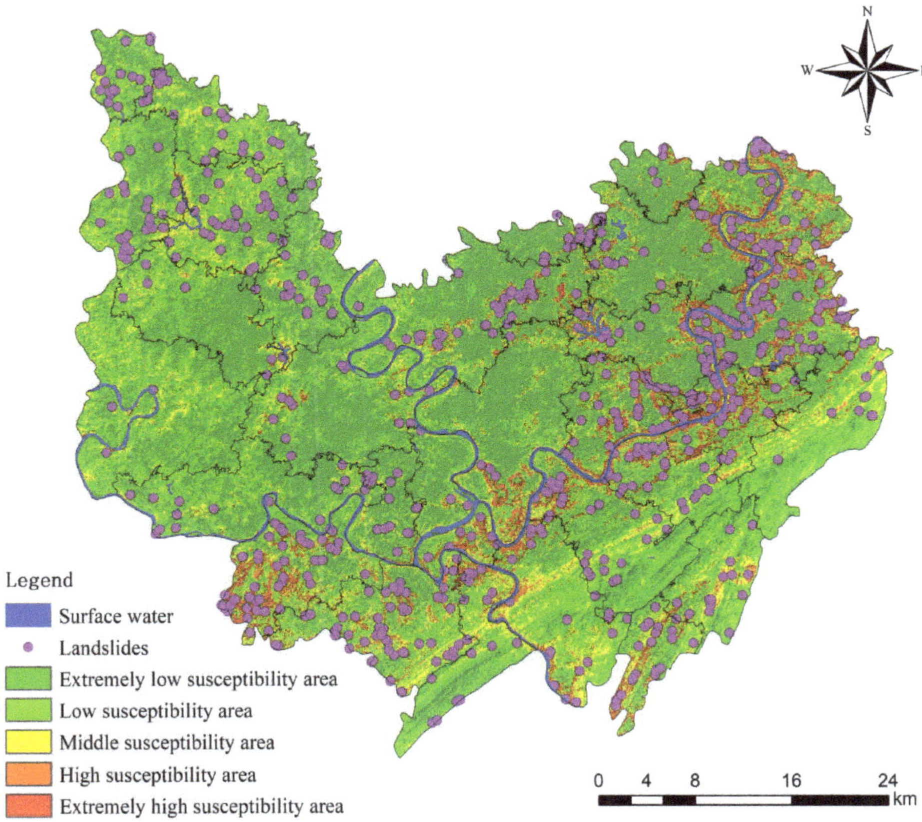

Figure 5. Landslide susceptibility zoning in Hechuan District.

Table 7. Statistical table of landslide susceptibility classification.

Probability of Landslide	Susceptibility Level	Grids	Area Ratio %	Landslides	Landslide Ratio %	Landslide Density/(Pcs/km^2)
<0.06	Extremely low	1,175,389	45.91	41	5.44	0.039
0.06~0.14	Low	791,083	30.67	74	9.81	0.104
0.14~0.24	Medium	373,736	14.41	80	10.61	0.238
0.24~0.38	High	174,718	6.53	91	12.07	0.579
>0.38	Extremely high	63,634	2.47	468	62.07	8.172

The majority of the regions in Hechuan District were found to have an extremely low or low landslide risk level. High-susceptibility zones are primarily situated in the northeast and near water systems. Landslides are rare in hilly basins with gentle terrain, and historical landslide areas correlate with the landslide susceptibility zones. With an enhancement in landslide susceptibility, the proportion of areas at each level, except the extremely high level, decreased. The number of landslides increased gradually, with the density strengthened, and there were a total of 753 landslide spots. The combined region of low and very low susceptibility accounts for 76.58% of Hechuan District's land area. The total landslides occurred in 15.25% of the total area. Landslides were possible in 74.14% of the land area, but regions of high and extremely high susceptibility accounted for only 9.01% of the land area.

As a crucial measure, extremely high-susceptibility areas should be largely concentrated along river valleys and mountains, which, to a considerable extent, affect urban development. The area of high or extremely high susceptibility spans 208.47 km^2, accounting for 9.01% of the total area, mainly distributed along the Qujiang and Jialing Rivers, including Xianglong Town, Shuanghuai Town, Xiaomian Town, Shitan Town, Guandu Town, Yunmen Subdistrict, Shuangfeng Town, Tongxi Town, Yanjing Street and Laitan Town. Such areas are strongly affected by surface water and rainfall. The rise and fall of river levels can result in landslide disasters during heavy rain. Landslide disasters in such areas induce changes in the courses of rivers; endanger infrastructure, residential areas and arable land; and have significant social impacts. The area of medium susceptibility spanned 333.62 km^2, accounting for 14.41% of the total area. Here, landslide disasters endanger infrastructure, residential areas and arable land, in addition to producing significant social impacts. The low-susceptibility area occupies 709.95 km^2, accounting for 30.67% of the total area, whereas the extremely low-susceptibility area spans 1062.88 km^2, accounting for 45.91% of the total area. Landslide disasters in such areas mainly threaten general facilities, residential areas and cultivated land, with a low level of risk. Differing from high- and extremely high =0susceptibility areas, low- and extremely low-susceptibility zones are extensively spread at lower altitudes in riverbank basins and around central metropolitan areas.

4.2. *Suitability Evaluation of Construction Lands*

4.2.1. Analysis of Evaluation Results

Given the complexity of mountainous areas, there are difficulties associated with determining the weights of evaluation factors using a quantitative assignment method. Although the judgement of expert experience has a certain degree of flexibility, the subjective assignment method can effectively adjust the weights for the land conditions of different regions, making the regional evaluation results more relevant and reliable. The safety factor in this research had a considerable impact in mountainous areas and was delineated as 33.377%. The natural factor, as the resource endowment of mountainous towns, involved more factors and had the highest weight. The social factor gradually emerged as a significant factor for evaluation. A veto system was adopted for the influence range indicator of the ecological factor, and the area to which it belongs was directly classified as a non-construction zone. Therefore, only the weight was calculated with the indicators using the hierarchical analysis method. The weighting values for the indicators of suitability of land for construction in mountainous areas were obtained with reference to previous studies

There are two types of indices for a judgement matrix: an index of consistency (CI) and a random consistency index (RI). The value ratios of the suitability, nature and society factors of construction land with respect to the consistency of the matrix were calculated to be 0.052, 0.097 and 0.000, respectively. All of the values were less than 0.1, thereby passing the consistency test and demonstrating that the results of the evaluation index weighting were reasonable (Tables 8–11).

Table 8. Construction land suitability index factor weights.

Index	Weight
Landslide	0.33377
Distance to built-up areas	0.09437
Distance to rural settlements	0.04719
Slope	0.09307
Distance from rivers	0.09232
Distance to roads	0.12069
NDVI	0.08149
Land cover	0.06794
Relief degree	0.02743
Aspect	0.02493
Elevation	0.01682

Table 9. Results of hierarchy analysis of suitability of construction land.

Item	Eigenvector	Weight Value	Maximum Eigenvalue	CI Value
Safety	1.001	33.377%	3.054	0.027
Nature	1.574	52.468%		
Society	0.425	14.156%		

Table 10. AHP results for social factors.

Item	Eigenvector	Weight Value	Maximum Eigenvalue	CI Value
Elevation	0.257	3.206%	8.959	0.137
Relief degree	0.418	5.227%		
Aspect	0.38	4.751%		
Distance from rivers	1.408	17.596%		
Land cover	1.036	12.949%		
NDVI	1.242	15.531%		
Distance to roads	1.84	23.002%		
Slope	1.419	17.738%		

Table 11. AHP results for natural factors.

Items	Eigenvector	Weight Value	Maximum Eigenvalue	CI Value
Distance to rural settlements	0.667	33.333%	2	0
Distance to built-up areas	1.333	66.667%		

According to the calculated weight results, as well as the classification and assignment of each index, ArcGIS was applied to superimpose a raster layer of each factor and to remove ecological red line space, thereby allowing for five levels of suitability for construction land in Hechuan District to be obtained: the most suitable area (1), more suitable area (2), basically suitable area (3), unsuitable area (4) and prohibited construction area (5) (Figure 6).

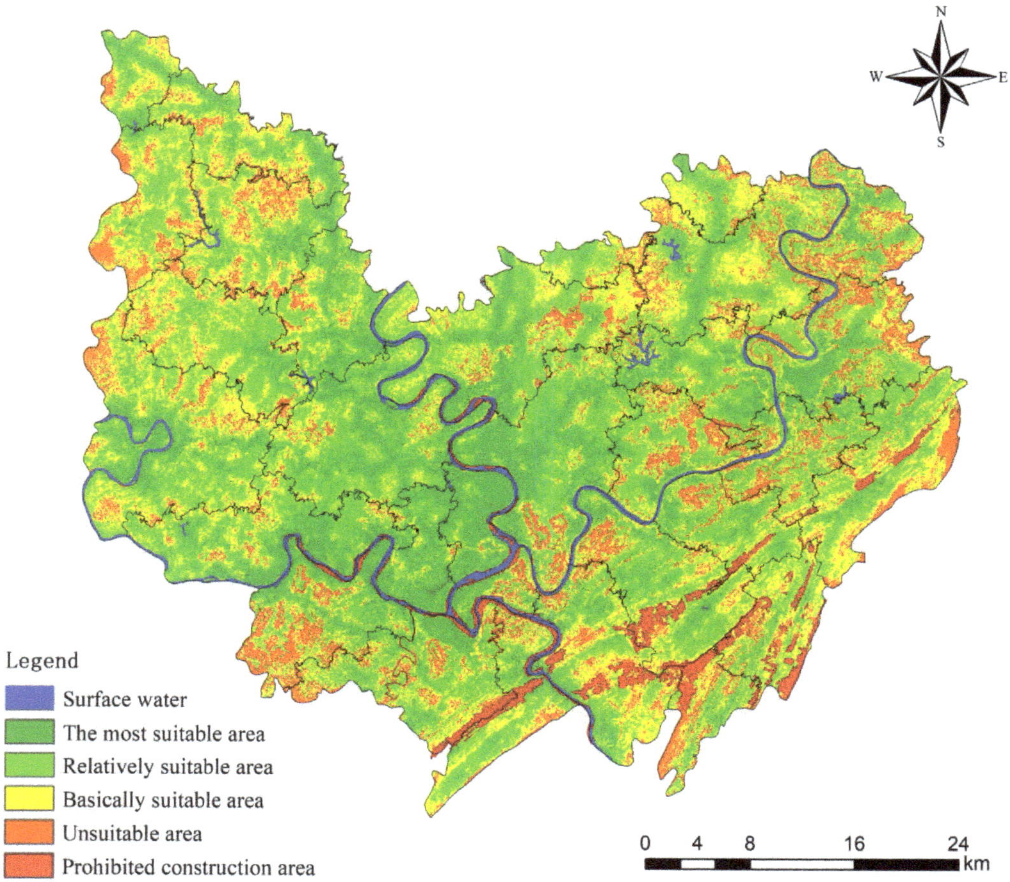

Figure 6. Suitability evaluation map of construction land in Hechuan District.

4.2.2. Suitability Zoning of Construction Lands

Using the ArcGIS10.6 platform, the aforementioned landslide susceptibility results and natural, social and ecological factors were superimposed and calculated according to Equation (3), with the results divided into the following zones according to the natural breakpoint method: prohibited construction, unsuitable, basically suitable, more suitable and the most suitable [2,30]. The ecological red line space was superimposed and divided into prohibited construction zones, and the final results of the suitability assessment of construction land in Hechuan District were obtained.

The suitability for construction was found to be good in Hechuan District. The suitable area spanned 2002.07 km^2, accounting for 84.28% of the total area, and the most suitable land was distributed in urban areas, where the three rivers meet or around small towns. The most suitable area spanned 637.18 km^2, accounting for 26.82% of the total area, and was mainly distributed in valley areas along the Jialing and Fujiang Rivers. Located at the confluence of many rivers, Yunmen Subdistrict has a low elevation and flat terrain, with sufficient water sources and convenient transportation. The more suitable area spanned 812.25 km^2, representing 34.19% of the total area. Such areas were divided into tracts in Zhongyunmen Subdistrict, Qiantang Town, Dashi Subdistrict, Heyangcheng Subdistrict and wide, hilly areas in Nanjin Town and Caojie Subdistrict. Most of the areas that cover such land are 220 m–350 m above sea level. The urban construction lands with basic suitability occupied 552.64 km^2, accounting for 23.26% of the total area, and were primarily found in the hilly regions outside the Huaying and Longduo Mountains. The less suitable areas spanned 241.90 km^2, accounting for 10.18% of the total area, scattered in the mountains with high elevations and were evenly distributed between Sanmiao Town, Yanwo Town, Sanlang Town, Longfeng Town, Taihe Town, Shayu Town, Guandu Town, Xianglong Town and Shuanghuai Town. The prohibited construction land covered an area of 131.65 km^2, accounting for 5.54% of the total area, and mainly distributed in the southeast area, the Huaying mountainous area and in regions in the ecological red line area, where there is the highest forest coverage.

5. Discussion and Conclusions

5.1. Discussion

5.1.1. Assessment of Significant Factors under Regional Characteristics

An assessment of the suitability of land for construction in mountainous areas is particularly significant for the development of towns and cities. Ying [31], Yi [32] and Peng [19] considered the impact of geological hazards on construction land and the safety of human life in the assessment of construction suitability in mountainous areas. Figure 7 shows the ranking of the importance of the factors in the suitability assessment of building sites (Figure 7).The top three factors for construction land suitability in this research area were identified using the AHP method, namely landslide susceptibility, distance from roads and distance from built-up areas (Figure 7). Among such factors, landslide susceptibility, as a geological hazard, is a significant indicator for assessment of the suitability of construction land and for the assessment of fragile ecological and geological environments, which is of considerable significance for the safety of regional residents, economic development and ecological protection. In this research, landslide susceptibility was incorporated into the construction land suitability assessment system from the perspective of hazards. Using the average Gini index module of the "randomForest" package, an importance ranking of the 22 factors of landslide susceptibility was achieved (Figure 8). According to the observation results, the three factors of multiyear average rainfall, elevation and lithology had the greatest influence on landslides (Figure 8). Landslides occur more often when rainfall is in the range of 1160 mm to 1266 mm. Several studies have shown [30,31,33] that the influence of rainfall on landslides depends largely on the amount of rainfall, in addition to the length of rainfall, and that landslides are more likely to occur in areas of prolonged heavy rainfall. Such parameters were all higher in the study area and were basically consistent. Landslides are also more likely to occur when the elevation is between 241 m and 461 m, which can be attributed to such an elevation range being conducive to human survival and a large number of human activities leading to changes in the geological environment, thereby increasing the probability of landslides [33]. The influence of the lithology of the strata on landslides mainly depends on the hardness, weathering and permeability of the rocks therein. The distance from the road and the distance from built-up areas show the significant role of geographical location with respect to the suitability of land for construction. Hechuan District is situated in a typical mountainous region with hillsides and limited land resources; thus, the convenience of road access becomes a significant factor that affects the development of mountainous areas. The distance from the road indicates road accessibility, with a greater road accessibility indicating a higher grade of construction land suitability. Built-up areas, such as mature town development areas, have relatively complete public service facilities and infrastructure, the economic benefits of which radiate to the surrounding areas; thus. the distance from built-up areas is a significant indicator in assessing the land value of construction land. As such, the importance of such factors can serve as the foundation for the development and construction of mountainous areas. However, at present, the safety level delineated by the random forest model is still subject to limitations in terms of data, and the misjudgments caused by the over-representation of non-landslides cannot yet be excluded [34]. As the survey research becomes more in-depth and the quality of data improves, a spatial database of multiple hazards, such as landslides, debris flows and floods, can be constructed. Furthermore, the zoning of such hazards should be refined in the future to clarify the inter-relationships between them to obtain more comprehensive safety zone assessment results and to improve the accuracy of the assessment of the suitability of land for construction in mountainous areas.

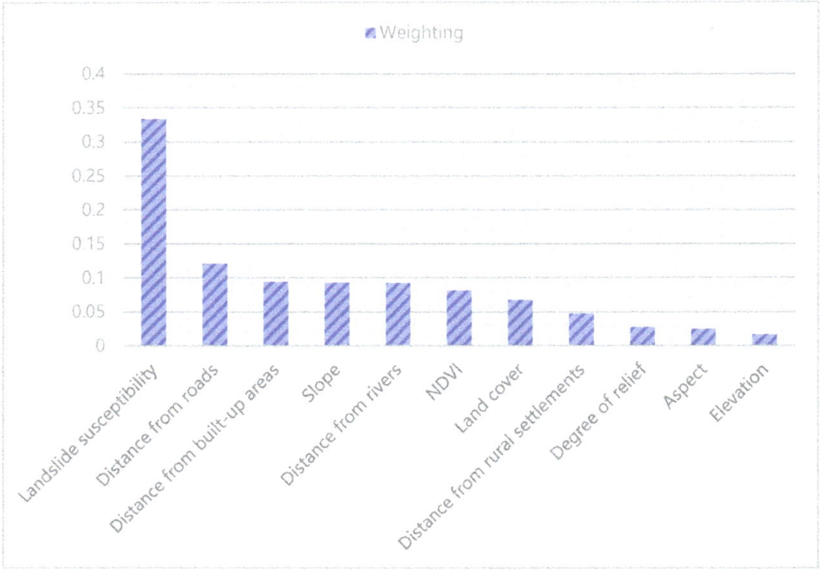

Figure 7. Ranking the importance of factors influencing the suitability of construction.

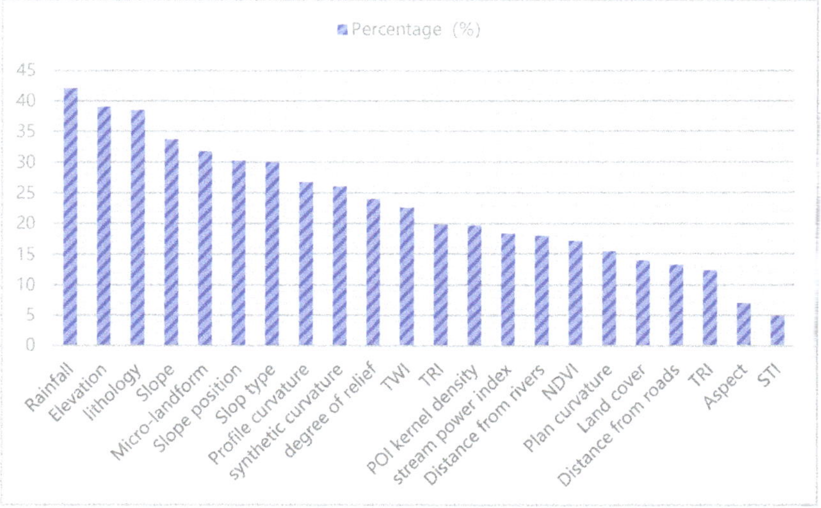

Figure 8. Ranking the importance of factors influencing landslide susceptibility.

5.1.2. Optimization of Disaster Prediction Models

Previous studies on the suitability of land for construction have mostly included statistical listings of existing disaster prevention results, whereas the integration of planning with geological hazard prevention and control has been neglected, thereby diminishing the guiding value of such studies in practice. Due to management problems and a lack of data, there is a scarcity of disaster prevention and control research. The results of previous studies mainly apply to geological hazard prevention and control, land-use planning and disaster prevention and mitigation. In this study, the random forest model was applied to the case study area. Moreover, the results demonstrate that the AUC value of the ROC curve in the training dataset was the highest, followed by the regional simulation and the validation dataset, at 0.999, 0.756 and 0.989, respectively. The RF model exhibited high reliability and stability. By selecting the optimal samples through 10-fold cross-validation and screening analysis of dominant condition factors, a more efficient and accurate random forest landslide susceptibility assessment model could be built with fewer dominant condition factors. The construction of a landslide susceptibility prediction model through the random forest model was the focus of the present study. The results show that the introduction of the RF model could improve the accuracy and precision of hazard prediction and that the RF model could be used in regions or countries with the same topography and geological conditions. Furthermore, the RF model could contribute to the hazard assessment of construction sites in mountainous areas, thereby reducing the workload and improving efficiency in practice.

5.1.3. Suitability of Construction Land vs. Non-Construction Land

Construction land and non-construction land, as two types of land, have a significant influence on the development of towns and cities. A suitability assessment of construction land is used to classify and secure land requirements for urban development and to balance the development costs. A random forest model was utilized in this research to construct five levels of landslide susceptibility, to determine the safety class of the study area and to explore the suitability of construction land on the basis of hazard assessment. The high- and very high-susceptibility areas for landslide prediction were found to be significant components of non-construction land, primarily belonging to the production and mostly ecological areas. To maintain the overall ecological safety of the area, the regional scope of the non-construction land that needs to be protected is clarified from a "counter-planning" perspective [35]. By analyzing and investigating ecological processes, Yu [36] used an ecological safety pattern approach to calculate the spatial extent of non-construction land in order to delineate the ecological safety pattern level of the study area. The prohibited areas and the unsuitable areas in the suitability assessment of construction land in this research were mostly non-construction areas, whereas the basically suitable areas could be incorporated into construction land or non-construction land, which needs to be comprehensively delineated according to geological conditions. The more suitable areas and the most suitable areas were found to be the main components of construction land. The basic starting points for determining the suitability of construction land and non-construction land are considerably different. The suitability of non-construction land is assessed from the perspectives of ecological safety and economy, whereas the suitability of construction land is assessed from the viewpoints of urban development, the impact of natural disasters and social and economic influences on the land.

6. Conclusions

In this research, the suitability of land for construction in mountainous areas was evaluated based on landslide susceptibility, and an indicator system was constructed that considers the four dimensions of safety, nature, society and ecology. In response to the drawbacks of existing methods, an attempt was made to identity the factors of landslide susceptibility using machine algorithms based on the number and spatial location of each indicator. Through such means, the rating of suitability of land for construction in mountainous areas was explored. The case study of the foregoing evaluation framework and method was conducted in the Hechuan District of Chongqing. The research results were as follows:

(1) The average accuracy of the tenfold cross-validation training set landslide data reached 0.978; the accuracy of the test set reached 0.913; the accuracy of the confusion matrix reached 97.2%; and the AUC values of the training test and all samples were 0.999, 0.756 and 0.989 respectively. The historical landslide sites in Hechuan District were mostly concentrated in highly susceptible areas, where the spatial areas of land with high landslide susceptibility and very high landslide susceptibility were 1.98 km^2 and 2.22 km^2, respectively, accounting for 2.47‰ and 6.53‰ of the study area. The areas with high landslide susceptibility were mainly concentrated in the south and southeast valleys and near the water system, whereas landslides were less frequent in the gentle hilly basin.
(2) The suitability of land for construction in mountainous areas was found to be most influenced by landslide susceptibility, the distance from roads and the distance from built-up areas. Furthermore, the annual average rainfall, elevation and lithological factors were significant factors influencing landslides in such areas. The suitability of land for construction in mountainous areas near the main city was promoted by locational advantages and restricted by disasters.
(3) Under the constraints of landslide susceptibility, the Hechuan District has considerable potential land reserves for construction in terms of more suitable areas and the most suitable areas (accounting for 61.01% of the study area) for construction. In terms of space, the more suitable and most suitable areas for construction were mainly distributed in the urban area, where the three rivers converge and the surrounding areas of small towns, showing a spatial distribution pattern characterized by a high central part and two low sides. The basically suitable areas for construction were mainly distributed at the buffer space on the periphery of the more suitable areas for construction.

Compared with existing research, the proposed evaluation indicator system and method with respect to the suitability of land for construction represent clear academic concepts, reflecting the essence and practical value of the suitability of land for construction in mountainous areas. The indicator system is simple and clearly structured with complete coverage, providing a basis for research and practice concerning the suitability of land for construction in other mountainous areas. The evaluation method is precise, easy, flexible and practical. In exploring a more accurate and convenient evaluation framework and method and extending the application scope of suitability evaluation in mountainous areas, the present study overcomes the issues encountered in previous research related to the suitability of land for construction in mountainous areas based on the perspective of disasters. However, the presented indicator system and evaluation method are only applicable to towns in mountainous areas with frequent disasters, and the validity thereof in other types of land and areas should be further verified.

Author Contributions: Conceptualization: J.Z.; Data curation: L.L. and X.C.; Formal analysis: R.L. Funding acquisition: L.L. and D.S.; Investigation: L.L.; Methodology: J.Z.; Resources: L.L. Software R.L.; Validation: D.S.; Visualization: X.C.; Writing—original draft: L.L. and X.C.; Writing—review and editing: J.Z. All authors have read and agreed to the published version of the manuscript.

Funding: This study was supported by the Science and Technology Research Program of Chongqing Municipal Education Commission (Grant No. KJQN202000525); Chongqing Graduate Research Innovation Project, Project Approval Number (Grant No. CYB22264); Natural Science Foundation of Chongqing (Grant No. CSTB2022NSCQ-MSX0594); Chongqing Natural Science Foundation (Grant No.cstc2020jcyj-msxmX0841).

Data Availability Statement: Not applicable.

Conflicts of Interest: The authors declare no conflict of interest.

References

1. Pu, J.; Zhao, X.-Q.; Miao, P.-P.; Li, S.-N.; Tan, K.; Wang, Q.; Tang, W. Integrating multisource RS data and GIS techniques to assist the evaluation of resource-environment carrying capacity in karst mountainous area. *J. Mt. Sci.* **2020**, *17*, 2528–2547. [CrossRef]
2. Cui, X.; Zhou, T.; Xiong, X.; Xiong, J.; Zhang, J.; Jiang, Y. Farmland Suitability Evaluation Oriented by Non-Agriculturalization Sensitivity: A Case Study of Hubei Province, China. *Land* **2022**, *11*, 488. [CrossRef]
3. Tan, M.; Li, X.; Xie, H.; Lu, C. Urban land expansion and arable land loss in China—A case study of Beijing–Tianjin–Hebei region. *Land Use Policy* **2005**, *22*, 187–196. [CrossRef]
4. Bagheri, M.; Zaiton Ibrahim, Z.; Mansor, S.; Manaf, L.A.; Akhir, M.F.; Talaat, W.I.A.W.; Beiranvand Pour, A. Land-Use Suitability Assessment Using Delphi and Analytical Hierarchy Process (D-AHP) Hybrid Model for Coastal City Management: Kuala Terengganu, Peninsular Malaysia. *ISPRS Int. J. Geo-Inf.* **2021**, *10*, 621. [CrossRef]
5. Ustaoglu, E.; Aydınoglu, A.C. Suitability evaluation of urban construction land in Pendik district of Istanbul, Turkey. *Land Use Policy* **2020**, *99*, 104783. [CrossRef]
6. Zhang, Y.; Chen, X. A Study on the Choices of Construction Land Suitability Evaluation of Ecological Index. *Procedia Comput. Sci.* **2016**, *91*, 180–183.
7. Zhang, Y.; Wen, H.; Xie, P.; Hu, D.; Zhang, J.; Zhang, W. Hybrid-optimized logistic regression model of landslide susceptibility along mountain highway. *Bull. Eng. Geol. Environ.* **2021**, *80*, 7385–7401. [CrossRef]
8. Sun, D.; Xu, J.; Wen, H.; Wang, Y. An Optimized Random Forest Model and Its Generalization Ability in Landslide Susceptibility Mapping: Application in Two Areas of Three Gorges Reservoir, China. *J. Earth Sci.* **2020**, *31*, 1068–1086. [CrossRef]
9. Miao, F.; Zhao, F.; Wu, Y.; Li, L.; Xue, Y.; Meng, J. A novel seepage device and ring-shear test on slip zone soils of landslide in the Three Gorges Reservoir area. *Eng. Geol.* **2022**, *307*, 106779. [CrossRef]
10. Miao, F.; Wu, Y.; Xie, Y.; Li, Y. Prediction of landslide displacement with step-like behavior based on multialgorithm optimization and a support vector regression model. *Landslides* **2017**, *15*, 475–488. [CrossRef]
11. Taalab, K.; Cheng, T.; Zhang, Y. Mapping landslide susceptibility and types using Random Forest. *Big Earth Data* **2018**, *2*, 159–178. [CrossRef]
12. Xie, P.; Wen, H.; Ma, C.; Baise, L.G.; Zhang, J. Application and comparison of logistic regression model and neural network model in earthquake-induced landslides susceptibility mapping at mountainous region, China. *Geomat. Nat. Hazards Risk* **2018**, *9*, 501–523. [CrossRef]
13. Xu, K.; Kong, C.; Li, J.; Zhang, L.; Wu, C. Suitability evaluation of urban construction land based on geo-environmental factors of Hangzhou, China. *Comput. Geosci.* **2011**, *37*, 992–1002. [CrossRef]
14. Sun, D.; Wen, H.; Xu, J.; Zhang, Y.; Wang, D.; Zhang, J. Improving Geospatial Agreement by Hybrid Optimization in Logistic Regression-Based Landslide Susceptibility Modelling. *Front. Earth Sci.* **2021**, *9*, 686. [CrossRef]
15. Wang, Y.; Wen, H.; Sun, D.; Li, Y. Quantitative Assessment of Landslide Risk Based on Susceptibility Mapping Using Random Forest and GeoDetector. *Remote Sens.* **2021**, *13*, 2625. [CrossRef]
16. Li, C.X.; Hong, Z.; Deliang, S.; Haijia, W.; Shuang, C. Evaluation of the suitability of rural settlement sites based on random forest algorithm: The case of Hechuan District, Chongqing. *J. Chongqing Norm. Univ.* **2022**, *39*, 93–104.
17. Chongqing Hechuan District Bureau of Statistics. *Hechuan District Statistical Yearbook*; China Yinshi Press: Beijing, China, 2020; pp. 1–364.
18. Liu, X.; Wang, Y.; Li, M. How to Identify Future Priority Areas for Urban Development: An Approach of Urban Construction Land Suitability in Ecological Sensitive Areas. *Int. J. Environ. Res. Public Health* **2021**, *18*, 4252. [CrossRef]
19. Peng, J.; Ma, J.; Du, Y.; Zhang, L.; Hu, X. Ecological suitability evaluation for mountainous area development based on conceptual model of landscape structure, function, and dynamics. *Ecol. Indic.* **2016**, *61*, 500–511. [CrossRef]
20. Huang, Y.; Zhao, L. Review on landslide susceptibility mapping using support vector machines. *Catena* **2018**, *165*, 520–529. [CrossRef]
21. Reichenbach, P.; Rossi, M.; Malamud, B.D.; Mihir, M.; Guzzetti, F.J.E. A review of statistically-based landslide susceptibility models. *Earth-Sci. Rev.* **2018**, *180*, 60–91. [CrossRef]
22. Chen, D.; Pan, Y.; Jin, X.; Du, H.; Li, M.; Jiang, P.J. The delineation of ecological redline area for catchment sustainable management from the perspective of ecosystem services and social needs: A case study of the Xiangjiang watershed, China. *Ecol. Indic.* **2021**, *121*, 107130. [CrossRef]
23. Sun, D.; Wen, H.; Wang, D.; Xu, J. A random forest model of landslide susceptibility mapping based on hyperparameter optimization using Bayes algorithm. *Geomorphology* **2020**, *362*, 107201. [CrossRef]
24. Xu, W.; Yu, W.; Jing, S.; Zhang, G.; Huang, J. Debris flow susceptibility assessment by GIS and information value model in a large-scale region, Sichuan Province (China). *Nat. Hazards* **2013**, *65*, 1379–1392. [CrossRef]
25. Zhang, X.; Fang, C.; Wang, Z.; Ma, H. Urban construction land suitability evaluation based on improved multi-criteria evaluation based on GIS (MCE-GIS): Case of New Hefei City, China. *Chin. Geogr. Sci.* **2013**, *23*, 740–753. [CrossRef]
26. Schapire, R.E. Random forests. *Mach. Learn.* **2001**, *45*, 5–32.
27. Sun, D.; Xu, J.; Wen, H.; Wang, D. Assessment of landslide susceptibility mapping based on Bayesian hyperparameter optimization: A comparison between logistic regression and random forest. *Eng. Geol.* **2021**, *281*, 105972. [CrossRef]
28. Saaty, T.L. Decision making with the analytic hierarchy process. *Int. J. Serv. Sci.* **2008**, *1*, 83–98. [CrossRef]

29. Saaty, T.L. What is the analytic hierarchy process? In *Mathematical Models for Decision Support*; Springer: Berlin/Heidelberg, Germany, 1988; pp. 109–121.
30. Yan, Y.; Zhang, Y.; Sharma, A.; Al-Amri, J.F. Evaluation of suitability of urban land using GIS technology. *Sustainability* **2021**, *13*, 10521. [CrossRef]
31. Liu, Y.; Deng, W.; Peng, L. The coupling mechanism between the suitable space and rural settlements considering the effect of mountain hazards in the upper Minjiang River basin. *J. Mt. Sci.* **2020**, *17*, 2774–2783. [CrossRef]
32. Wang, Y.; Jin, C.; Lu, M.; Lu, Y. Assessing the suitability of regional human settlements environment from a different preferences perspective: A case study of Zhejiang Province, China. *Habitat Int.* **2017**, *70*, 1–12. [CrossRef]
33. Zhou, X.; Wen, H.; Li, Z.; Zhang, H.; Zhang, W. An interpretable model for the susceptibility of rainfall-induced shallow landslides based on SHAP and XGBoost. *Geocarto Int.* 2022, 1–27, *(just-accepted)*.
34. Zhou, X.; Wen, H.; Zhang, Y.; Xu, J.; Zhang, W. Landslide susceptibility mapping using hybrid random forest with GeoDetector and RFE for factor optimization. *Geosci. Front.* **2021**, *12*, 101211. [CrossRef]
35. Zhang, R.; Wang, D.; Ai, D.; Sun, W.; Hou, Z. Suitability evaluation of urban land development based on niche and anti-planning. *Trans. Chin. Soc. Agric. Eng.* **2018**, *34*, 258–264.
36. Yu, K.; Li, D.; Li, C. Ecological Security Patterns and Urban Growth Scenarios in Beijing. *Acta Ecol. Sin.* **2009**, *29*, 1189–1204.

forests

MDPI

Article

A Comparative Study of Shallow Machine Learning Models and Deep Learning Models for Landslide Susceptibility Assessment Based on Imbalanced Data

Shiluo Xu [1], Yingxu Song [2,*] and Xiulan Hao [1]

1 School of Information Engineering, Huzhou University, Huzhou 313000, China
2 School of Information Engineering, East China University of Technology, Nanchang 330013, China
* Correspondence: yxsong@ecut.edu.cn

Abstract: A landslide is a type of geological disaster that poses a threat to human lives and property. Landslide susceptibility assessment (LSA) is a crucial tool for landslide prevention. This paper's primary objective is to compare the performances of conventional shallow machine learning methods and deep learning methods in LSA based on imbalanced data to evaluate the applicability of the two types of LSA models when class-weighted strategies are applied. In this article, logistic regression (LR), random forest (RF), deep fully connected neural network (DFCNN), and long short-term memory (LSTM) neural networks were employed for modeling in the Zigui-Badong area of the Three Gorges Reservoir area, China. Eighteen landslide influence factors were introduced to compare the performance of four models under a class balanced strategy versus a class imbalanced strategy. The Spearman rank correlation coefficient (SRCC) was applied for factor correlation analysis. The results reveal that the elevation and distance to rivers play a dominant role in LSA tasks. It was observed that DFCNN (AUC = 0.87, F1-score = 0.60) and LSTM (AUC = 0.89, F1-score = 0.61) significantly outperformed LR (AUC = 0.89, F1-score = 0.50) and RF (AUC = 0.88, F1-score = 0.50) under the class imbalanced strategy. The RF model achieved comparable outcomes (AUC = 0.90, F1-score = 0.61) to deep learning models under the class balanced strategy and ran at a faster training speed (up to 63 times faster than deep learning models). The LR model performance was inferior to that of the other three models under the balanced strategy. Meanwhile, the deep learning models and the shallow machine learning models showed significant differences in susceptibility spatial patterns. This paper's findings will aid researchers in selecting appropriate LSA models. It is also valuable for land management policy making and disaster prevention and mitigation.

Keywords: landslide susceptibility assessment; deep learning; machine learning; Three Gorges Reservoir area

Citation: Xu, S.; Song, Y.; Hao, X. A Comparative Study of Shallow Machine Learning Models and Deep Learning Models for Landslide Susceptibility Assessment Based on Imbalanced Data. *Forests* **2022**, *13*, 1908. https://doi.org/10.3390/f13111908

Academic Editors: Víctor Resco de Dios, Haijia Wen, Chong Xu, Weile Li and Hiromu Daimaru

Received: 12 October 2022
Accepted: 11 November 2022
Published: 14 November 2022

1. Introduction

A landslide is one of the most destructive geological disasters around the world. Landslides are widely distributed in mountainous and reservoir bank areas, which seriously threaten people's lives and property safety. Landslide susceptibility assessment (LSA) evaluates potential landslides spatially, which is an important tool for landslide prevention. LSA selects a series of landslide influence factors and estimates the probability of landslide occurrence. The LSA models typically work with the geographic information system (GIS) frameworks [1,2] and are grouped into two categories: model-driven and data-driven models. Model-driven LSA models can be expressed by mathematical formulas and are driven by physical theories; for instance, the shallow landslide stability model (SHALTAB) model assumes that the intensity of rainfalls remains constant and the rain seeps into the ground completely [3,4]. For model-driven LSA models, specific geotechnical parameters are necessary [5–7].

According to the literature review, data-driven models are the most common methods for LSA. The data-driven models are classified into two groups: knowledge-based models and machine learning models. For knowledge-based models [8], domain expertise is requested from the experts; they need to assign weights to landslide influence factors based on their expertise or available literature [9].

Machine learning techniques play a crucial role in data-driven models, and they are the most widely used approaches for LSA [10]. In general, machine learning approaches consist of two phases: training and testing. During the training phase, the input features and the targets are sent into the models, and the inner parameters of the models are fine-tuned according to some specified rules. The targets are predicted based on the trained models during the testing phase. The representative machine learning approaches involve frequency ratio [11], logistic regression (LR) [12,13], support vector machine (SVM) [14], Bayesian network (BN) [15], random forest (RF) [16,17], back propagation network (BP) [18,19], and ensemble learning techniques [20,21]. These models are the so-called shallow machine learning methods, which have the ability to handle more complex data than knowledge-based approaches [22]. These models can be integrated with other models to provide better performance [23]. Furthermore, it is observed that combining data-driven machine learning methods with qualitative analysis can improve the reliability of LSA models [24].

In recent years, deep learning approaches have demonstrated surprising feature extraction and data fitting capabilities and they are applied in the fields of computer vision [25], natural language processing [26], autopiloting [27], and intelligent medicine [28]. Recent literature describes deep learning approaches as a potent tool for LSA, attracting the interest of numerous researchers [29–32]. Prior research has demonstrated that LSA models based on deep learning techniques outperform LSA models based on shallow machine learning techniques [33,34]. The convolutional neural network (CNN) is the most widely used deep learning algorithm [33]. The authors of [35] applied CNN to Jiuzhaigou, Sichuan province, China, and verified that CNN achieved better performance compared with multi-layer perceptron (MLP). The authors of [36] developed a 1D-CNN with a high dropout rate to evaluate landslide susceptibility in South Korea. The results showed that 1D-CNN outperformed the artificial neural networks (ANN) and SVM because of its sophisticated architecture. Other deep learning approaches, including deep belief networks (DBN) and recurrent neural networks (RNNs), were also applied for LSA and achieved promising performances [30,32].

In the real world, the number of non-landslide samples is far more than the number of landslide samples. Many researchers have examined various LSA models, and most of them have utilized a balanced sampling technique in the training stage, which involves selecting an equal number of data from both landslide and non-landslide occurrences at random [37]. Without a doubt, this sampling technique is quite useful and effective [38,39]. However, the sampling technique introduces additional complexity to LSA models. A class-weighted strategy is a simple and effective solution to address the problem of imbalanced data in LSA [40]. The class-weighted strategies are mainly discussed in conventional machine learning LSA models and rarely reported in deep learning LSA models. Are the class-weighted strategies still effective for the deep learning LSA models? Additionally, in the scenario of extremely imbalanced data, do the deep learning LSA models outperform conventional shallow machine learning LSA models? To bridge this gap, the real-world matched imbalanced data were used as the training dataset to compare the landslide and the non-landslide evaluation performances based on shallow LSA models and deep LSA models. The Three Gorges reservoir area was chosen as the study area in this paper, which then employed the SRCC approach to assess the correlation between various landslide influence factors, and finally selected the informative factors to build LSA models. Two conventional shallow machine learning models, LR and RF, and two typical deep learning models, DFCNN and LSTM, were employed as LSA models. The performances of these models were assessed using the area under curve (AUC) and the

F1-score for various landslide to non-landslide class-weight ratios. Finally, this research concludes with some insightful recommendations for the selection of LSA models based on the experimental findings.

2. Materials

2.1. Study Area

The study area is located in the Zigui-Badong section of the Three Gorges Reservoir area, Hubei Province, China, with a longitude between 110°15′51″ E and 110°52′33″ E and a latitude between 30°51′21″ N and 31°5′1″ N (Figure 1). There are a large number of mountains, valleys, and hills in the study area, with a total area of 662.671 km^2 and a maximum altitude of 2004 m. The main stream of the Yangtze River flows through the whole area. Landslides are mainly distributed on both sides of the main stream and tributaries of the Yangtze River. According to the field survey data, there have been 332 identified slides (stable and unstable), with a total area of about 4210 m^2. These historical landslides include soil landslides, rock landslides, rock and soil mixed landslides, and other types of landslides, which accounted for 53%, 37%, 2%, and 8% of the total area, respectively.

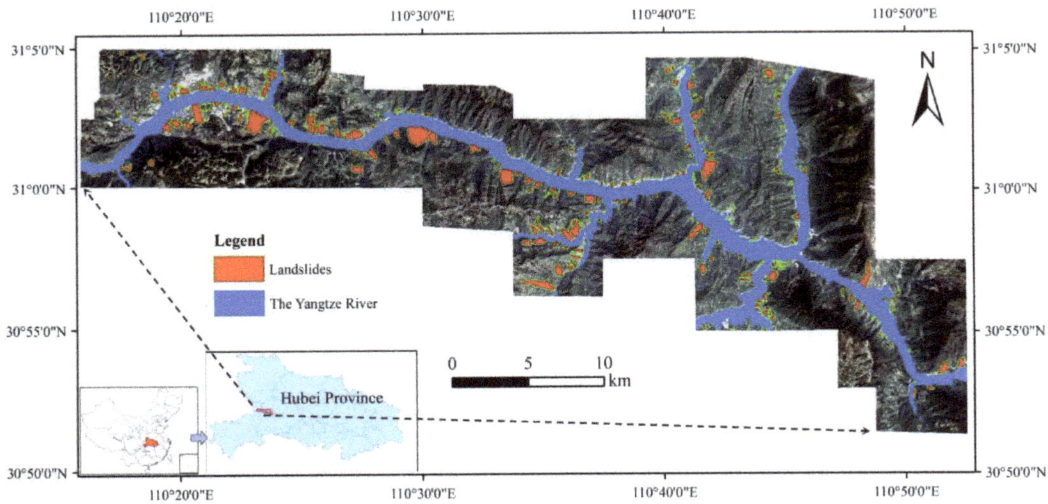

Figure 1. Location of the study area and the landslides' distribution.

The study area is mainly located in the pre-Nanhua metamorphic basement area at the core of the Huangling dome and the surrounding sedimentary cover area in the south, which belongs to the Yangtze Craton core area in South China. The strata in the study area are well developed and gradually new from east to west. The strata are mainly composed of the Badong Formation. The principal composition of the Badong Formation is an argillaceous rock with low mechanical strength and weak weathering resistance, which makes the formation prone to geological disasters. The Badong Formation in the study area belongs to the middle Triassic (Figure 2). The overlying strata are the Xujiahe Formation, Jiuligang Formation, and Shazhenxi Formation of the Upper Triassic, and the underlying strata are the Jialingjiang Formation of the Lower Triassic.

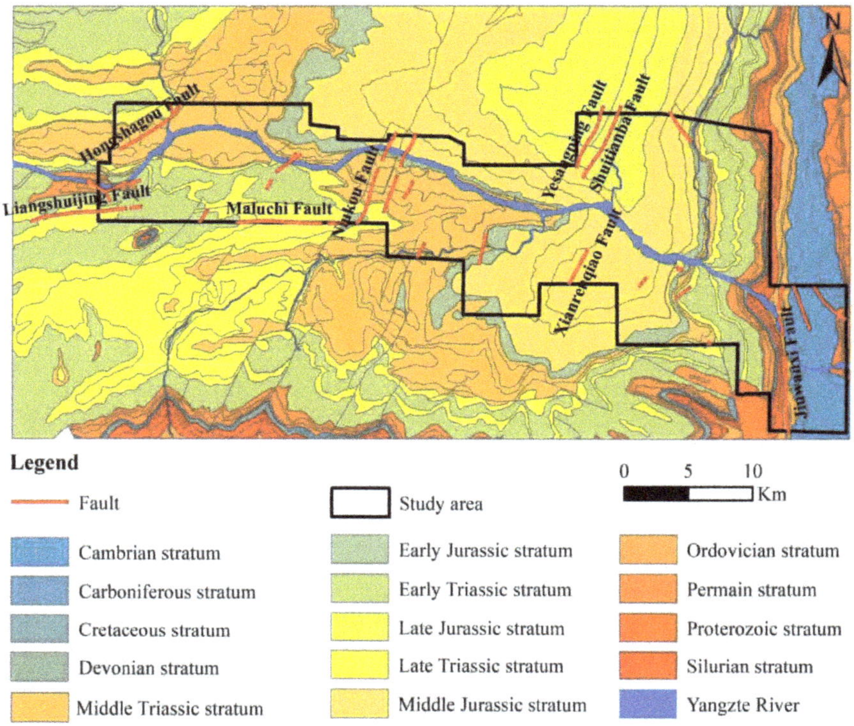

Figure 2. Geological map of the study area.

The study area has abundant rainfall, and the monthly average rainfall can exceed 1000 mm. Heavy rainfall leads to a considerable increase of water content in soil, which is a key influence factor inducing landslides. Meanwhile, the Xiannvshan fault, the Niukou fault, and other small faults are also located in the study area. The rock stratum along the fault zone is squeezed and stretched, and the stratum instability increases, which controls the generation of landslides around the fault zone.

2.2. Data Source

In this research, Landsat 8 OLI remote sensing images (2013) were used to extract the land cover factor, and a 1:50,000 geological map was utilized to derive the geological and hydrological related factors. Elevation data were derived from an ASTER GDEM V2 image and rainfall data were collected from Climate Hazards Group InfraRed Precipitation with Station data (CHIRPS) (2013). Seismic data were collected from 1978 to 2013. In addition, landslide inventory data from field surveys were available. On the basis of the original data, all influence factors were vectorized, rasterized, and converted into raster layers prior to model computation. All the influence raster layers were resampled to 30 × 30 m using the nearest neighbor sampling technique to match the resolution of elevation data and Landsat images.

3. Methodology

In the present study, the LSA modeling involves four principal steps (Figure 3): (1) collecting the raw data, (2) data preparation, which includes landslide influence factors generation, factors standardization, factor correlation analysis and data splitting, (3) modeling based on two shallow machine learning models and two deep learning models, (4) performance evaluation and landslide susceptibility zoning of the study areas.

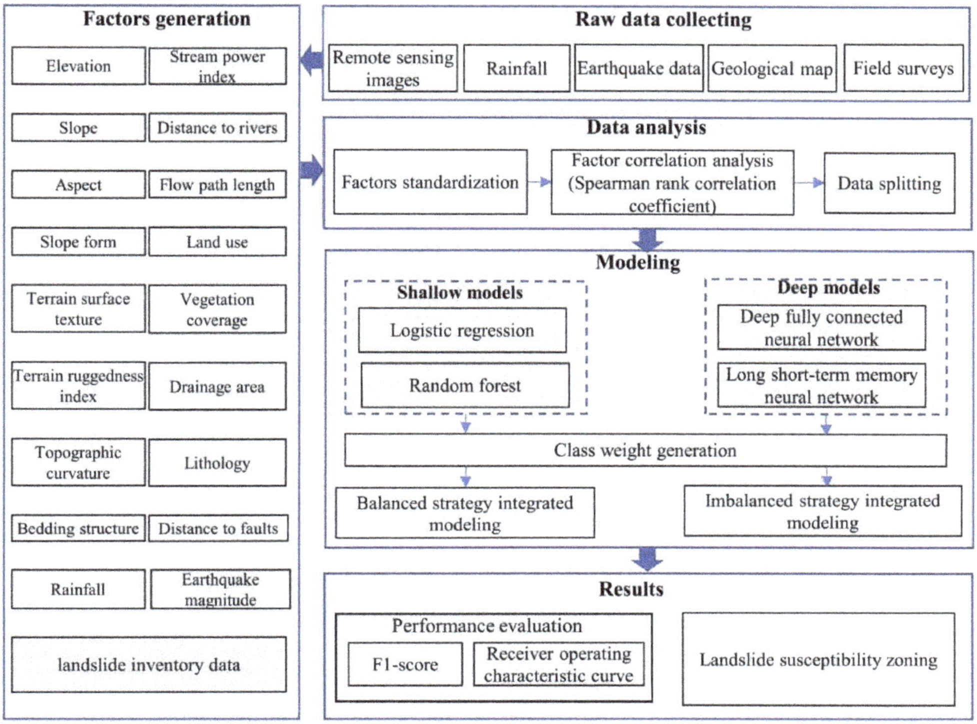

Figure 3. Methodological flowchart of the study.

3.1. Landslide Susceptibility Models

3.1.1. Logistic Regression

Logistic regression (LR) has been widely used in landslide susceptibility assessment [41–43]. In general, $y = w_1x_1 + w_2x_2 + \ldots + w_nx_n + b$ is used for the linear fitting of data. Landslide susceptibility assessment is a nonlinear binary classification problem. Thus, a nonlinear mapping is required to be applied to y. The sigmoid function is generally used to add nonlinear characteristics to y. The sigmoid function is defined as Equation (1) [44].

$$g(y) = \frac{1}{1+e^{-y}} = \frac{1}{1+e^{-(w_1x_1+w_2x_2+\ldots+w_nx_n)}} \tag{1}$$

The above equation will make the model's output range from 0 to 1. An output value close to 0 means samples are classified as non-landslide. An output value close to 1 means samples are classified as landslides. To facilitate calculation, the logarithm of the above equation is usually taken.

3.1.2. Random Forest

Random forest (RF) [45–47] is an ensemble learning model that integrates multiple weak decision tree models to form a robust classification model. RF picks the input samples and features randomly, so it can avoid over-fitting problems to a certain extent and has good anti-noise abilities. The RF construction processes are summarized as following steps:

(1) Random sample selection. Assume that the total number of samples is N and there are m decision trees in RF; to create subset Ns, select s samples from N samples randomly; a total of m sample subsets are constructed.

(2) Random feature selection. Assume that the total number of features is F, and h features are selected from the F to form a feature subset F_h. The optimal feature is generated from the feature subset F_h when the decision tree splits each time.
(3) Classifier voting. m decision trees produce a total number of m classification results. The class with the highest votes is used as the final prediction class.

3.1.3. Deep Fully Connected Neural Network

Conventional shallow neural networks generally consist of three layers, which are the input layer, the hidden layer, and the output layer [48]. There are connections between the nodes of the previous layer and those of the following layer. The data fitting ability of a three-layer neural network is limited due to its shallow structure in practice. The deep fully connected neural network (DFCNN) builds a deeper structure by adding extra hidden layers to the basic three-layer structure. The structure of the DFCNN is shown in Figure 4.

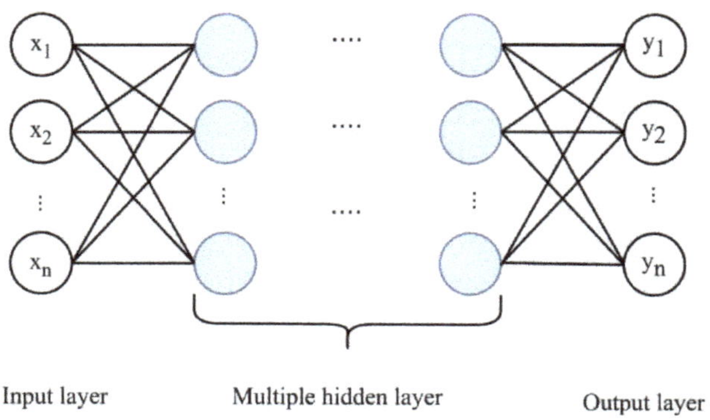

Figure 4. Architecture of DFCNN.

The input of the network is $[x_1, x_2, \cdots, x_n]$, and the weights w and the bias b are generally applied to the inputs. To make the neurons have nonlinear mapping capabilities, it is common to send the outputs of the neurons to an activation function, σ. The process can be expressed by the following Equation (2) [49].

$$h = \delta\left(\sum_{i=1}^{n} w_i x_i + b\right) \quad (2)$$

The commonly used activation functions are the sigmoid function, the tanh function, and the rectified linear units (ReLU) [50] function. Since the sigmoid function and tanh function have the problem of gradient vanishing, the ReLU function was adopted in this article. The derivative of ReLU is 1 while the inputs are greater than 0, which maintains the gradient without decay. The ReLU function alleviates the gradient vanishing problem to a certain extent. The definition of the ReLU function is indicated by Equation (3).

$$f(x) = \begin{cases} x, & x \geqslant 0 \\ 0, & x < 0 \end{cases} \quad (3)$$

Since landslide susceptibility assessment is a binary classification problem, the last output layer is connected to a sigmoid function that outputs the probability value for each class.

3.1.4. Long Short-Term Memory Neural Network

A long short-term memory (LSTM) neural network [51] is a special recurrent neural network (RNN) [52]. An RNN has a loop in the hidden layer, which sends the information from the previous hidden layer into the current hidden layer. It represents that there exists a connection between the nodes at different time points. However, for long sequences, a RNN still has the gradient vanishing problem. To address this problem, LSTM adopts a gating mechanism to control the state of information flow, which is called "memory block". There are three types of gating units in the memory block: input gate, forget gate, and output gate. The input gate selectively retains the essential information, the forget gate filters the irrelevant or interfering information, and the output gate integrates information together and outputs them. The architecture of the memory block is indicated in Figure 5.

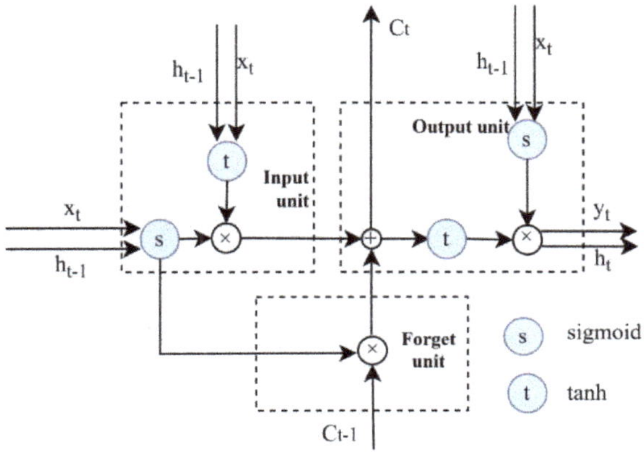

Figure 5. Architecture of the memory block at time step *t*.

In Figure 5, x_t represents the original input at time step t, h_{t-1} is the hidden layer state transmitted from time step $t-1$, and C_{t-1} is the memory block state transmitted at time step $t-1$. After passing through the memory block, three results, y_t, h_t and C_t, are generated, which are the class output at time step t, the hidden layer output at time step t, and the memory block state output at time step t, respectively. Sigmoid and tanh activation functions are used in memory blocks to obtain nonlinear mapping abilities.

3.2. Landslide Influence Factors

This paper adopted six types of landslide influence factors based on prior research [53,54] and the real situation of the study area. These include land cover, geography, hydrology, geology, earthquakes, and rainfall. The extraction of factors was conducted using GDAL, SAGA, QGIS, and ArcGIS (Table 1).

- Land cover includes land use and vegetation coverage in this study. Land use is an indicator of the intensity of human activity. In regions where human activities are violent, slope instability rises. Land uses are exacted by pixel classifications from Landsat 8 OLI images and they are then validated by a public dataset [55]. The overall validation accuracy of land use is 89.8%. To make the land use results more reliable, visual interpretation is used to correct misclassifications. Plant roots hold soil and rock in place, and absorb water in the soil at the same time. It usually exerts a positive effect on landslides [56]. The vegetation coverage is computed based on the dimidiate pixel model and normalized difference vegetation index (NDVI).
- Topography is an essential condition to control the development of landslides [56]. The slope, aspect, slope form, terrain surface texture, terrain ruggedness index, and topographic curvature are derived from the elevation. It must be noted that the bedding structures are a combination of topographic conditions and geological conditions. The bedding structures are classified into six types according to the previous literature [57].
- Hydrological conditions are also an important factor influencing the occurrence of landslides in the study area [58]. The rivers erode the riverbed on both sides, making the slope unstable. Thus, the landslides in the study area are mainly distributed on

both sides of the river. Drainage area, flow path length, stream power index, and distance to rivers are used for LSA modeling.

- Geological conditions play a decisive role in the development of landslides and are important internal controlling factors. The faults destroy rock formations and affect their stability [59]. The closer to the faults, the greater impact on the rock formation. The geological factors including the lithology and the faults are mainly exacted from geological maps by vectorization.
- Earthquakes squeeze and stretch the formation, causing huge deformations of the formation and resulting in drastic changes in the geological environment, and finally affect the stability of the slope [60]. The earthquake data were collected from earthquake monitoring sites and then imported into GIS software for further processing.
- Rainfall affects the water content in the soil and it is one of the main factors that trigger landslides [61,62]. Short-term heavy rainfall may lead to soil erosion, an increase in surface runoff, and a reduction in the soil's and rock's absorption capacity. Simultaneously, the rainfall causes fluctuations in reservoir water levels, and the static and dynamic water pressures vary accordingly, aggravating the slope further.

Different factors, regardless of whether the original data source is a raster layer or a vector layer, were converted into 30 × 30 m raster layers eventually. The influence raster layers are shown in Figure 6. The river area is masked from the original layers. The influence factor layers and the landslide inventory layer were aligned and stacked by geographical location in GIS software. It must be noted that each landslide influence factor with continuous values was reclassified into several classes in most of the previous literature [16,63]. However, the continuous landslide influence factors were not reclassified in this study and they were directly fed to the LSA models. Thus, the reclassification processes of factors were performed implicitly by the LSA models.

Table 1. Landslide influence factors, the exaction tools used, and their variable types.

Data Source	Type	Exaction Tools	Variable Type	Influence Factor
Landsat images	Land cover	Preprocessing, vegetation coverage estimation based on dimidiate pixel models, and implemented using python in ArcGIS	Continuous	Vegetation coverage
		Preprocessing, pixel classification in QGIS	Discrete	Land use
DEM	Topography	Direct utilization	Continuous	Elevation
		Model calculation based on the terrain analysis tool of GDAL	Continuous	Slope
		Model calculation based on the terrain analysis tool of GDAL	Continuous	Aspect
		Model calculation using curvature tools of SAGA	Discrete	Slope form
		Model calculation using the terrain analysis tool of SAGA	Continuous	Terrain surface texture
		Model calculation using the terrain analysis tool of SAGA	Continuous	Terrain ruggedness index
		Model calculation using curvature tools of SAGA	Continuous	Topographic curvature
		Vectorization, rasterization, and calculation based on slope and aspect in QGIS	Discrete	Bedding structure

Table 1. *Cont.*

Data Source	Type	Exaction Tools	Variable Type	Influence Factor
DEM Landsat image	Hydrological	Model calculation using terrain analysis—hydrology tool of SAGA	Continuous	Drainage area
		Model calculation using terrain analysis—hydrology tool of SAGA	Continuous	Flow path length
		Model calculation using terrain analysis—hydrology tool of SAGA	Continuous	Stream power index
		Model calculation based on proximity (Raster distance) tool of QGIS	Continuous	Distance to rivers
Geological map	Geological	Vectorization, rasterization, and reclassifying in QGIS	Discrete	Lithology
		Vectorization, rasterization, and model calculation based on the proximity (raster distance) tool of QGIS	Continuous	Distance to faults
Earthquake monitoring sites	Earthquake	Model calculation using TIN interpolation in QGIS	Continuous	Earthquake magnitude
Rainfall data	Rainfall	Direct utilization	Continuous	Rainfall

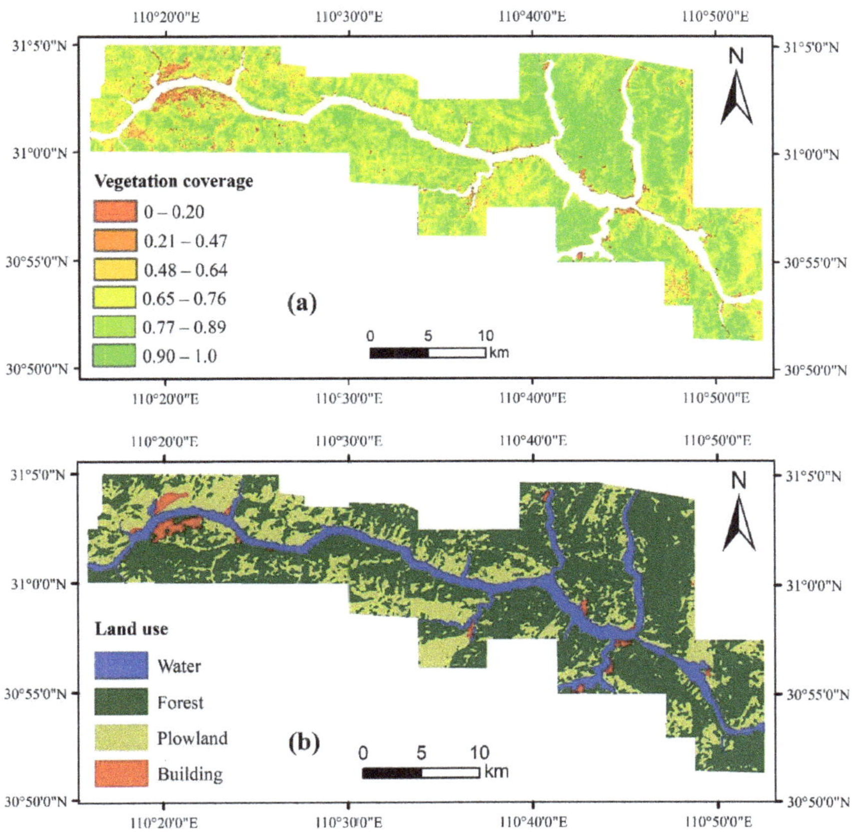

Figure 6. *Cont.*

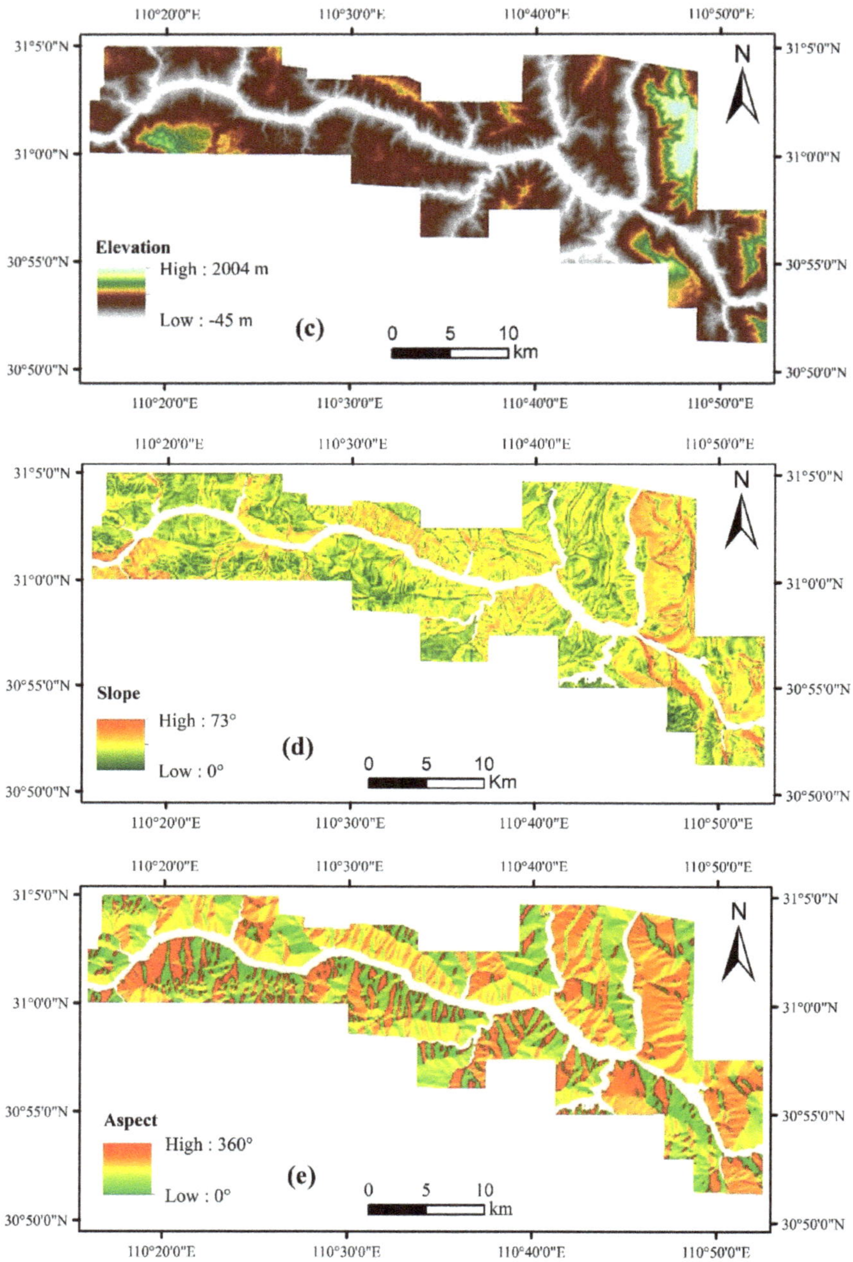

Figure 6. *Cont.*

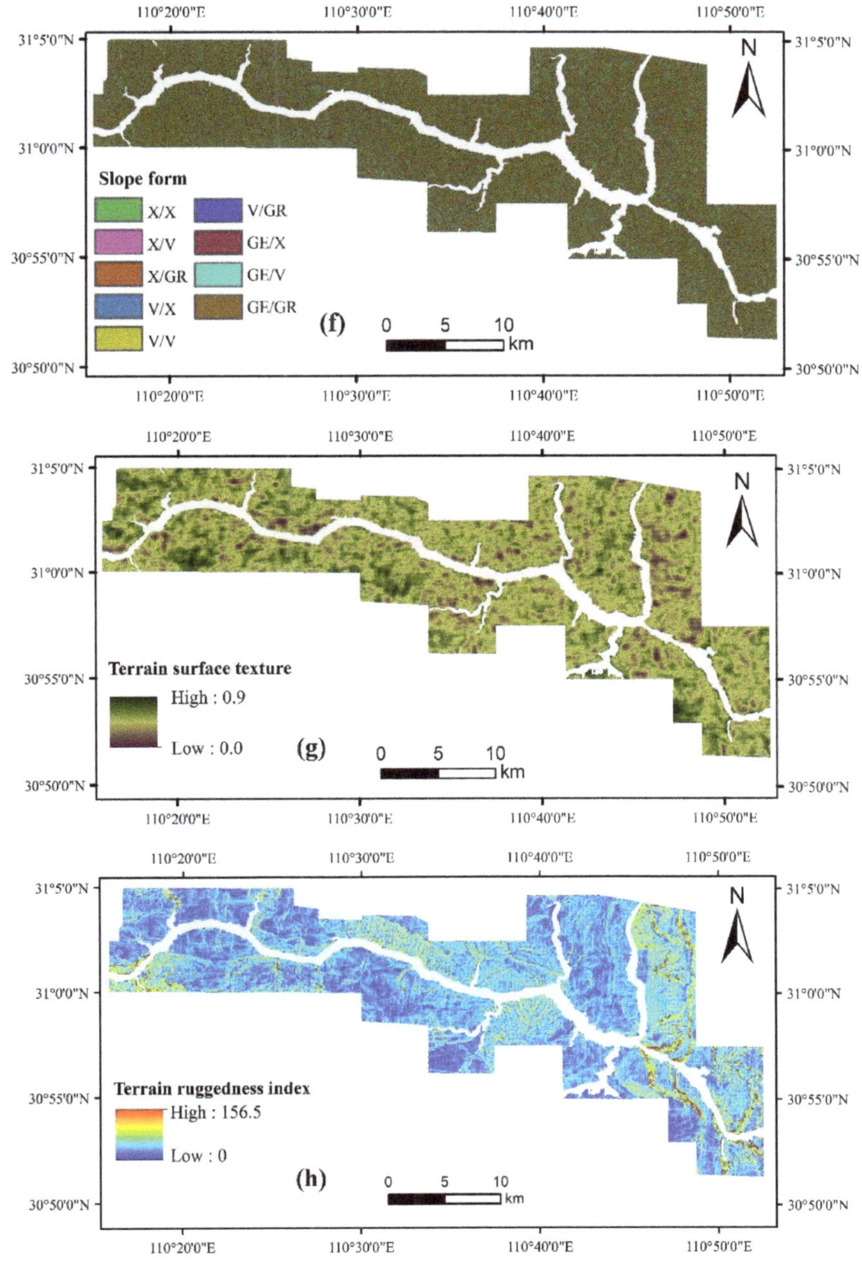

Figure 6. *Cont.*

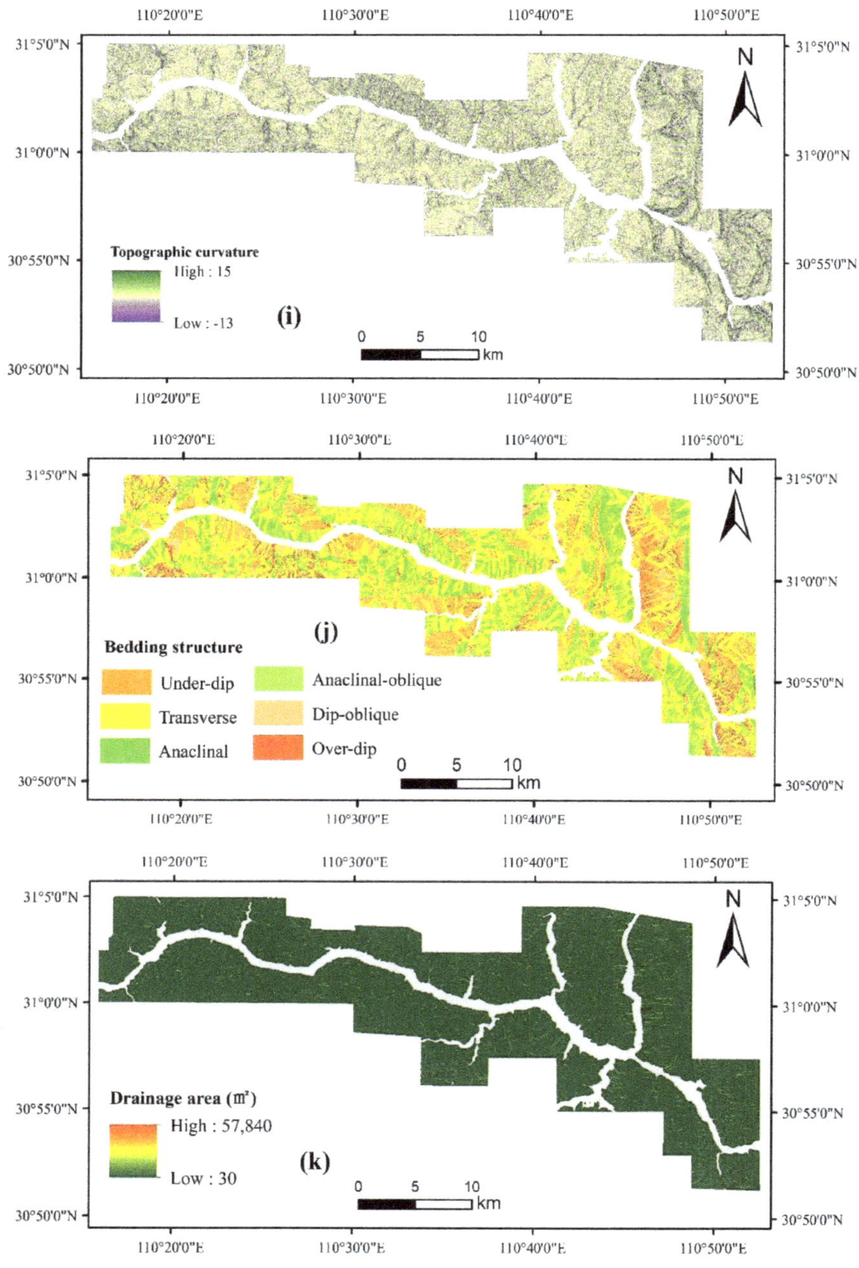

Figure 6. *Cont.*

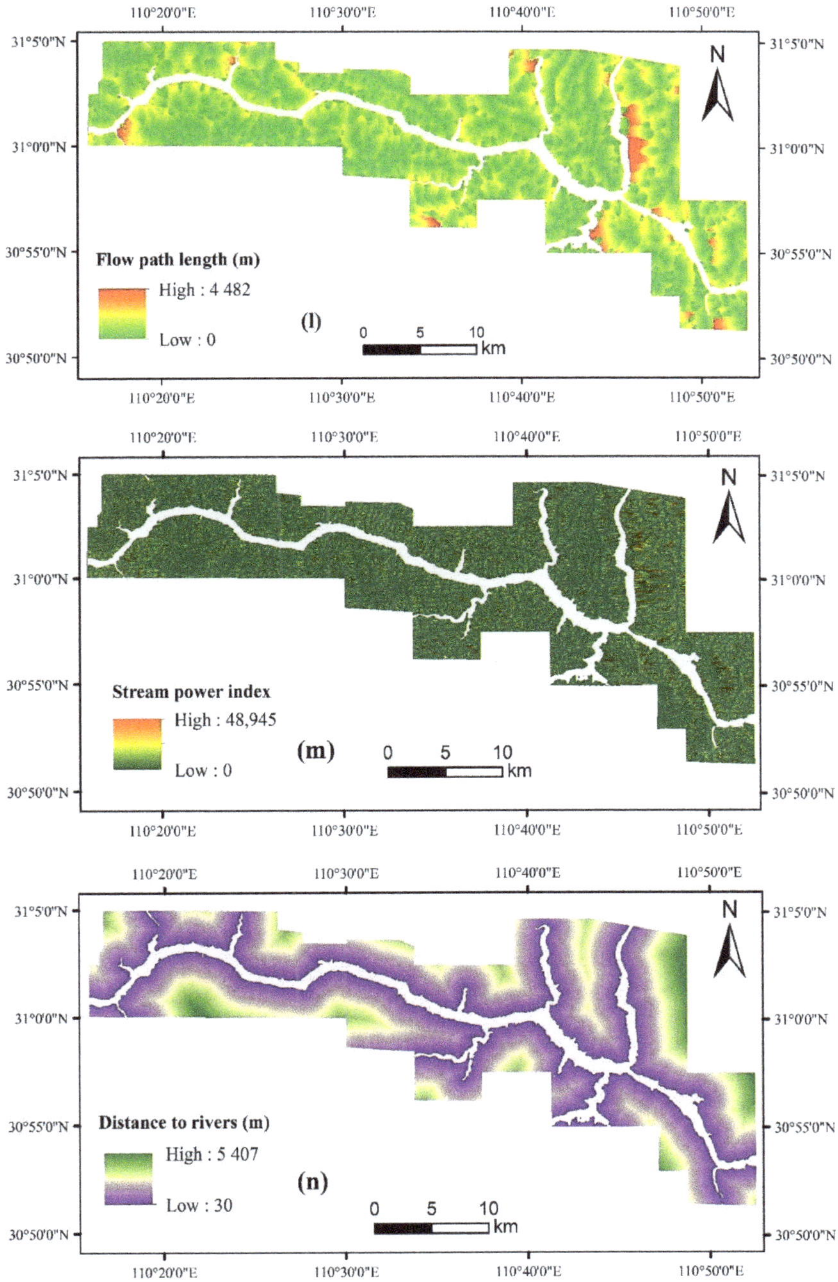

Figure 6. *Cont.*

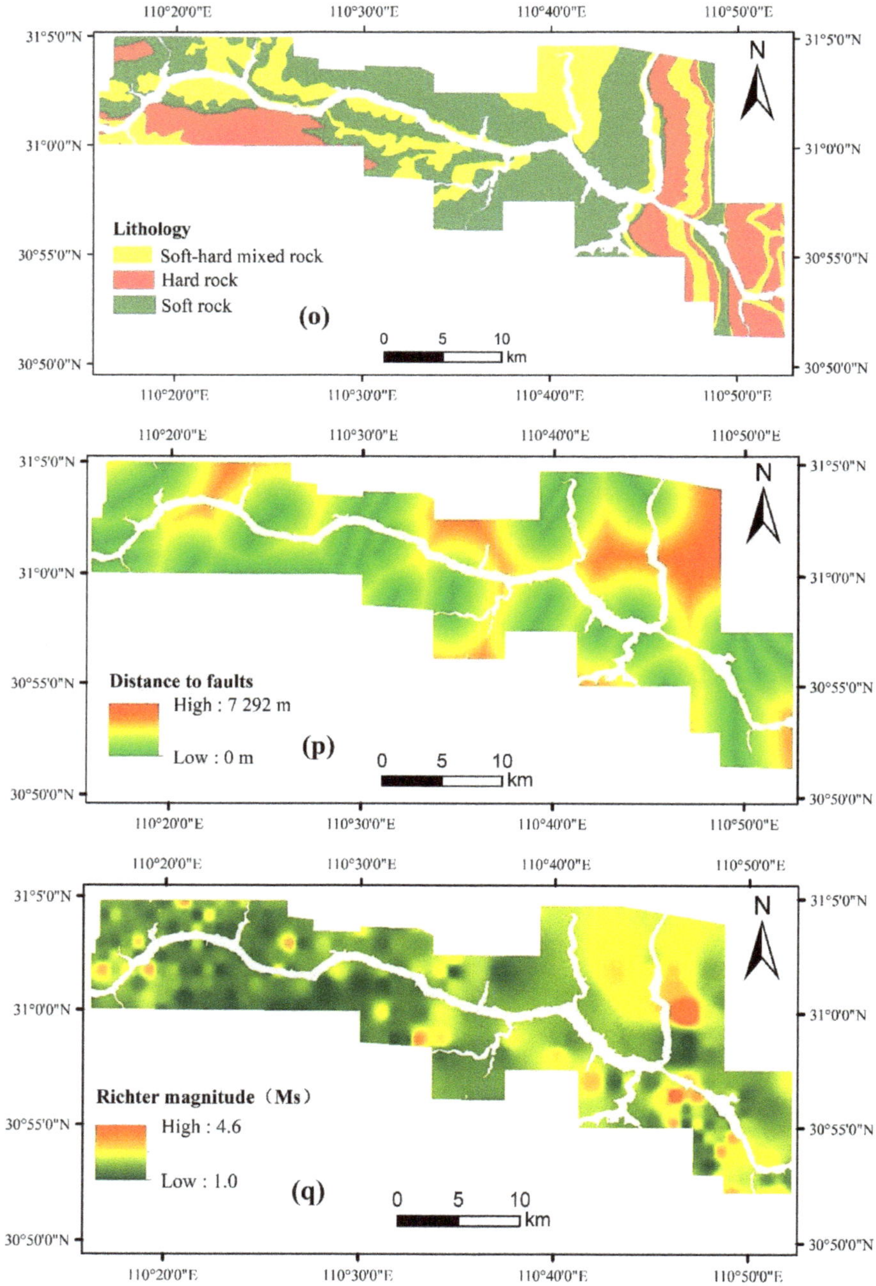

Figure 6. *Cont.*

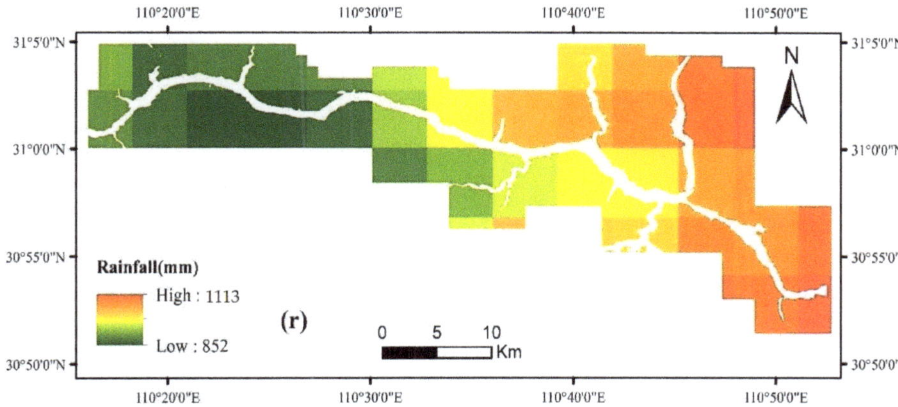

Figure 6. Landslide influence factors: (**a**) vegetation coverage, (**b**) land use, (**c**) elevation, (**d**) slope, (**e**) aspect, (**f**) slope form, (**g**) terrain surface texture, (**h**) terrain ruggedness index, (**i**) topographic curvature, (**j**) bedding structure, (**k**) drainage area, (**l**) flow path length, (**m**) stream power index, (**n**) distance to rivers, (**o**) lithology, (**p**) distance to faults, (**q**) earthquake magnitude, (**r**) rainfall.

3.3. Spearman Rank Correlation Coefficient

The Pearson correlation coefficient (PCC) is usually used to measure the linear correlations of different datasets [64]. However, landslide influence factors have intricate nonlinear relationships with each other. To eliminate redundant features and reduce noises, the Spearman rank correlation coefficient (SRCC) is employed to select the influence factors with strong predictive ability in LSA. SRCC can be applied to both continuous factors and discrete factors. The computation of SRCC is similar to the computation of PCC. The difference is that the SRCC calculations use rank factors rather than raw factors [65]. The SRCC is defined as Equation (4).

$$\mathrm{SRCC}_{X,Y} = \frac{cov(R(X), R(Y))}{\sigma_{R(X)}\sigma_{R(Y)}} = \frac{E\left[\left(R(X) - \mu_{R(X)}\right)\left(R(Y) - \mu_{R(Y)}\right)\right]}{\sigma_{R(X)}\sigma_{R(Y)}} \tag{4}$$

where the $cov(R(X), R(Y))$ is the covariance of the two variables, the $\sigma_{R(X)}$ and $\sigma_{R(Y)}$ are the standard deviations of the two variables, whereas $\mu_{R(X)}$ and $\mu_{R(Y)}$ are the mean values of the two variables. The absolute value of SRCC ranges from 0 to 1, whereas 0 corresponds to a weak linear correlation and 1 corresponds to a strong linear correlation.

3.4. Performance Evaluation

Since the proportion of landslide area was relatively small compared with the whole study area, the landslide susceptibility assessment is a typical class imbalanced classification task. The non-landslide area was much larger than the landslide area. The ratio of non-landslide to landslide in the study area was up to 15:1. Consequently, using the accuracy to evaluate the landslide susceptibility performance would result in the model preferring to identify the samples as non-landslide zones. Non-landslide samples were defined as the negative class, and landslide samples were defined as the positive class. Four types of prediction results were defined for landslide and non-landslide events. TP (true positive) denotes the landslide sample is classified as the landslide, and the prediction is correct. FP (false positive) denotes the non-landslide sample is classified as the landslide, and the prediction is incorrect. TN (true negative) denotes the non-landslide sample is classified as the non-landslide, and the prediction is correct. FN (false negative) denotes the landslide sample is classified as the non-landslide, and the prediction is incorrect.

1. F1-score

$$\text{Recall} = \frac{\text{TP}}{\text{TP} + \text{FN}} \tag{5}$$

$$\text{Precision} = \frac{\text{TP}}{\text{TP} + \text{FP}} \tag{6}$$

$$\text{F}_1 = 2 \times \frac{\text{Precision} \times \text{Recall}}{\text{Precision} + \text{Recall}} \tag{7}$$

Recall is the proportion of correctly predicted landslides relative to the actual landslides (Equation (5) [66]). Precision is the proportion of correctly predicted landslides among all the data predicted as landslides (Equation (6) [66]). The cost caused by the misclassification of landslides is relatively larger than that of the misclassification of non-landslide. Thus, the recall value can be used to measure whether the landslide prediction is comprehensive. However, recall as the evaluation criterion will make the model tend to predict more samples as landslides, resulting in a high false positives rate. The F1-score is the harmonic average of recall and precision to avoid the model being completely inclined to a certain class, as shown in Equation (7) [67]. Therefore, the F1-scores were used as the evaluation criteria for LSA results.

2. Receiver operating characteristic curve

The x-axis of the receiver operating characteristic curve (ROC) [68] is the false positive rate (FPR), and the y-axis is the true positive rate (TPR). FPR and TPR are defined in Equations (8) and (9), respectively.

$$\text{FPR} = \frac{\text{FP}}{\text{FP} + \text{TN}} \tag{8}$$

$$\text{TPR} = \frac{\text{TP}}{\text{TP} + \text{FN}} \tag{9}$$

When the sample distribution changes, the ROC curve remains stable. Therefore, it is objective to measure the results of landslide susceptibility with dramatic changes in sample distributions. To quantify the ROC curve, the area under curve (AUC) is used to evaluate the performance of the LSA models. The value of AUC ranged from 0 to 1. A high AUC value corresponds to superior model performance, whereas a low AUC value corresponds to inferior model performance. In general, the AUC value of the model was greater than 0.5. The prediction result of the model was worse than that of a random guess when the AUC value was less than 0.5.

3.5. Assessment Units

The common assessment units for LSA can be summarized into three types: raster pixel units, grid units, and slope units [69,70]. In this paper, raster pixel units were chosen as the assessment units. Each raster pixel unit represents a vector $[x_1, x_2, \cdots, x_n, y]$, where x_i denotes the value of a landslide influence factor and y denotes the target value at a specified pixel, respectively. Since raster layers cannot be fed directly to a model, the combined raster layer (including the influence factor layers and landslide inventory layer) was exported to a two-dimensional table. The conversion process was as follows: (1) Create a polygon and convert it into a raster layer (called L_s) with the same extent as the study area. (2) Convert the L_s into a point layer L_p. (3) Exact the pixel values of the combined raster layer (called L_c) into the attribute table of the point layer L_p. The rows of the attribute table represent landslide samples, and the columns of the attribute table represent different landslide features.

3.6. Factors Standardization

Different factors have different dimensions, and their values vary widely. Some influence factors are discrete values, whereas other influence factors are continuous values. The discrete factors were mapped to numeric values. In order to reduce the impacts of dimensions on model performances, all factors were normalized by Equation (10) [71].

$$z = \frac{x - \overline{x}}{\sigma} \tag{10}$$

where x is the value of the given factor, $\overline{x}$ is the mean value of the given factor and σ is the standard deviation of the given factor.

3.7. Class Weighted Strategy

For the class imbalanced problem, the LSA models tended to predict the minority class (non-landslide) as the majority class (non-landslide) to obtain better overall accuracy. This dramatically diminished the reliability of the LSA models. The class-weighted strategy assigns different penalty weights for the majority class and the minority class, respectively [40]. For example, the class weight ratio (non-landslides vs. landslides) of 1:4 means that the penalty intensity for incorrectly predicting landslides is four times higher than that for incorrectly predicting non-landslides. This mechanism can correct the prediction preferences of the LSA model. Since there was a high ratio of non-landslide samples to landslide samples in the study area, two training strategies were employed: (1) A class balanced strategy; that is, the weights of landslides were greater than the weights of non-landslides. A high weight ratio means the model tends to classify more samples as landslide-prone zones. A low weight ratio leads to the number of misclassified landslide samples increasing. In order to find out the best class weight, the weight ratio of landslides to non-landslides was set at a range of 1 to 15. (2) A class imbalanced strategy; that is, the weights of landslide and non-landslide were handled by the model itself. Generally, the weight ratio was 1:1, which indicated the majority class (non-landslide) and the minority class (landslide) had identical class weights. The model tended to predict more samples as non-landslide to achieve better accuracy. However, it was easy to miss detecting some landslide areas. The two training strategies were compared using the evaluation criteria.

3.8. Experiment Setup

A total of 636,190 pixel-based units were generated and they were used as the entire dataset. The study area was divided into ten parts. Parts 1~6 was the training dataset, parts 7~8 was the validation dataset, and the rest of parts 9 ~10 was the testing dataset (Figure 7). The experiment adopted the class balanced strategy and the class imbalanced strategy. The training device was a personal computer with an AMD 5800 CPU, 32GB of memory, and GeForce RTX3060 12GB GPU. LR and RF were trained on the CPU with parallel settings. For DFCNN and LSTM, the GPU was also used to accelerate the training process. The LR and RF were implemented using the scikit-learn 0.24.2. LR was implemented using LogisticRegression that was built in scikit-learn and newton-cg was used to optimize the algorithm parameters. The maximum iteration of LR was set to 100. RF was implemented using RandomForestClassifier that was built in scikit-learn. Since the distribution of the testing data and the training data may have been quite dissimilar and large tree depths are prone to overfitting, a tiny subset of the study area with significant changes was employed in the pre-testing. The pre-testing results demonstrated that a small tree depth could effectively avoid overfitting and achieve better performances. Thus, the maximum depth of the RF trees was set to 10 based on the outcomes of the preliminary test. The DFCNN was implemented with 4 hidden layers using Keras 2.6.0. Each hidden layer consisted of 8 nodes and was connected to the ReLU activation function. LSTM consisted of 6 LSTM layers and a fully connected layer. The ReLU activation function also was employed for the hidden layer. Both LSTM and DFCNN were trained 200 times.

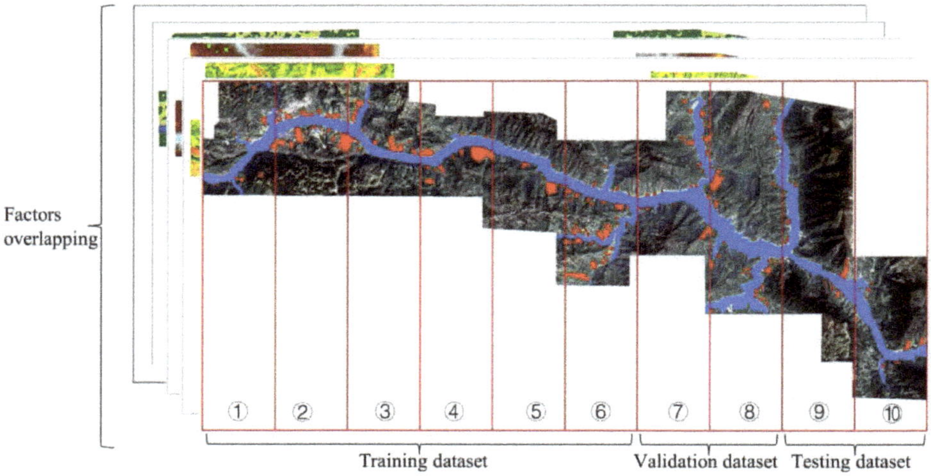

Figure 7. Diagram of dataset splitting.

4. Results

4.1. Performances of Different Class Weights

Different class weights lead to changes in model performances. Figure 8 denotes that with the change of class weights, the AUC and F1-score of deep learning models (DFCNN and LSTM) fluctuated without any obvious upward or downward trend. The AUC values of LR remained stable and nearly constant under different class weights. The F1-scores of LR achieved the best performance at class weights of 1:4 and 1:5, and then the performances decreased substantially and gradually. The AUC values of RF fluctuated slightly with different class weights, while F1-scores climbed gradually with increasing class weights, reaching a maximum at class weights of 1:4 and 1:5, and then maintaining small fluctuations. The class weight of 1:4 was chosen to construct the prediction model under the balanced strategy, while the class weight of 1:1 was used to build the prediction model under the imbalanced strategy.

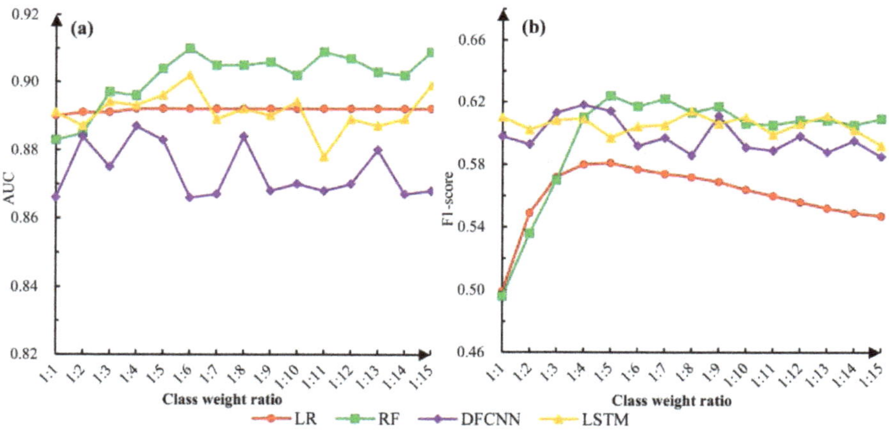

Figure 8. Performances with different class weight ratios: (**a**) AUC performances, (**b**) F1-score performances.

4.2. Correlations of Influence Factors

SRCC was employed to establish the most related landslide influence factors. Figure 9 suggests that the stream power index and the drainage area (SRCC = 0.92), the stream power index and the slope form (SRCC = −0.51), the slope form and the drainage area (SRCC = −0.58), the distance to rivers and the elevation (SRCC = 0.80), the terrain ruggedness index and the slope (SRCC = 0.67), the rainfall and the earthquake magnitude (SRCC = 0.61) were highly correlated, with the absolute SRCC values greater than 0.5. This implies that there may have existed redundant information in these paired landslide influence factors. To explore whether these highly correlated landslide influence factors would affect the performances of the LSA models, these paired influences were first removed separately and then removed simultaneously. Thus, there were 15 scenarios in total, which are shown in Table 2.

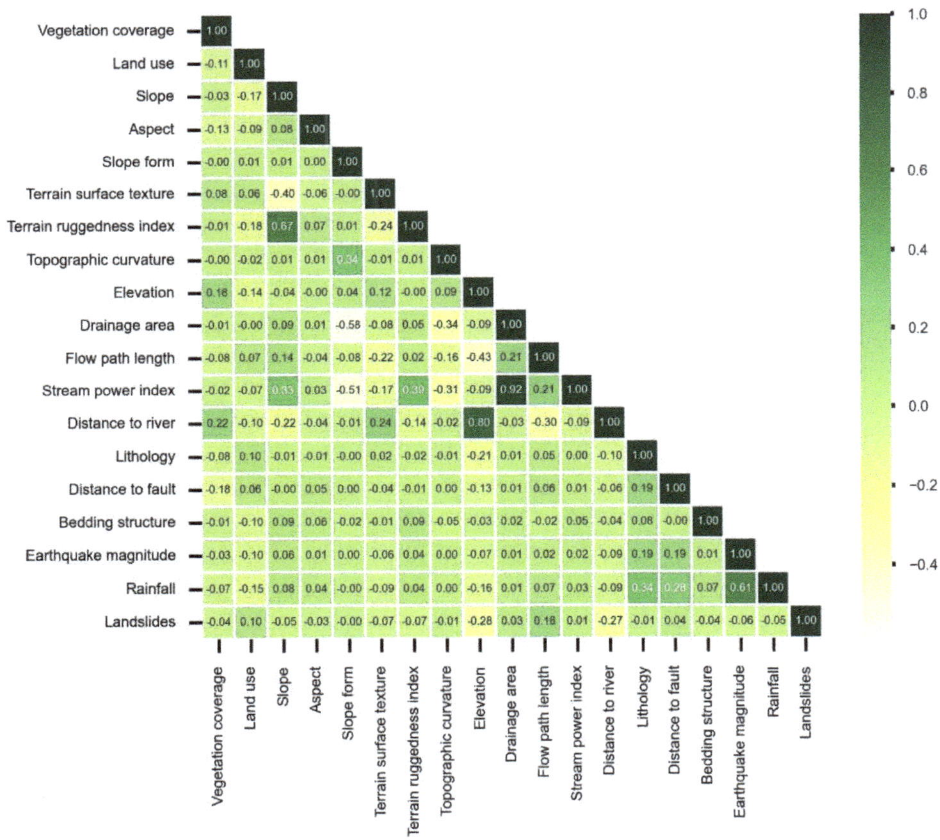

Figure 9. SRCC between landslide influence factors and the landslides.

Table 2. Scenarios of removing highly correlated influence factors.

Scenario	Influence Factors Removed
Base	All factors are used
A	Stream power index
A*	Drainage area
A-	Stream power index, drainage area
B	Slope form
B-	Slope form, drainage area
C	Distance to rivers
C*	Elevation
C-	Distance to rivers, elevation
D	Terrain ruggedness index
D*	Slope
D-	Terrain ruggedness index, slope
E	Rainfall
E*	Earthquake magnitude
E-	Rainfall, earthquake magnitude

Figure 10 illustrates the performance of these 15 scenarios under a class balanced strategy (class weight ratio is 1:1) and a class imbalanced strategy (class weight ratio is 1:4). In most scenarios, removing some influence factors in the highly correlated factor pairs did not result in significant losses of model performances. However, there were some exceptions. In scenario C-, the AUC values of the four models decreased significantly. Similarly, the F1-scores of four models also dropped dramatically in scenario C-. This demonstrates that the influence factors involved in scenario C- exerted a considerable influence on the occurrence of landslides. They were indispensable in the modeling process.

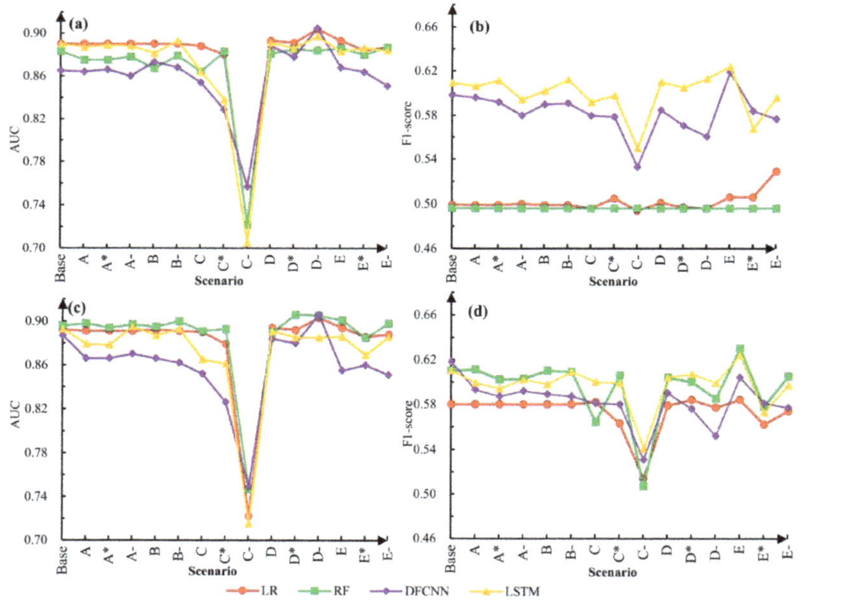

Figure 10. Performances of removing landslide influence factors: (**a**) AUC under class imbalanced strategy, (**b**) F1-score under class imbalanced strategy, (**c**) AUC under class balanced strategy, (**d**) F1-score under class balanced strategy.

4.3. Landslide Susceptibility Results

The above experimental results show that the most stable results could be achieved by using the original influence factors. Therefore, the original landslide influence factors were finally used to compare the performances of the four models under the balanced strategy and the imbalanced strategy. Their performances are shown in Table 3. All models achieved high AUC values under two distinct strategies in Figure 11. Therefore, it was not reliable to evaluate the performances of LSA models using the AUC values alone. Under the class imbalanced strategy, the F1-scores of the deep learning models were significantly better than that of the shallow models, while no significant performance differences were observed between the two shallow models. Similarly, no noticeable performance differences were observed between the two deep learning models. Compared with the performances under the class imbalanced strategy, the AUC and F1-score performances of the deep learning models under the class balanced strategy did not change significantly, while the F1-score performances of the shallow models had huge improvements. The improvement of the RF model was more pronounced relative to the LR (up to 0.11).

Table 3. Performances of four models under the imbalanced strategy and balanced strategy.

Model	AUC		F1-Score	
	Imbalanced	Balanced	Imbalanced	Balanced
LR	0.89	0.89	0.50	0.58
RF	0.88	0.90	0.50	0.61
DFCNN	0.87	0.89	0.60	0.62
LSTM	0.89	0.89	0.61	0.61

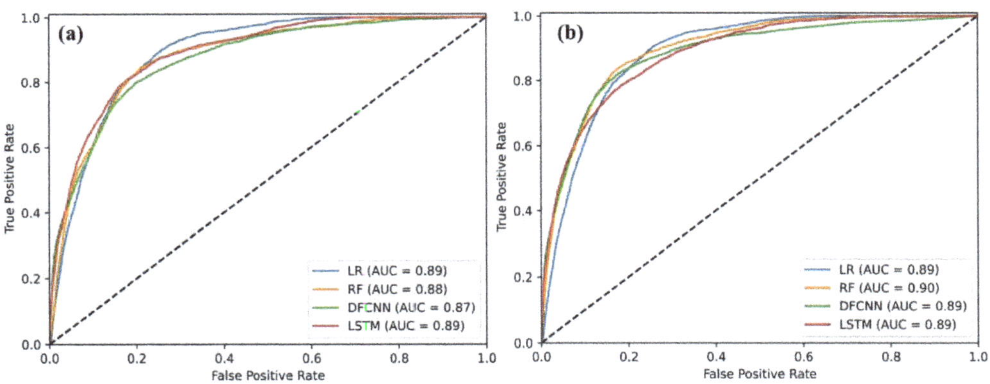

Figure 11. ROC curves of four models: (**a**) ROC curves under class imbalanced strategy, (**b**) ROC curves under class balanced strategy.

The evaluation results of the models were classified into four susceptibility classes using an equal intervals scheme; namely, very low (<0.25), low (0.25~0.5), moderate (0.5~0.75), and high (>0.75). The experimental results of the four models were connected to the raster pixel unit layer to produce susceptibility images and were compared with historical landslide data for validation. The results of the landslide susceptibility based on the testing dataset are indicated in Figures 12 and 13. The high-risk areas predicted by RF, DFCNN, and LSTM models were distributed on both banks of the river, which is consistent with reality. For the shallow machine learning models, the area of high-risk and moderate-risk regions predicted under the class balanced strategy was significantly higher than those predicted under the class imbalanced strategy. For the deep learning models, the area of high-risk and moderate-risk regions predicted under two different class weighting strategies was nearly identical. Compared with the landslide distribution predicted by deep learning models, the shallow machine learning models predicted wider distributions of landslides and more differentiated hazard hierarchies. It can be found that there was a significant difference in the susceptibility patterns between the deep learning models and shallow machine learning models.

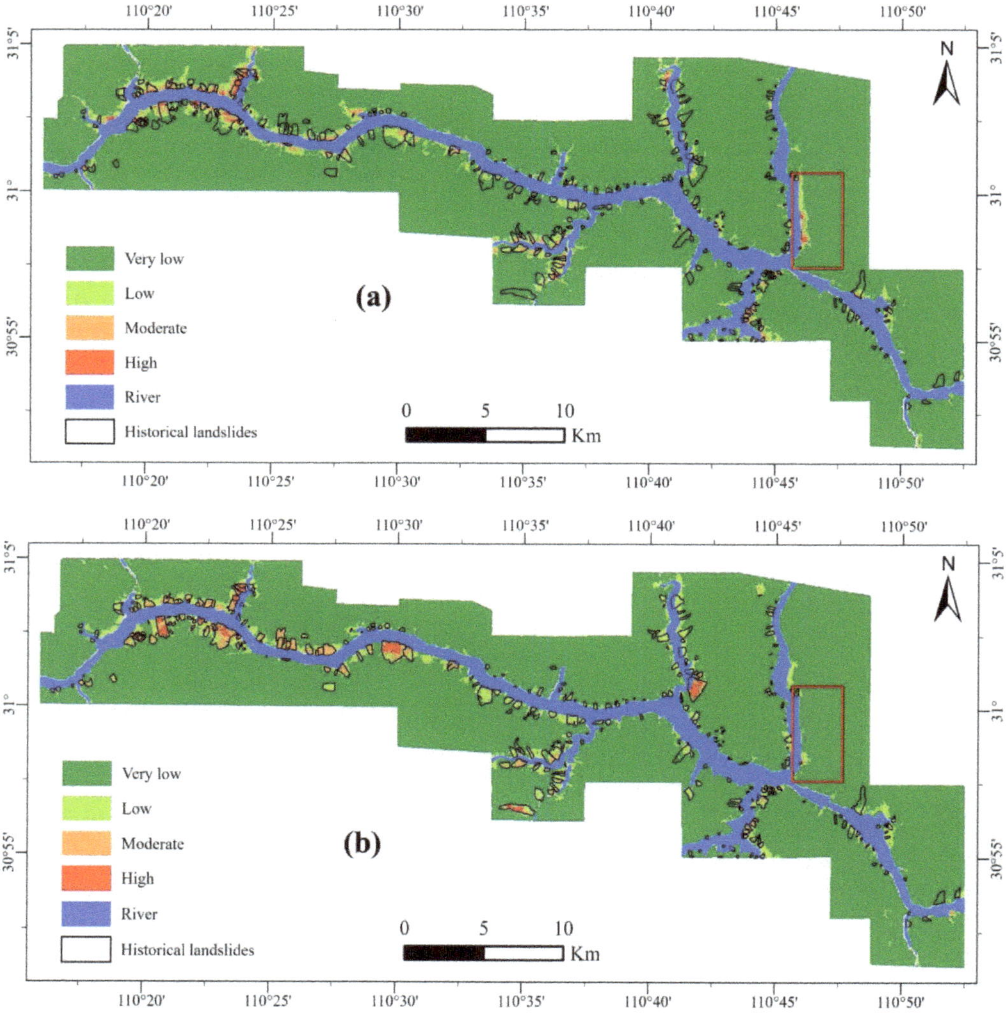

Figure 12. *Cont.*

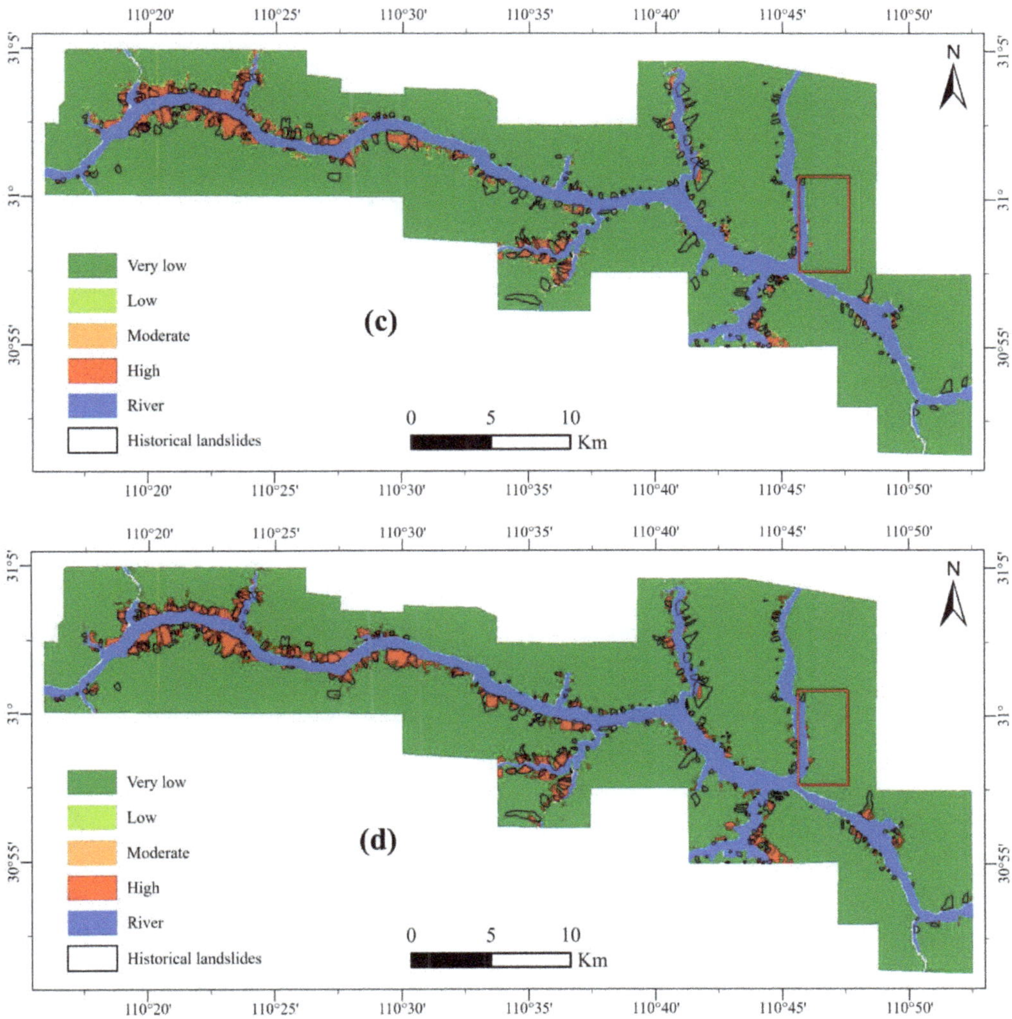

Figure 12. LSA results under the class imbalanced strategy: (**a**) result of LR, (**b**) result of RF, (**c**) result of DFCNN, (**d**) result of LSTM.

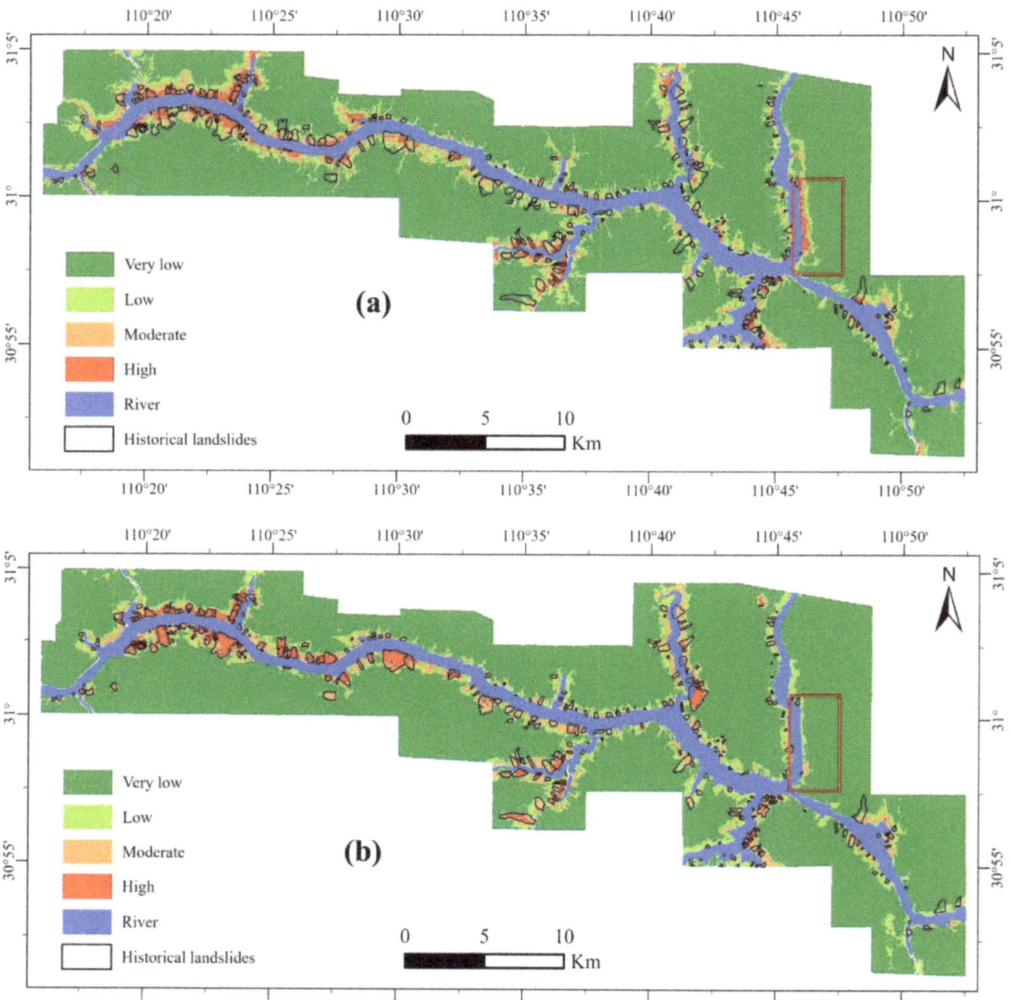

Figure 13. *Cont.*

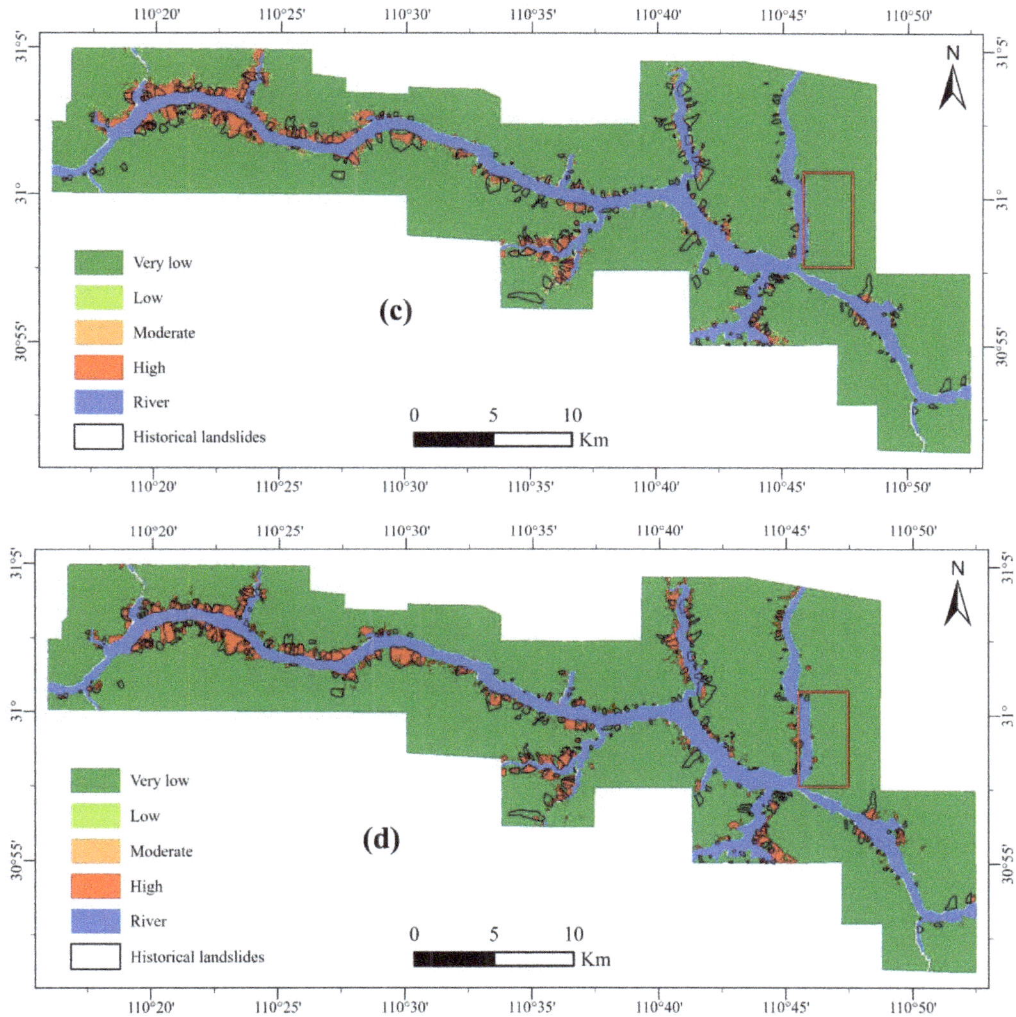

Figure 13. LSA results under the class balanced strategy: (**a**) result of LR, (**b**) result of RF, (**c**) result of DFCNN, (**d**) result of LSTM.

5. Discussion

5.1. Comparison of Performances

Figure 12 demonstrates the high-risk areas predicted by the deep models had a higher overlap with the historical landslides, and this is consistent with those demonstrated by the performance evaluation metrics in Table 3. The AUC values of the shallow models slightly outperformed the deep learning model under the class imbalanced strategy. However, the F1-scores of the deep learning models outperformed the shallow machine learning models by an increase of 0.10 to 0.11. These results have some commonalities with those reported in previous studies. Deep learning models do not always outperform shallow models in AUC performances; this is especially the case with the RF model among the various shallow models, which is reported to be superior to the deep learning models in terms of AUC values in some cases [72]. Meanwhile, the deep learning models generally achieve better F1-score performances compared to the shallow models [73]. Thus, the deep learning models still outperformed shallow models in terms of the overall performances.

It is worth mentioning that in previous research, sampling techniques were often used to obtain roughly equal numbers of landslide and non-landslide samples [74]. In this study, no sampling techniques were used and the performances of deep learning models were still similar to those in previous studies. It might be required to investigate the necessity of sampling techniques in deep learning LSA models in the future.

When the class balanced strategy was applied, the F1-scores of LR and RF rose rapidly, which indicates a substantial improvement in predicting high-risk regions. The proportion of moderate-risk predicted by LR increased from 1.57% to 6.81%, and the proportion of high-risk predicted by LR increased from 0.23% to 2.32%. Meanwhile, the proportion of low-risk decreased significantly from 92.10% to 78.26% (Figure 14). Similar trends were observed in the results of the RF. These phenomena verify the effectiveness of the class-weighted strategy for shallow machine learning models. When the class balanced strategy was adopted, the proportion of very low-risk, low-risk, and moderate-risk predicted by DFCNN and LSTM varied slightly and the overall performances of deep learning models were not significantly improved compared with the class imbalanced strategy. This phenomenon reveals that the two deep learning models were more robust to varying class weights than the conventional shallow machine learning methods. This is because deep learning models have more intricate connections between nodes in different layers, which can represent more complex mapping relations, and have stronger anti-interference abilities. However, the deep learning model also has some shortcomings. It is evident that the deep learning model predicted a small area of low-risk zones and moderate-risk zones, and had a weak predictive power for potential landslides, which may lead it to miss detection of landslides and cause problems for disaster management and decision making. Compared with deep learning models, LR and RF had powerful abilities to explore unknown regions while reliably predicting high-risk areas.

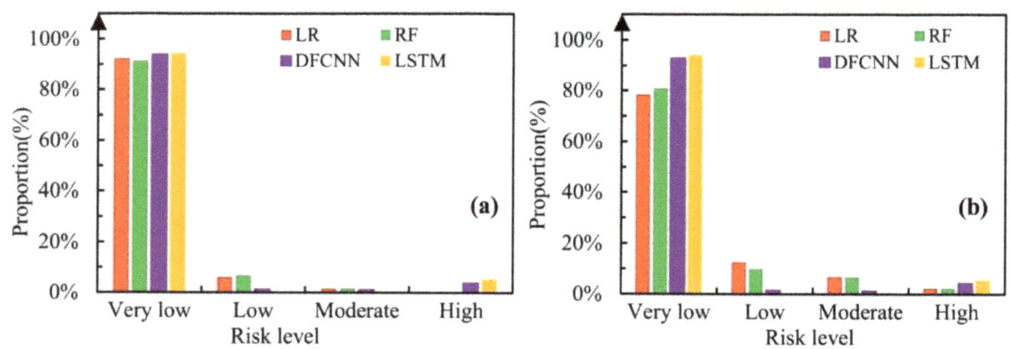

	LR	RF	DFCNN	LSTM
Very low	92.10%	91.16%	93.69%	94.23%
Low	6.11%	6.85%	1.31%	0.14%
Moderate	1.57%	1.66%	1.16%	0.15%
High	0.23%	0.32%	3.83%	5.49%

	LR	RF	DFCNN	LSTM
Very low	78.26%	80.85%	92.81%	94.08%
Low	12.61%	10.05%	1.55%	0.15%
Moderate	6.81%	6.74%	1.30%	0.16%
High	2.32%	2.36%	4.33%	5.61%

Figure 14. Proportions of different landslide risk levels of four models: (**a**) class imbalanced strategy; (**b**) class balanced strategy.

5.2. Susceptibility Spatial Patterns

The susceptibility spatial patterns of shallow models (LR and RF) and deep learning models (DFCNN and LSTM) were different. In the cases of deep learning models, high-risk areas were surrounded by pure very low-risk areas, and the proportion of the moderate-risk regions was quite low. However, in the case of shallow models, high-risk areas were surrounded by a much larger proportion of moderate-risk and low-risk areas compared to the deep learning models, and the deep learning models obtained more accurate results in high-risk landslide areas. The results of shallow models had richer hierarchies than deep learning models. This indicates the shallow models are suitable for scenarios requiring hierarchical emergency response. The deep learning models generated more accurate predictions of high-risk landslide areas, which represents that deep learning models are more suitable for scenarios that need precise investigations and specified treatments. However, neither the deep learning models nor the shallow models can completely predicted all landslide zones, and all of these models missed some potentially dangerous landslides. It is worth noting that there were also some regions (red rectangle in Figures 12 and 13) that were predicted as risk zones by two models (LR and RF), but these regions are not marked in the historical landslide survey data. These parts may be future high-risk zones that require some precautions by decision makers.

5.3. Influences of Class Weights

Since LSA is a problem of imbalanced data classification, it is usually considered that class weights can improve the reliability of the evaluation results. Figure 8 illustrates that class weights had a negligible impact on deep learning models. In contrast, class weights had a tremendous impact on the shallow models and can substantially improve the balanced performances between landslides and non-landslides. However, with the class weights increase of the minority class, the F1-score values of the shallow models (LR and RF) first increased and then decreased. This is due to the fact that with the change of class weights, the shallow model can predict landslides more accurately. When a certain limit is exceeded, it leads to a decrease in the performance of the predicted non-landslides, thus reducing the overall performance. It can be concluded that determining the appropriate class weights is critical for shallow models when faced with LSA problems in unknown zones. The optimal class weight reported by [40] differs from this study. This implies that the optimal class weights vary with the dataset and the class weighted strategy usually exerts a positive effect on the shallow machine learning models. Typically, the optimal class weight is determined by conducting numerous tests on existing historical data. However, this way of obtaining the optimal class weights becomes less reliable when the disparities between the known area and the evaluated area are too large. This difficulty is an important reason that prevents the widespread use of class-weighted strategies in LSA. In scenarios where the training data and test data are very different, it is preferable to employ a well-trained deep learning model.

5.4. Model Evaluation and Factors Processing

The AUC value of each model did not vary tremendously under the two strategies. This suggests that AUC is not feasible as a single evaluation metric for LSA models. It is necessary to use multiple model evaluation metrics. We employed the F1-score as the main metric for evaluating the model performances. Nevertheless, high F1-score values prefer to predict more samples as landslide areas, which may raise the cost of landslide prevention. Determining the appropriate F1-score is a tricky problem to be solved, which is related to our practical needs and the cost we are willing to pay to prevent landslides.

Figure 10 illustrates that removing the distance to rivers and elevation greatly affected the performances of the models while removing the distance to rivers or elevation alone did not significantly reduce the performance. This suggests that these two factors are the most critical factors affecting LSA modeling. However, the most important landslide influence factors vary across the literature [19,75–77]. It is important to note that various studies in the literature adopt diverse factor analysis methods, which may lead to different key factors being observed. However, even if the same factor analysis method is used, the importance rankings of landslide influence factors observed in different study areas are different [33]. This indicates that there is no constant pattern in the relationship between landslide influence factors and landslide occurrences. Thus, it is necessary to perform factor analysis to select the appropriate landslide influence factor in data-driven LSA modeling. It also reveals that the relationships between influencing factors are complicated, and the joint action of the influence factors affects the final prediction ability of the models.

5.5. Training Time Consumed

The average time consumed by shallow machine learning methods and deep learning methods in the training stage is illustrated in Table 4. The training time required by deep learning models was up to 765 times longer than that required by shallow machine learning models. In the shallow models, RF achieved pretty good performances with a very low time overhead. DFCNN and LSTM achieved the best performances and stable results at the cost of the enormous time overhead. As it is reported in [78], the training time consumed is a necessary consideration when choosing a suitable LSA model. Deep learning models require faster computing devices, and the model building processes are more complicated [33]. Since deep learning models generally require GPUs to accelerate the computation, in the absence of high-performance computing cards, rapid landslide susceptibility mapping cannot be accomplished using deep learning methods. The model structures and hyperparameters need to be fine-tuned in the training stage [22] and these activities are generally not counted in the consumption time. Nevertheless, the time cost of designing network structures and adjusting hyperparameters cannot be ignored in actual disaster prevention. Therefore, RF is still an attractive option to build the LSA model quickly and obtain acceptable prediction results. Deep learning methods are preferable for achieving more consistent results.

Table 4. Average time consumed by different LSA models.

Model	Average Time Consumed (s)	
	Class Balanced	Class Imbalanced
LR	1	1
RF	12	11
DFCNN	418	405
LSTM	765	761

6. Conclusions

In this study, eighteen landslide influence factors were selected for landslide susceptibility assessment. LR, RF, DFCNN, and LSTM were employed as the assessment models and their performances were compared. In general, the AUC values of the four models were greater than 0.8 under the class balanced strategy and the class imbalanced strategy which implies they could be used as credible LSA models. RF performed best in shallow machine learning models, which can balance the training time and the performance. The deep learning models (DFCNN and LSTM) achieved better overall performances and greater robustness than shallow models at the cost of the more significant time overhead. The shallow models were more sensitive to class weights, while the deep learning models were relatively insensitive. It is verified that removing a single or a few landslide influence factors did not substantially affect the accuracy of LSA results except distance to rivers and elevation in this study. The evaluation results of the shallow models had richer landslide susceptibility hierarchies, whereas the deep learning models identified the high-risk landslide zones better. However, whether most shallow machine learning models and deep learning models have the same benefits and drawbacks requires further experimental validation. To ensure the LSA models' stability across various imbalanced landslide datasets, more robust methods for estimating class weights must be developed in the future.

In conclusion, the class balanced strategy must be adopted when using LR and RF for LSA modeling. The deep learning models (DFCNN and LSTM) have superior adaptability and can be used in an imbalanced dataset. The findings presented in this research can serve as a reference for model selection in LSA and further provide decision support for land management and disaster prevention and mitigation.

Author Contributions: Conceptualization, Y.S. and S.X.; methodology, S.X. and Y.S.; software, S.X.; validation, X.H. and S.X; formal analysis, S.X.; investigation, Y.S.; data curation, S.X. and X.H.; writing—original draft preparation, S.X.; writing—review and editing, Y.S and X.H. All authors have read and agreed to the published version of the manuscript.

Funding: This research was funded by the Zhejiang Province Key Laboratory of Smart Management & Application of Modern Agricultural Resources (Grant No. 2020E10017), Science and Technology Research Project of Jiangxi Provincial Department of Education (Grant No. GJJ200748) and Jiangxi Provincial Natural Science Foundation (Grant No. 20202BAB204035).

Institutional Review Board Statement: Not applicable.

Informed Consent Statement: Not applicable.

Data Availability Statement: The elevation data are available at https://www.earthdata.nasa.gov/ (accessed on 10 June 2021). The Landsat images presented in this study are obtained from USGS earth explorer and they are openly available at https://earthexplorer.usgs.gov/ (accessed on 10 June 2021). Rainfall data are collected from Climate Hazards Group InfraRed Precipitation with Station data at https://data.chc.ucsb.edu/products/CHIRPS-2.0/ (accessed on 3 November 2022). Seismic data and landslide inventory data are available upon request from the corresponding author.

Conflicts of Interest: The authors declare that they have no known competing financial interests or personal relationships that could have appeared to influence the work reported in this paper.

References

1. Argyriou, A.V.; Teeuw, R.M.; Rust, D.; Sarris, A. GIS multi-criteria decision analysis for assessment and mapping of neotectonic landscape deformation: A case study from Crete. *Geomorphology* **2016**, *253*, 262–274. [CrossRef]
2. Vahidnia, M.H.; Alesheikh, A.A.; Alimohammadi, A.; Hosseinali, F. A GIS-based neuro-fuzzy procedure for integrating knowledge and data in landslide susceptibility mapping. *Comput. Geosci.* **2010**, *36*, 1101–1114. [CrossRef]
3. Cabral, V.C.; Reis, F.A.G.V. Assessment of Shallow Landslides Susceptibility Using SHALSTAB and SINMAP at Serra Do Mar, Brazil. In *Understanding and Reducing Landslide Disaster Risk*; Guzzetti, F., Mihalić Arbanas, S., Reichenbach, P., Sassa, K., Bobrowsky, P.T., Takara, K., Eds.; Springer: Cham, Switzerland, 2021; Volume 2, pp. 257–265. [CrossRef]
4. Fernandes, N.F.; Guimarães, R.F.; Gomes, R.A.T.; Vieira, B.C.; Montgomery, D.R.; Greenberg, H. Topographic controls of landslides in Rio de Janeiro: Field evidence and modeling. *Catena* **2004**, *55*, 163–181. [CrossRef]
5. Ciurleo, M.; Cascini, L.; Calvello, M. A comparison of statistical and deterministic methods for shallow landslide susceptibility zoning in clayey soils. *Eng. Geol.* **2017**, *223*, 71–81. [CrossRef]
6. Lin, W.; Yin, K.; Wang, N.; Xu, Y.; Guo, Z.; Li, Y. Landslide hazard assessment of rainfall-induced landslide based on the CF-SINMAP model: A case study from Wuling Mountain in Hunan Province, China. *Nat. Hazards* **2021**, *106*, 679–700. [CrossRef]
7. Melo, C.M.; Kobiyama, M.; Michel, G.P.; de Brito, M.M. The Relevance of Geotechnical-Unit Characterization for Landslide-Susceptibility Mapping with SHALSTAB. *GeoHazards* **2021**, *2*, 383–397. [CrossRef]
8. Chen, W.; Li, W.; Chai, H.; Hou, E.; Li, X.; Ding, X. GIS-based landslide susceptibility mapping using analytical hierarchy process (AHP) and certainty factor (CF) models for the Baozhong region of Baoji City, China. *Environ. Earth Sci.* **2016**, *75*, 1–14. [CrossRef]
9. Vilasan, R.T.; Kapse, V.S. Evaluation of the prediction capability of AHP and F-AHP methods in flood susceptibility mapping of Ernakulam district (India). *Nat. Hazards* **2022**, *112*, 1767–1793. [CrossRef]
10. Merghadi, A.; Yunus, A.P.; Dou, J.; Whiteley, J.; ThaiPham, B.; Bui, D.T.; Avtar, R.; Abderrahmane, B. Machine learning methods for landslide susceptibility studies: A comparative overview of algorithm performance. *Earth-Sci. Rev.* **2020**, *207*, 103225. [CrossRef]
11. Wang, L.; Guo, M.; Sawada, K.; Lin, J.; Zhang, J. A comparative study of landslide susceptibility maps using logistic regression, frequency ratio, decision tree, weights of evidence and artificial neural network. *Geosci. J.* **2016**, *20*, 117–136. [CrossRef]
12. Wang, Q.; Wang, Y.; Niu, R.; Peng, L. Integration of Information Theory, K-Means Cluster Analysis and the Logistic Regression Model for Landslide Susceptibility Mapping in the Three Gorges Area, China. *Remote Sens.* **2017**, *9*, 938. [CrossRef]
13. Zhou, X.; Wu, W.; Qin, Y.; Fu, X. Geoinformation-based landslide susceptibility mapping in subtropical area. *Sci. Rep.* **2021**, *11*, 1–16. [CrossRef] [PubMed]
14. Yu, X.; Wang, Y.; Niu, R.; Hu, Y. A Combination of Geographically Weighted Regression, Particle Swarm Optimization and Support Vector Machine for Landslide Susceptibility Mapping: A Case Study at Wanzhou in the Three Gorges Area, China. *Int. J. Environ. Res. Public Health* **2016**, *13*, 487. [CrossRef] [PubMed]
15. Kainthura, P.; Sharma, N. Machine learning driven landslide susceptibility prediction for the Uttarkashi region of Uttarakhand in India. *Georisk Assess. Manag. Risk Eng. Syst. Geohazards* **2021**, *16*, 570–583. [CrossRef]

16. Chen, W.; Xie, X.; Wang, J.; Pradhan, B.; Hong, H.; Bui, D.T.; Duan, Z.; Ma, J. A comparative study of logistic model tree, random forest, and classification and regression tree models for spatial prediction of landslide susceptibility. *Catena* **2017**, *151*, 147–160. [CrossRef]
17. Zhou, X.; Wen, H.; Zhang, Y.; Xu, J.; Zhang, W. Landslide susceptibility mapping using hybrid random forest with GeoDetector and RFE for factor optimization. *Geosci. Front.* **2021**, *12*, 101211. [CrossRef]
18. Dou, J.; Yamagishi, H.; Pourghasemi, H.R.; Yunus, A.P.; Song, X.; Xu, Y.; Zhu, Z. An integrated artificial neural network model for the landslide susceptibility assessment of Osado Island, Japan. *Nat. Hazards* **2015**, *78*, 1749–1776. [CrossRef]
19. Pham, B.T.; Van Dao, D.; Acharya, T.D.; Van Phong, T.; Costache, R.; Van Le, H.; Nguyen, H.B.T.; Prakash, I. Performance assessment of artificial neural network using chi-square and backward elimination feature selection methods for landslide susceptibility analysis. *Environ. Earth Sci.* **2021**, *80*, 686. [CrossRef]
20. Zhou, X.; Wen, H.; Li, Z.; Zhang, H.; Zhang, W. An interpretable model for the susceptibility of rainfall-induced shallow landslides based on SHAP and XGBoost. *Geocarto Int.* **2022**, 1–32. [CrossRef]
21. Saha, S.; Roy, J.; Pradhan, B.; Hembram, T.K. Hybrid ensemble machine learning approaches for landslide susceptibility mapping using different sampling ratios at East Sikkim Himalayan, India. *Adv. Space Res.* **2021**, *68*, 2819–2840. [CrossRef]
22. Bui, D.T.; Tsangaratos, P.; Nguyen, V.; Liem, N.V.; Trinh, P.T. Comparing the prediction performance of a Deep Learning Neural Network model with conventional machine learning models in landslide susceptibility assessment. *Catena* **2020**, *188*, 104426. [CrossRef]
23. Sun, D.; Gu, Q.; Wen, H.; Shi, S.; Mi, C.; Zhang, F. A Hybrid Landslide Warning Model Coupling Susceptibility Zoning and Precipitation. *Forests* **2022**, *13*, 827. [CrossRef]
24. Zhang, W.; Liu, S.; Wang, L.; Samui, P.; Chwała, M.; He, Y. Landslide Susceptibility Research Combining Qualitative Analysis and Quantitative Evaluation: A Case Study of Yunyang County in Chongqing, China. *Forests* **2022**, *13*, 1055. [CrossRef]
25. Chai, J.; Zeng, H.; Li, A.; Ngai, E.W.T. Deep learning in computer vision: A critical review of emerging techniques and application scenarios. *Mach. Learn. Appl.* **2021**, *6*, 100134. [CrossRef]
26. Otter, D.W.; Medina, J.R.; Kalita, J.K. A Survey of the Usages of Deep Learning for Natural Language Processing. *IEEE Trans. Neural Netw. Learn. Syst.* **2021**, *32*, 604–624. [CrossRef]
27. Grigorescu, S.; Trasnea, B.; Cocias, T.; Macesanu, G. A survey of deep learning techniques for autonomous driving. *J. Field Robot.* **2020**, *37*, 362–386. [CrossRef]
28. Esteva, A.; Chou, K.; Yeung, S.; Naik, N.; Madani, A.; Mottaghi, A.; Liu, Y.; Topol, E.; Dean, J.; Socher, R. Deep learning-enabled medical computer vision. *NPJ Digit. Med.* **2021**, *4*, 5. [CrossRef] [PubMed]
29. Bera, S.; Upadhyay, V.K.; Guru, B.; Oommen, T. Landslide inventory and susceptibility models considering the landslide typology using deep learning: Himalayas, India. *Nat. Hazards* **2021**, *108*, 1257–1289. [CrossRef]
30. Thi Ngo, P.T.; Panahi, M.; Khosravi, K.; Ghorbanzadeh, O.; Kariminejad, N.; Cerda, A.; Lee, S. Evaluation of deep learning algorithms for national scale landslide susceptibility mapping of Iran. *Geosci. Front.* **2021**, *12*, 505–519. [CrossRef]
31. Wang, H.; Zhang, L.; Luo, H.; He, J.; Cheung, R.W.M. AI-powered landslide susceptibility assessment in Hong Kong. *Eng. Geol.* **2021**, *288*, 106103. [CrossRef]
32. Wang, W.; He, Z.; Han, Z.; Li, Y.; Dou, J.; Huang, J. Mapping the susceptibility to landslides based on the deep belief network: A case study in Sichuan Province, China. *Nat. Hazards* **2020**, *103*, 3239–3261. [CrossRef]
33. Liu, R.; Yang, X.; Xu, C.; Wei, L.; Zeng, X. Comparative Study of Convolutional Neural Network and Conventional Machine Learning Methods for Landslide Susceptibility Mapping. *Remote Sens.* **2022**, *14*, 321. [CrossRef]
34. Azarafza, M.; Azarafza, M.; Akgün, H.; Atkinson, P.M.; Derakhshani, R. Deep learning-based landslide susceptibility mapping. *Sci. Rep.* **2021**, *11*, 24112. [CrossRef] [PubMed]
35. Yi, Y.; Zhang, Z.; Zhang, W.; Jia, H.; Zhang, J. Landslide susceptibility mapping using multiscale sampling strategy and convolutional neural network: A case study in Jiuzhaigou region. *Catena* **2020**, *195*, 104851. [CrossRef]
36. Sameen, M.I.; Pradhan, B.; Lee, S. Application of convolutional neural networks featuring Bayesian optimization for landslide susceptibility assessment. *Catena* **2020**, *186*, 104249. [CrossRef]
37. Habumugisha, J.M.; Chen, N.; Rahman, M.; Islam, M.M.; Ahmad, H.; Elbeltagi, A.; Sharma, G.; Liza, S.N.; Dewan, A.M. Landslide Susceptibility Mapping with Deep Learning Algorithms. *Sustainability* **2022**, *14*, 1734. [CrossRef]
38. Wu, B.; Qiu, W.; Jia, J.; Liu, N. Landslide Susceptibility Modeling Using Bagging-Based Positive-Unlabeled Learning. *IEEE Geosci. Remote Sens. Lett.* **2021**, *18*, 766–770. [CrossRef]
39. Yao, J.; Qin, S.; Qiao, S.; Liu, X.; Zhang, L.; Chen, J. Application of a two-step sampling strategy based on deep neural network for landslide susceptibility mapping. *Bull. Eng. Geol. Environ.* **2022**, *81*, 148. [CrossRef]
40. Zhang, H.; Song, Y.; Xu, S.; He, Y.; Li, Z.; Yu, X.; Liang, Y.; Wu, W.; Wang, Y. Combining a class-weighted algorithm and machine learning models in landslide susceptibility mapping: A case study of Wanzhou section of the Three Gorges Reservoir, China. *Comput. Geosci.* **2022**, *158*, 104966. [CrossRef]
41. Süzen, M.L.; Kaya, B.Ş. Evaluation of environmental parameters in logistic regression models for landslide susceptibility mapping. *Int. J. Digit. Earth* **2012**, *5*, 338–355. [CrossRef]
42. Zhang, M.; Cao, X.; Peng, L.; Niu, R. Landslide susceptibility mapping based on global and local logistic regression models in Three Gorges Reservoir area, China. *Environ. Earth Sci.* **2016**, *75*, 958. [CrossRef]

43. Akinci, H.; Zeybek, M. Comparing classical statistic and machine learning models in landslide susceptibility mapping in Ardanuc (Artvin), Turkey. *Nat. Hazards* **2021**, *108*, 1515–1543. [CrossRef]
44. Mira, J.; Sandoval, F. *From Natural to Artificial Neural Computation: International Workshop on Artificial Neural Networks, Malaga-Torremolinos, Spain, 7–9 June 1995: Proceedings*; Springer-Verlag: New York, USA, 1995; Volume 930, pp. 195–201.
45. Chen, W.; Sun, Z.; Zhao, X.; Lei, X.; Shirzadi, A.; Shahabi, H. Performance Evaluation and Comparison of Bivariate Statistical-Based Artificial Intelligence Algorithms for Spatial Prediction of Landslides. *ISPRS Int. J. Geo-Inf.* **2020**, *9*, 696. [CrossRef]
46. Akinci, H.; Kilicoglu, C.; Dogan, S. Random Forest-Based Landslide Susceptibility Mapping in Coastal Regions of Artvin, Turkey. *ISPRS Int. J. Geo-Inf.* **2020**, *9*, 553. [CrossRef]
47. Akinci, H. Assessment of rainfall-induced landslide susceptibility in Artvin, Turkey using machine learning techniques. *J. Afr. Earth Sci.* **2022**, *191*, 104535. [CrossRef]
48. Li, J.; Cheng, J.; Shi, J.; Huang, F. Brief Introduction of Back Propagation (BP) Neural Network Algorithm and Its Improvement. In *Advances in Intelligent and Soft Computing, Proceedings of the Advances in Computer Science and Information Engineering*; Jin, D., Lin, S., Eds.; Springer-Verlag: Berlin/Heidelberg, Germany, 2012; Volume 169, pp. 553–558.
49. Wang, S.-C. Artificial Neural Network. In *Interdisciplinary Computing in Java Programming*; Wang, S.-C., Ed.; The Springer International Series in Engineering and Computer Science; Springer US: Boston, MA, USA, 2003; Volume 743, pp. 81–100.
50. Nair, V.; Hinton, G.E. Rectified Linear Units Improve Restricted Boltzmann Machines. In Proceedings of the 27th International Conference on Machine Learning, Haifa, Israel, 21–24 June 2010; ICML: Haifa, Israel, 2010; pp. 807–814.
51. Hochreiter, S.; Schmidhuber, J. Long short-term memory. *Neural Comput.* **1997**, *9*, 1735–1780. [CrossRef]
52. Wang, Y.; Fang, Z.; Wang, M.; Peng, L.; Hong, H. Comparative study of landslide susceptibility mapping with different recurrent neural networks. *Comput. Geosci.* **2020**, *138*, 104445. [CrossRef]
53. Moosavi, V.; Niazi, Y. Development of hybrid wavelet packet-statistical models (WP-SM) for landslide susceptibility mapping. *Landslides* **2016**, *13*, 97–114. [CrossRef]
54. Zhao, C.; Chen, W.; Wang, Q.; Wu, Y.; Yang, B. A comparative study of statistical index and certainty factor models in landslide susceptibility mapping: A case study for the Shangzhou District, Shaanxi Province, China. *Arab. J. Geosci.* **2015**, *8*, 9079–9088. [CrossRef]
55. Yang, J.; Huang, X. The 30 m annual land cover dataset and its dynamics in China from 1990 to 2019. *Earth Syst. Sci. Data* **2021**, *13*, 3907–3925. [CrossRef]
56. Bourenane, H.; Bouhadad, Y.; Guettouche, M.S.; Braham, M. GIS-based landslide susceptibility zonation using bivariate statistical and expert approaches in the city of Constantine (Northeast Algeria). *Bull. Eng. Geol. Environ.* **2015**, *74*, 337–355. [CrossRef]
57. Peng, L.; Niu, R.; Huang, B.; Wu, X.; Zhao, Y.; Ye, R. Landslide susceptibility mapping based on rough set theory and support vector machines: A case of the Three Gorges area, China. *Geomorphology* **2014**, *204*, 287–301. [CrossRef]
58. Pourghasemi, H.R.; Rahmati, O. Prediction of the landslide susceptibility: Which algorithm, which precision? *Catena* **2018**, *162*, 177–192. [CrossRef]
59. Regmi, N.R.; Giardino, J.R.; McDonald, E.V.; Vitek, J.D. A comparison of logistic regression-based models of susceptibility to landslides in western Colorado, USA. *Landslides* **2014**, *11*, 247–262. [CrossRef]
60. Tang, C.; Ma, G.; Chang, M.; Li, W.; Zhang, D.; Jia, T.; Zhou, Z. Landslides triggered by the 20 April 2013 Lushan earthquake, Sichuan Province, China. *Eng. Geol.* **2015**, *187*, 45–55. [CrossRef]
61. Fang, Z.; Wang, Y.; Peng, L.; Hong, H. Integration of convolutional neural network and conventional machine learning classifiers for landslide susceptibility mapping. *Comput. Geosci.* **2020**, *139*, 104470. [CrossRef]
62. Melillo, M.; Brunetti, M.T.; Peruccacci, S.; Gariano, S.L.; Guzzetti, F. Rainfall thresholds for the possible landslide occurrence in Sicily (Southern Italy) based on the automatic reconstruction of rainfall events. *Landslides* **2016**, *13*, 165–172. [CrossRef]
63. Fallah-Zazuli, M.; Vafaeinejad, A.; Alesheykh, A.A.; Modiri, M.; Aghamohammadi, H. Mapping landslide susceptibility in the Zagros Mountains, Iran: A comparative study of different data mining models. *Earth Sci. Inform.* **2019**, *12*, 615–628. [CrossRef]
64. Rodgers, J.L.; Nicewander, W.A. Thirteen Ways to Look at the Correlation Coefficient. *Am. Stat.* **1988**, *42*, 59–66. [CrossRef]
65. Gauthier, T. Detecting Trends Using Spearman's Rank Correlation Coefficient. *Environ. Forensics* **2001**, *2*, 359–362. [CrossRef]
66. Yu, X.; Zhang, K.; Song, Y.; Jiang, W.; Zhou, J. Study on landslide susceptibility mapping based on rock–soil characteristic factors. *Sci. Rep.* **2021**, *11*, 15476. [CrossRef] [PubMed]
67. Flach, P.; Kull, M. Precision-recall-gain curves: PR analysis done right. In *Advances in Neural Information Processing Systems*; MIT Press: Cambridge, MA, USA, 2015; Volume 1, pp. 838–846.
68. Pontius, R.G.; Parmentier, B. Recommendations for using the relative operating characteristic (ROC). *Landsc. Ecol.* **2014**, *29*, 367–382. [CrossRef]
69. Erener, A.; Düzgün, H.S.B. Landslide susceptibility assessment: What are the effects of mapping unit and mapping method? *Environ. Earth Sci.* **2012**, *66*, 859–877. [CrossRef]
70. Hussin, H.Y.; Zumpano, V.; Reichenbach, P.; Sterlacchini, S.; Micu, M.; van Westen, C.; Bălteanu, D. Different landslide sampling strategies in a grid-based bi-variate statistical susceptibility model. *Geomorphology* **2016**, *253*, 508–523. [CrossRef]
71. Spiegel, M.R.; Schiller, J.J.; Srinivasan, R.A. *Schaum's Outline of Probability and Statistics*, 4th ed.; McGraw-Hill Education: Los Angeles, CA, USA, 2013; pp. 75–107.
72. Wang, H.; Zhang, L.; Yin, K.; Luo, H.; Li, J. Landslide identification using machine learning. *Geosci. Front.* **2021**, *12*, 351–364. [CrossRef]

73. Xiong, Y.; Zhou, Y.; Wang, F.; Wang, S.; Wang, Z.; Ji, J.; Wang, J.; Zou, W.; You, D.; Qin, G. A Novel Intelligent Method Based on the Gaussian Heatmap Sampling Technique and Convolutional Neural Network for Landslide Susceptibility Mapping. *Remote Sens.* **2022**, *14*, 2866. [CrossRef]
74. Fang, Z.; Wang, Y.; Duan, G.; Peng, L. Landslide Susceptibility Mapping Using Rotation Forest Ensemble Technique with Different Decision Trees in the Three Gorges Reservoir Area, China. *Remote Sens.* **2021**, *13*, 238. [CrossRef]
75. Wang, Y.; Song, C.; Lin, Q.; Li, J. Occurrence probability assessment of earthquake-triggered landslides with Newmark displacement values and logistic regression: The Wenchuan earthquake, China. *Geomorphology* **2016**, *258*, 108–119. [CrossRef]
76. Pineda, M.C.; Viloria, J.; Martínez-Casasnovas, J.A. Landslides susceptibility change over time according to terrain conditions in a mountain area of the tropic region. *Environ. Monit. Assess.* **2016**, *188*, 255. [CrossRef]
77. Hua, Y.; Wang, X.; Li, Y.; Xu, P.; Xia, W. Dynamic development of landslide susceptibility based on slope unit and deep neural networks. *Landslides* **2021**, *18*, 281–302. [CrossRef]
78. Dao, D.V.; Jaafari, A.; Bayat, M.; Mafi-Gholami, D.; Qi, C.; Moayedi, H.; Phong, T.V.; Ly, H.; Le, T.; Trinh, P.T.; et al. A spatially explicit deep learning neural network model for the prediction of landslide susceptibility. *Catena* **2020**, *188*, 104451. [CrossRef]

forests

MDPI

Technical Note

Landslide Susceptibility Research Combining Qualitative Analysis and Quantitative Evaluation: A Case Study of Yunyang County in Chongqing, China

Wengang Zhang [1], Songlin Liu [1], Luqi Wang [1,*], Pijush Samui [2], Marcin Chwała [3] and Yuwei He [1]

1 School of Civil Engineering, Chongqing University, Chongqing 400045, China; zhangwg@cqu.edu.cn (W.Z.); songlinl@cqu.edu.cn (S.L.); 20164985@cqu.edu.cn (Y.H.)
2 Department of Civil Engineering, National Institute of Technology Patna, Patna 800005, Bihar, India; pijush@nitp.ac.in
3 Department of Geotechnics and Hydrotechnics, Faculty of Civil Engineering, Wroclaw University of Science and Technology, Wybrzeże Wyspiańskiego 27, 50-370 Wrocław, Poland; marcin.chwala@pwr.edu.pl
* Correspondence: wlq93@cqu.edu.cn

Abstract: Machine learning-based methods are commonly used for landslide susceptibility mapping. Most of the recent publications focused on quantitative analysis, i.e., improving data processing methods, comparing and perfecting the data-driven model itself, but rarely taking the qualitative aspects of the local landslide occurrences into consideration and the further analysis of the key features was always lacking. This study aims to combine qualitative and quantitative analysis and examine its effect on mapping accuracy; based on the feature importance ranks and the related literature, the key features for identifying landslide/non-landslide points of different sub-zones were further analyzed. Before modeling, the study area Yunyang County, Chongqing City, China, was manually divided into four sub-zones based on the information from geological hazards exploration in Chongqing, including the mechanism of landslide formation and sliding failure and geomorphic unit characteristics. Upon the qualitative analysis basis, five grid searches tuned random forest models (one for the whole region and four for the sub-zones independently) were established by 1654 data points and 20 conditioning features. Compared with the conventional data-driven method, the integrated quantitative evaluation based on the qualitative analysis results showed higher reliability, which not only improved the mapping accuracy but also increased the AUC values of all four sub-models, which were 8.8%, 2.3%, 1.9% and 9.1% higher than that of the parent model. Moreover, the quantitative evaluation based on the qualitative analysis revealed the key factors affecting local landslide formation. Therefore, qualitative analysis is recommended in future landslide susceptibility modeling with the additional combination of data-driven methods.

Keywords: landslide susceptibility mapping; random forest model; qualitative analysis; quantitative evaluation; Yunyang County

Citation: Zhang, W.; Liu, S.; Wang, L.; Samui, P.; Chwała, M.; He, Y. Landslide Susceptibility Research Combining Qualitative Analysis and Quantitative Evaluation: A Case Study of Yunyang County in Chongqing, China. *Forests* **2022**, *13*, 1055. https://doi.org/10.3390/f13071055

Academic Editors: Chong Xu, Haijia Wen, Weile Li and Hiromu Daimaru

Received: 11 May 2022
Accepted: 28 June 2022
Published: 4 July 2022

1. Introduction

Landslides are one of the most destructive geological hazards, which not only cause enormous damage to houses and infrastructure, such as bridges and roads, but also lead to loss of life [1]. According to the World Health Organization, approximately 4.8 million people were affected, and more than 18,000 deaths were caused by landslides between 1998 and 2017. Specifically, as one of the countries with a high incidence of landslides, China suffered severe loss of life [2,3]. The China Statistical Yearbook indicates that during 2000 to 2015, 373,630 landslides occurred in this country, killing 10,996 people, which is approximately 690 landslide-related deaths per year [4]. To mitigate the serious social impact caused by landslides, constructive and productive activities should be avoided in areas with high susceptibility to landslides. Therefore, developing an efficient method to distinguish landslide-prone zones is an essential need for both local governments and research institutes [5]. Landslide susceptibility describes the likelihood that a landslide will occur in a certain area based on local terrain conditions [6]. Landslide susceptibility mapping (LSM) is one of the most widely used assessment methods, it visualizes the spatial distribution of zones with different probabilities of occurrence of landslides in a certain area.

Various methods such as probabilistic analysis, statistical analysis, analytic process, and weighted overlay were widely applied to LSM by researchers in the early stages. With the development of Artificial Intelligence (AI) and Geographic Information System (GIS), machine learning-based methods, with the capability of solving complex nonlinear problems, are becoming increasingly popular compared to opinion-driven models and statistical learning, making the accuracy and precision of susceptibility models evolve rapidly [7,8]. Huang et al. [9] adopted logic regression (LR), support vector machine (SVM), and random forest (RF) on LSM for model comparison. He et al. [10] used RF in the global assessment of earthquake-induced landslide susceptibility. Sun et al. [11] applied the Bayes algorithm to optimize the hyper-parameters of the RF model for LSM. Smith et al. [12] compared the effect of landslide inventories assembled by different methods on the performance of RF and LF for LSM. Lim et al. [13] applied the RF model to estimate the probability of a landslide. Nhu et al. [14] investigated and compared the Logistic Model Tree, LR, NBTress, Artificial Neural Network (ANN), and SVM in the shallow landslide susceptibility mapping for Bijar City in Kurdistan City. Zhang et al. [15] used the predictive performance of RF, XGBoost, SVM, and LR on landslide susceptibility mapping in Yunyang County. Hu et al. [16] compared the effect of different non-landslide sampling methods on the performance of SVM and NB for LSM. Zhou et al. [17] applied GeoDetector and RFE for factor optimization and then used the selected factors as inputs to train an RF model to obtain the LSM of Wuxi County. Sun et al. [18] proposed a hybrid landslide warning model based on RF susceptibility zoning and precipitation. Zhou et al. [19] constructed an interpretable model for the susceptibility to rainfall-induced shallow landslides based on SHAP and XGBoost.

Among those methods, RF is the most commonly used method in large-scale mapping and classification [20–22] due to its characteristics of low computational cost, low data requirement, convenience of hyper-parameters tuning, and robustness in solving complex nonlinear problems [23]. Previous work usually focused on quantitative analysis, such as the selection and improvement of models and input features, but rarely took into account the qualitative analysis of landslide areas. Actually, as one of the major geological hazards, landslides are highly area dependent, the mechanism of landslide formation and its corresponding triggering factors are undoubtedly different in distinct areas. The frequent fluctuation of reservoir water seriously reduces the stability of the slopes in the reservoir area, making them prone to landslides [24]. For mountainous areas, however, rainfall is the major triggering factor for the occurrence of landslides [25]. With increased population, human activities have become the major issue that accelerates landslide formation in areas with high population density. Therefore, manually dividing a relatively large region into different sub-zones according to the qualitative analysis of the landslide formation and geomorphic unit characteristics will theoretically improve mapping accuracy. This paper aims to use the 827 historical landslide data points in Yunyang County and the 20 conditioning factors to build 5 RF models, including an RF model (referred to as the parent model below) for the whole region and four RF models for the divided four sub-zones (referred to as sub-model one to sub-model four below). Then, the feature importance and the performances of the parent model and the four sub-models are analyzed and compared to verify the effectiveness of applying experience-based zonation before modeling.

2. Study Area

Chongqing City is located in the mountainous area around the eastern Sichuan basin and the slope area of the basin margin. It spans two tectonic units, namely the Yangtze quasi-platform and the Qinling fold system. The landscape of Chongqing City is mainly mountains and hills, which make up 92% of its total area. There are many adverse geological conditions accelerating the formation of landslides, dangerous rock collapse, ground collapse, debris flow, and other geological disasters, including developed surface water networks, strongly cut terrain, complex rock and soil structure, and geological structure, making it one of the cities with the highest geological disaster frequency in the country.

The spatial distribution of geological hazards in Chongqing City shows a certain degree of concentration and can be concluded as a striped distribution and vertical zonal distribution; moreover, its temporal distribution presents a seasonal cluster pattern. According to the statistics, there are currently 14,926 geological hazard-prone points in Chongqing City, of which 5776 (38.7%) are located in the 7 districts and counties of northeast Chongqing City (Wanzhou, Kaizhou, Chengkou, Wuxi, Wushan, Fengjie, Yunyang), 1864 (12.49%) are in the 5 districts and counties of southeastern Chongqing City (Wulong, Youyang, Qianjiang, Pengshui, Xiushan), and 1320 (8.84%) are in the 11 districts of the main city. Therefore, northeast Chongqing City is the key area with a high probability of potential geological disasters.

As one of the seven districts and counties in the northeast of Chongqing City (Figure 1), Yunyang County (spans 108°24′37″–109°14′47″ E and 30°34′59″–31°26′28″ N) is located in the middle of the Three Gorges Reservoir Project area, being the important hub of the ecological and economic zone along the Yangtze River. According to the announcement of the Chongqing Forest Bureau, while the forest area of Chongqing city reaches 54.5%, that of Yunyang County exceeds 58.5%, making it one of the greenest counties in China. Based on the Seventh National Census of China, there were 929,034 long-term residents (48% of them are urban residents) in this area in the year 2020. Yunyang County is crossed by twelve major folds, namely Changdianfang Syncline (1), Macaoba Anticline (2), Qvmahe Syncline (3), Tiefengshan Anticline (4), Yangliuwan Syncline (5), Dongcun Anticline (6), Xinchang Anticline (7), Huangpoxi Syncline (8), Guling Syncline (9), Fangdoushan Anticline (10), Ganchang Syncline (11), and Longjukan Syncline (12). Under the subtropical monsoon climate, Yunyang County has an average annual rainfall of 1123.7 to 1264.8 mm and an average annual temperature from 10.2 to 18.5 °C.

Mountainous areas are generally susceptible to mass movements due to preparatory and triggering causal factors [26]; not only the weathering effects but anthropogenic activities in the region also commonly accelerate the formation of unstable areas on both the earth material and on hill slopes [27]. As a part of Chongqing City, Yunyang County has always been a significant hotspot for landslide occurrences. There are a total of 836 historical landslides recorded in the dataset; 827 data points are left after data cleaning. A total of 28.2% of them are small landslides, 51.8% are medium landslides, and 20% are large landslides. Among them, trust-load-caused landslides accounted for 53.7%, and loosen-caused landslides and multi-caused landslides accounted for 14.5% and 31.8%, respectively. To build sub-models, we manually divided the study area into four different sub-zones (Figure 2) based on the information from the exploration of geological hazards in Chongqing City, such as the mechanism of landslide formation and sliding failure and the geomorphic unit characteristics. Among the four sub-zones, sub-zone II contains all the strip-distributing landslides along the mainstream of the Yangtze River, so it can also be called the Yangtze River mainstream zone. From a larger scope, a part of Yunyang County belongs to the low-hills section that crosses Yunyang, Fengjie, and Kaizhou; this area is classified as sub-zone IV. Sub-zone I (south of sub-zone II) is crossed by the main highway called S305, and the main area of sub-zone III (between sub-zone II and IV) is crossed by the S103 and S305. Similarly, the density of the road network is also at a high stage in the other two parts of sub-zone III. The landslides that occurred in these two sub-zones are found to be mainly along the roads (Figure 3). After zonation, 89 landslides are located in sub-zone I, 285 of them are in sub-zone II, sub-zone III contains 44 landslides, and with the largest area, 408 of the historical landslides occurred in sub-zone IV.

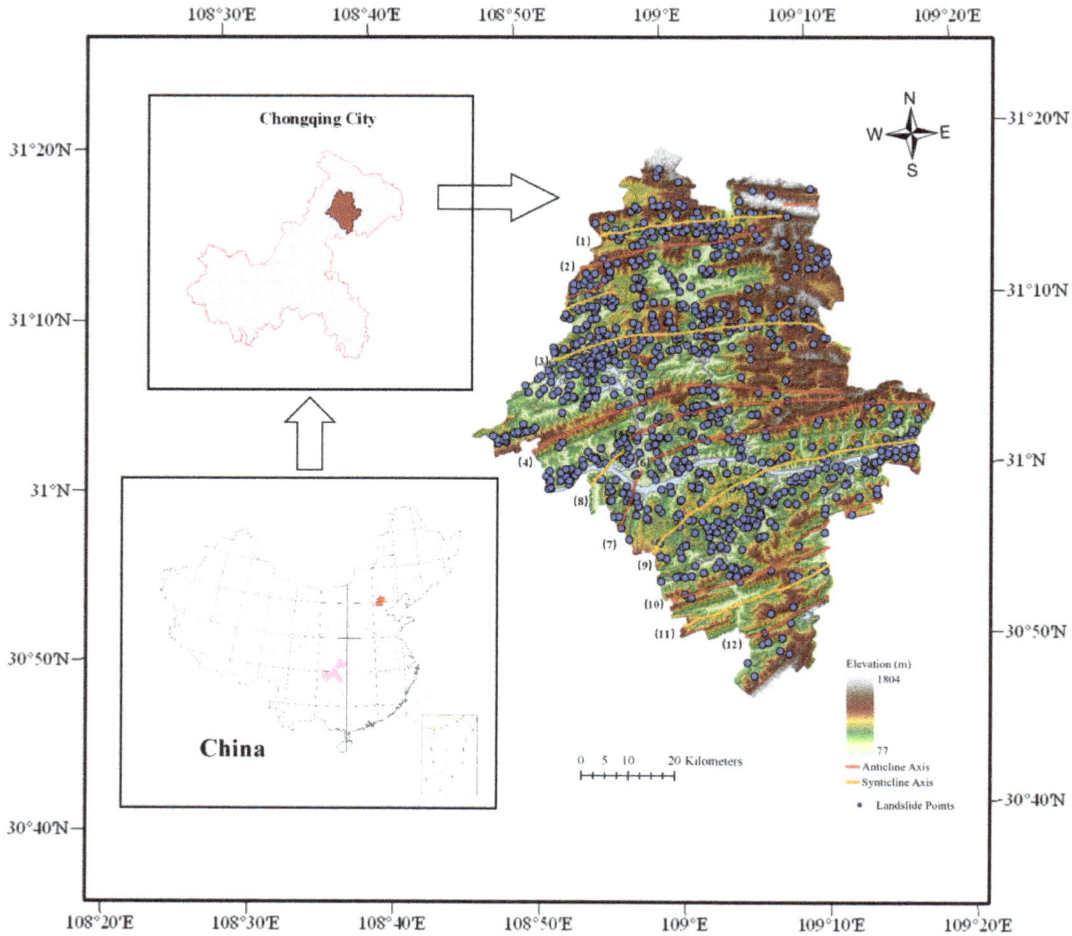

Figure 1. Location, landslide distribution and tectonic map of the study area. (1) Changdianfang Syncline, (2) Macaoba Anticline, (3) Qvmahe Syncline, (4) Tiefengshan Anticline, (5) Yangliuwan Syncline, (6) Dongcun Anticline, (7) Xinchang Anticline, (8) Huangpoxi Syncline, (9) Guling Syncline, (10) Fangdoushan Anticline, (11) Ganchang Syncline, and (12) Longjukan Syncline.

As one of the typical landslides in the Three Gorges Reservoir area, the Jiuxianping landslide (in sub-zone II) is located on the left bank of the Yangtze River (Figure 2b). After the Three Gorges Reservoir project, the fluctuation of the Three Gorges Reservoir water level restarted the displacement and deformation of the ancient landslide, making this area more prone to geological hazards. A subsidence of about one meter occurred on a roadway in the middle of the landslide body after heavy rain in 2003 and 2004, causing the roadway to be abandoned. With the impact of continuous heavy rain, landslides occurred in the back accumulation of Jiuxianping on 19 and 22 June 2007, causing the houses of the villagers to collapse, and the mountain body cracked. On 9 June 2009, the back-accumulation of Jiuxianping deformed again under the impact of heavy rain, causing cracks on both the accumulation body and the houses of the villagers. Recently, under the continuous effect of the Three Gorges Reservoir, this area has been in the overall creep deformation stage for years, especially the cliffs near the river, which often suffer from local collapse and damage.

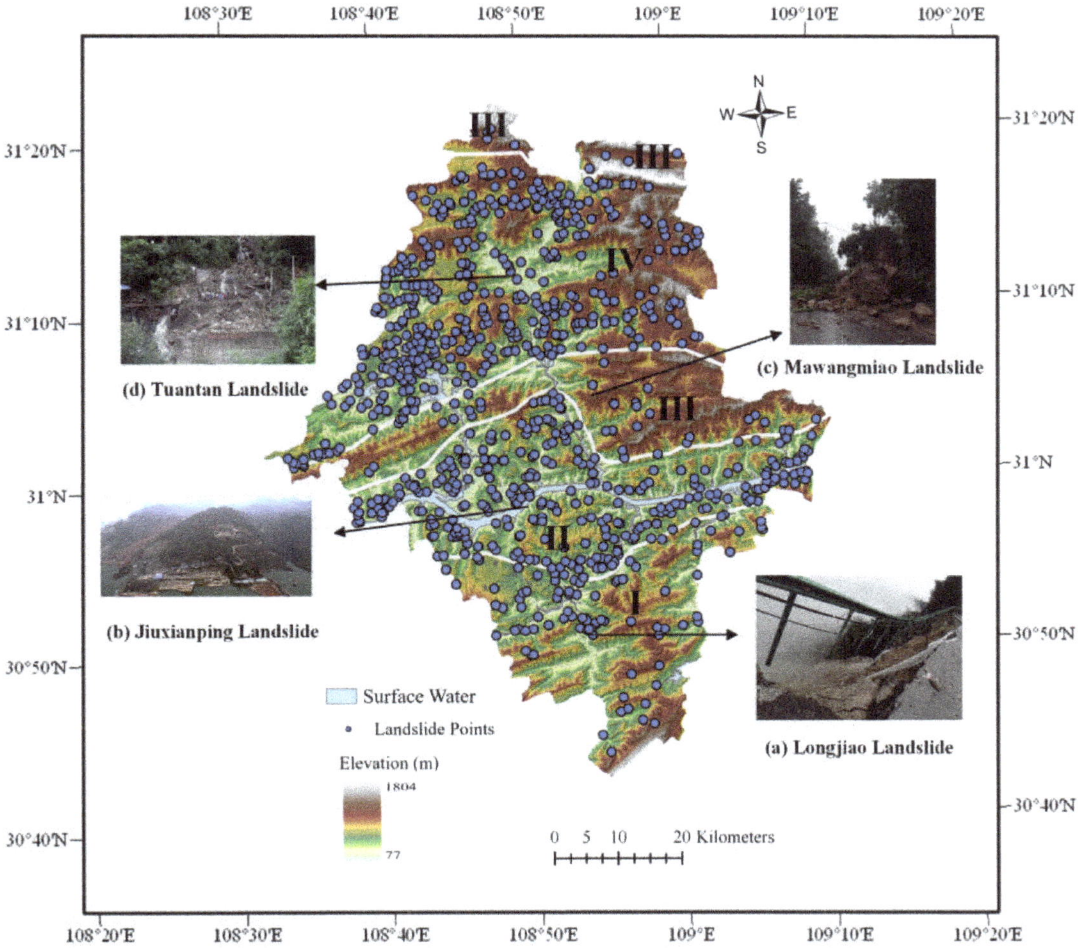

Figure 2. Zonation and typical landslides: (**a**) Longjiao Landslide, (**b**) Jiuxianping Landslide, (**c**) Mawangmiao Landslide, (**d**) Tuantan Landslide.

The continuous heavy rains from 30 August to 1 September 2014 made the accumulated rainfall in Jiangkou Town more than 300 mm. The day after that, the Tuantan landslide (Figure 2d) occurred on the back mountain and on the left side of the Yongfa Coal Mine staff dormitory in Tuantan village, Jiangkou Town, Yunyang County (in sub-zone IV). Although the employees were notified to evacuate from the area subjected to the massive landslide, twelve of them were buried on the spot. Unfortunately, only one of the twelve was saved.

Typically, in sub-zone I, under the impact of rainfall, a landslide occurred on the S202 Highway in the direction from Longjiao to Rucao (Figure 2a) on 13 July 2021. Similarly, there was a 10,000 cubic-meter landslide triggered by heavy rainfall in the area of Mawang Temple (Figure 2c), which trapped two four-wheel cars and a motorcycle, and blocked the highway section for five days.

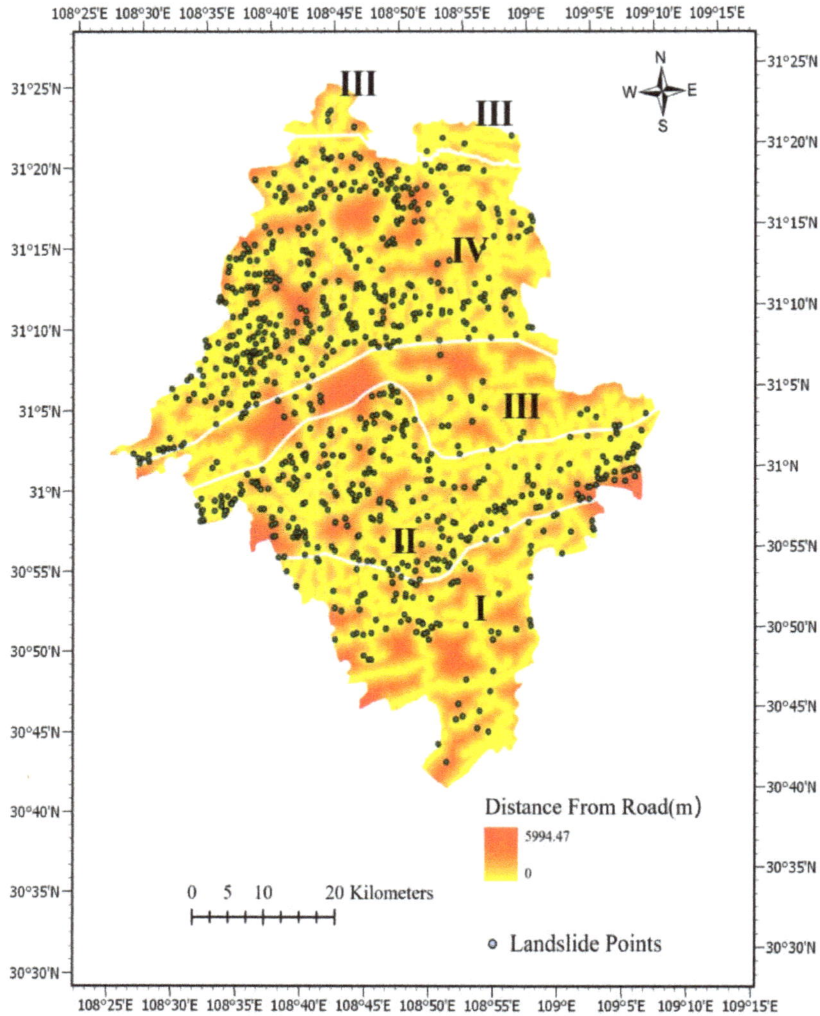

Figure 3. Landslides distribution versus distance from road in different sub-zones.

3. Method Explanation

3.1. Random Forest

As one of the most popular classification methods, RF was first proposed by Breiman [28] and Cutler [29]. During its training process, different data subsets obtained by random sampling are used to train multiple decision trees as independent estimators; each of them is only allowed to fit the data based on the part of the input features, and the final output of the RF model is based on the voting results of the constructed estimators. With the double randomness, those estimators will be trained as distinct ones, and such a special structure makes RF less sensitive to noise and outliers, less likely to overfit and has a lower dependency on feature selection, but more robust and can provide more accurate predictions compared with the other machine learning models.

The RF training process can be concluded as the flowing three simple steps (Figure 4):

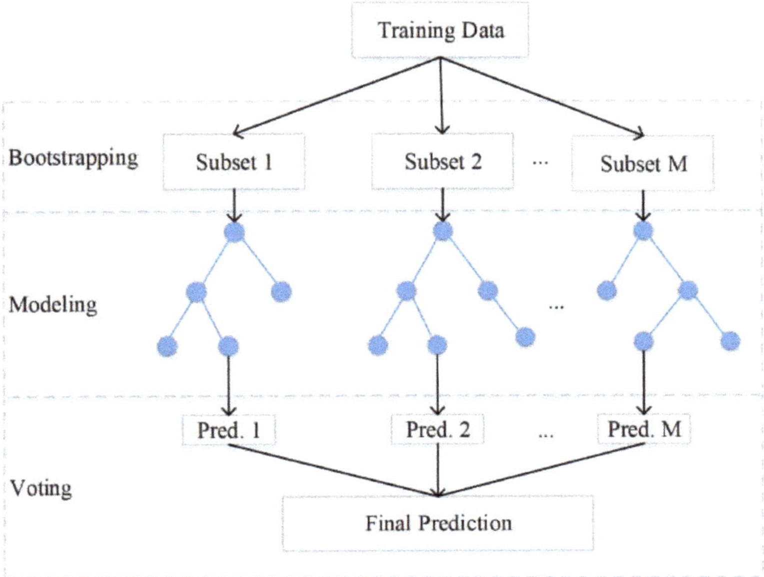

Figure 4. Random forest flowchart.

Bootstrapping: To build M decision trees (estimators), M subsets will be generated as training sets from the original dataset by sampling with replacement.

Modeling: Each subset obtained by bagging is used to train the corresponding estimator. Similarly, for the purpose of getting estimators as distinct as possible, the available features for each estimator are also randomly chosen from the entire feature list.

Voting: Each estimator will output a result independently, then the final result of the model will be generated based on the voting results (Equation (1)).

$$\mathrm{f(x)} = \frac{1}{M}\sum_{m=1}^{M} f_m(x) \quad (1)$$

where f(x) represents the final output, and $f_m(x)$ means the output of the mth tree.

3.2. Grid Search

As pre-set parameters, hyper-parameters play a crucial role in model performance, and there are rare algorithms that are hyper-parameter-free [30]. Therefore, proper hyper-parameter tuning is essential for improving model performance. Grid search, as one of the conventional automatic hyper-parameter tuning methods, is widely used because of its simple operation. Its basic idea is to choose the best hyper-parameter combination by enumerating and iterating over all possible combinations. Although it is computationally expensive because of the exhaustive search process, grid search suits random forest very well, as random forest is a hyper-parameter tuning friendly model; only two hyper-parameters are to be tuned in our case.

3.3. Performance Measure

For classification problems, accurate-oriented modeling sometimes ends up with "rabid" models that tend to classify all samples as a certain type, especially for unbalanced datasets. Therefore, the confusion matrix is introduced in this case (Table 1).

Table 1. Confusion matrix.

Predicted Values	Actual Values	
	Positive	Negative
Positive	True Positive (TP)	False Positive (FP)
Negative	False Negative (FN)	True Negative (TN)

According to Table 1, the true positive rate, also called the recall, is defined as TPR = TP/(TP + FN), and the false positive rate is defined as FPR = FP/(FP + TN). After model establishment, its corresponding TPR versus FPR at different cutoff values can be plotted, which is called as receiver operation characteristic (ROC) curve and can be used to represent the predictive ability of the model. The area under the ROC curve (AUC) is commonly used as an index that represents the true classification accuracy of the model.

4. Methodology

4.1. Data Collection and Preparation

Data collection and data cleaning are of the top importance in machine learning applications; high-quality data is the foundation of accurate prediction models. Model perfection also plays a vital role; choosing appropriate hyper-parameters will help models to extract useful information from input data more effectively and precisely. In this section, we will introduce the procedure for data preparation and model establishment. The analysis of the performance and the factor importance of different models will be presented in the following sections. The flowchart for this study is shown in Figure 5.

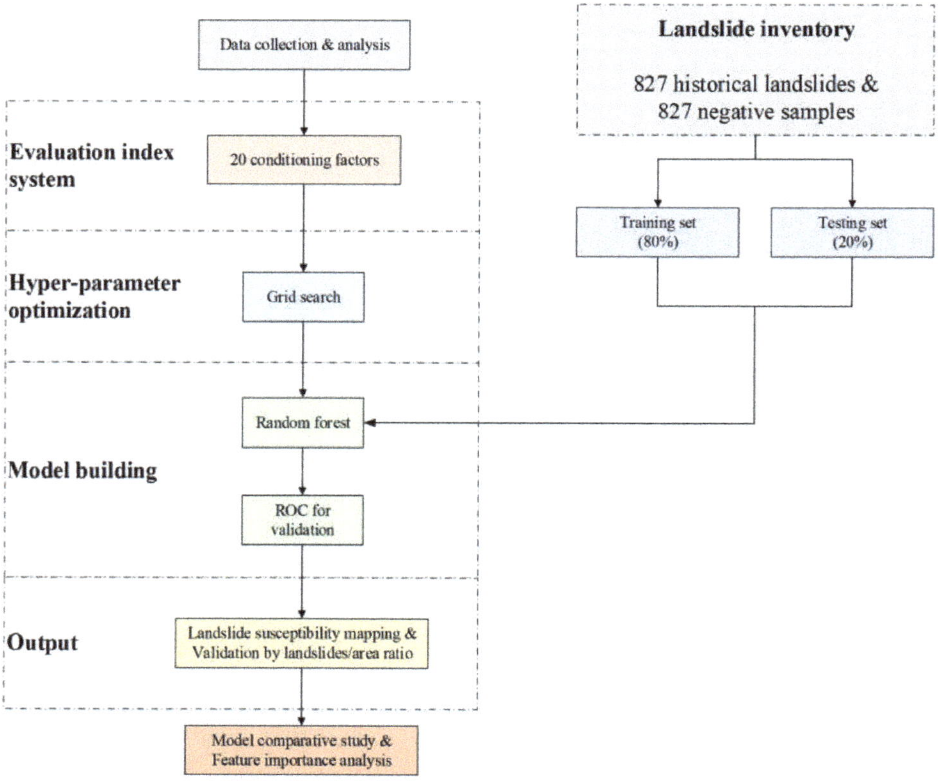

Figure 5. Flowchart of this study.

Triggered by the joint effects of natural factors and human activities, the mechanism of the occurrence of landslides is very complex [5]. Based on the assumption that future landslides will occur under the same conditions as past landslides [31], evaluating these factors and analyzing their relationships with historically recorded landslides can contribute to forming the basis for the prediction of future landslides in an area [32,33]. However, after reviewing the studies related to the evaluation of landslide sensitivity from 1983 to 2016, Reichenbach et al. [34] found that there was a total of 596 different factors that affected landslide formation, considering all of them will be laborious and time-consuming. According to previous articles, topography, hydrology, geology, land cover, and natural and human-related factors are generally used for landslide susceptibility analysis [14,17]. Therefore, in this study, topographical factors (aspect, elevation, plane curvature, profile curvature, relief amplitude, slope), hydrological factors (aridity, distance from rivers, index of moisture (IM), and topographic wetness index (TWI)), geological factors (lithology, distance from anticline axis, distance from syncline axis), factors related to land cover (namely land use and normalized difference vegetation index (NDVI)), human factors (such as human activity intensity of land surface (HAILS), distance from road, and population), and natural factors (average annual temperature, average annual rainfall) are all considered (Figure 6).

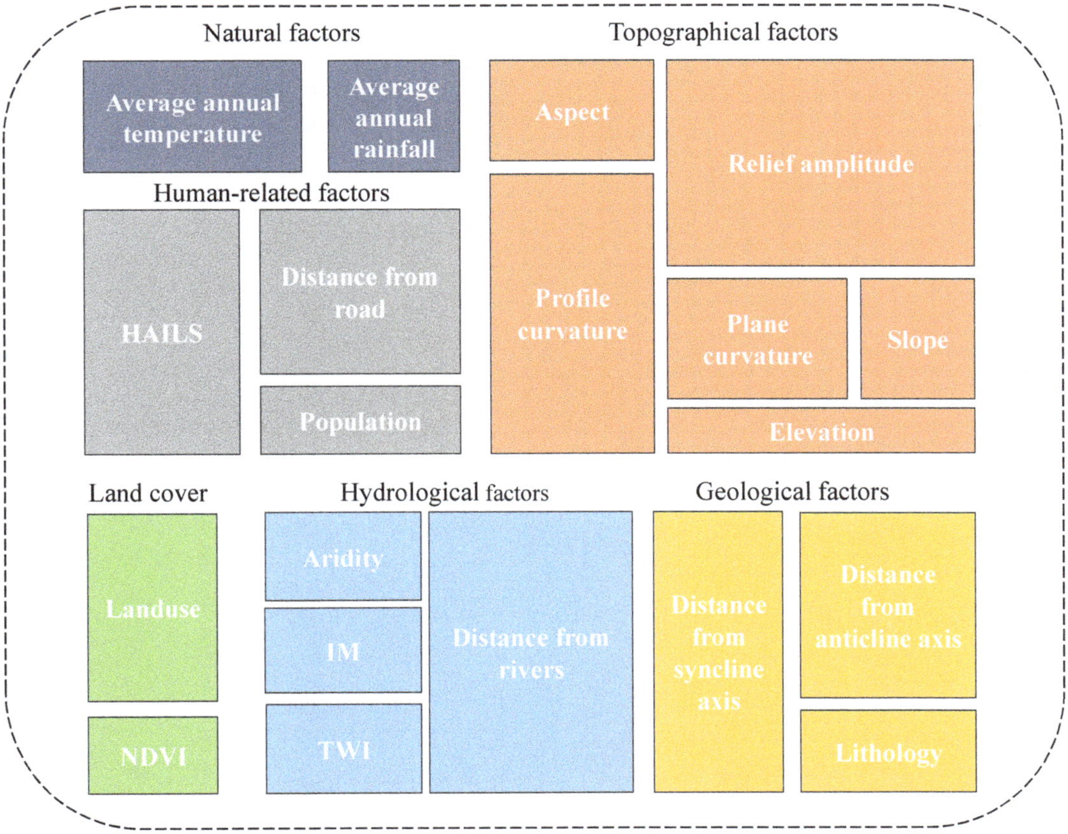

Figure 6. Conditioning factor types.

As the grid data layers cannot be directly used for model training, the Yunyang County fishnet with a cell size of 25 × 25 m was created for the purpose of data extraction for model training and measuring the distances from specific structures/natural sources, the total number of cells is 5,831,382. The secondary data includes aspect, plane curvature, profile curvature, relief amplitude, slope, distance from rivers, TWI, distance from anticline axis, distance from syncline axis, and distance from road. Among them, TWI, aspect, plane curvature, profile curvature, relief amplitude, and slope were obtained by ArcGIS processing of DEM. The characteristics associated with distances (i.e., distance from rivers, distance from anticline axis, and distance from syncline axis) were obtained by applying the near function of ArcGIS to the corresponding primary data (i.e., the river network of Chongqing City, the road network of Chongqing City, and the geological structure of Yunyang County) and the created fishnet of Yunyang County. The ranges and categories of the 20 characteristics are summarized in Table 2.

Table 2. (**a**) Categories of conditioning factors. (**b**) Ranges of conditioning factors.

Feature ID	*Feature Name*	*# of Classes*	*Class Name*
		(a)	
1	lithology	37	1. T_1d_3; 2. T_1d_4; 3. J_2s_2; 4. Ss; 5. T_1J_4; 6. T_1j_1; 7. T_1j_2; 8. J_2s_1; 9. J_2x; 10. J_1z_1da; 11. J_1z_1m; 12. J_1z_1d; 13. J_1z; 14. T_1j_3; 15. T_2b_2; 16. T_2b_3; 17. T_2b_2; 18. T_2b_1; 19. T_3xj; 20. J_1zlm; 21. J_1zld; 22. J_3sn; 23. J_3p_2; 24. J_3p; 25. T_3xj_3; 26. T_3xj_2; 27. T_3xj_1; 28. J_1zlda; 29. J_1z_1; 30. J_2sn; 31. T_1d_2; 32. T_1d; 33. P_3w_2; 34. P_3d; 35. $P_2m - g$; 36. P_3wl; 37. P_2q
2	landuse	15	1. Paddy filed; 2. Dry land; 3. Forestland; 4. Shrub land; 5. Sparse woods; 6. Other woods; 7. Grassland with high coverage; 8. Grassland with medium coverage; 9. Grassland with low coverage; 10. Cannel; 11. Reservoir; 12. Urban land use; 13. Rural settlements; 14. Other types of building; 15. Marshland

		(b)
Feature ID	*Feature Name*	**Value Range**
3	Slope (°)	0–67.0356
4	Average Annual Temperature (°C)	10.2–18.5
5	Average Annual Rainfall (mm)	1123.7–1264.8
6	Aridity	0.698–0.952
7	IM	14.5–53.47
8	Population	88.8744–695.201
9	HAILS	0–78.2
10	NDVI	−0.1372–0.8837
11	Distance from Syncline Axis (m)	0–19,026.1
12	Distance from Anticline Axis (m)	0–16,958.8
13	Distance from Road (m)	0–5994.47
14	Distance from Rivers (m)	0–17,236.8
15	Elevation (m)	76.0774–1794.29
16	Profile Curvature	0–44.2457
17	Aspect (°)	0–360
18	Plane Curvature	0–360
19	TWI	−0.82–30.1
20	Relief Amplitude	0–148.865

All 20 factors were visualized by ArcGIS with a resolution of 25 m (Figure 7). Then the corresponding classes and rating values of the 20 factors were assigned to each cell of the prepared fishnet.

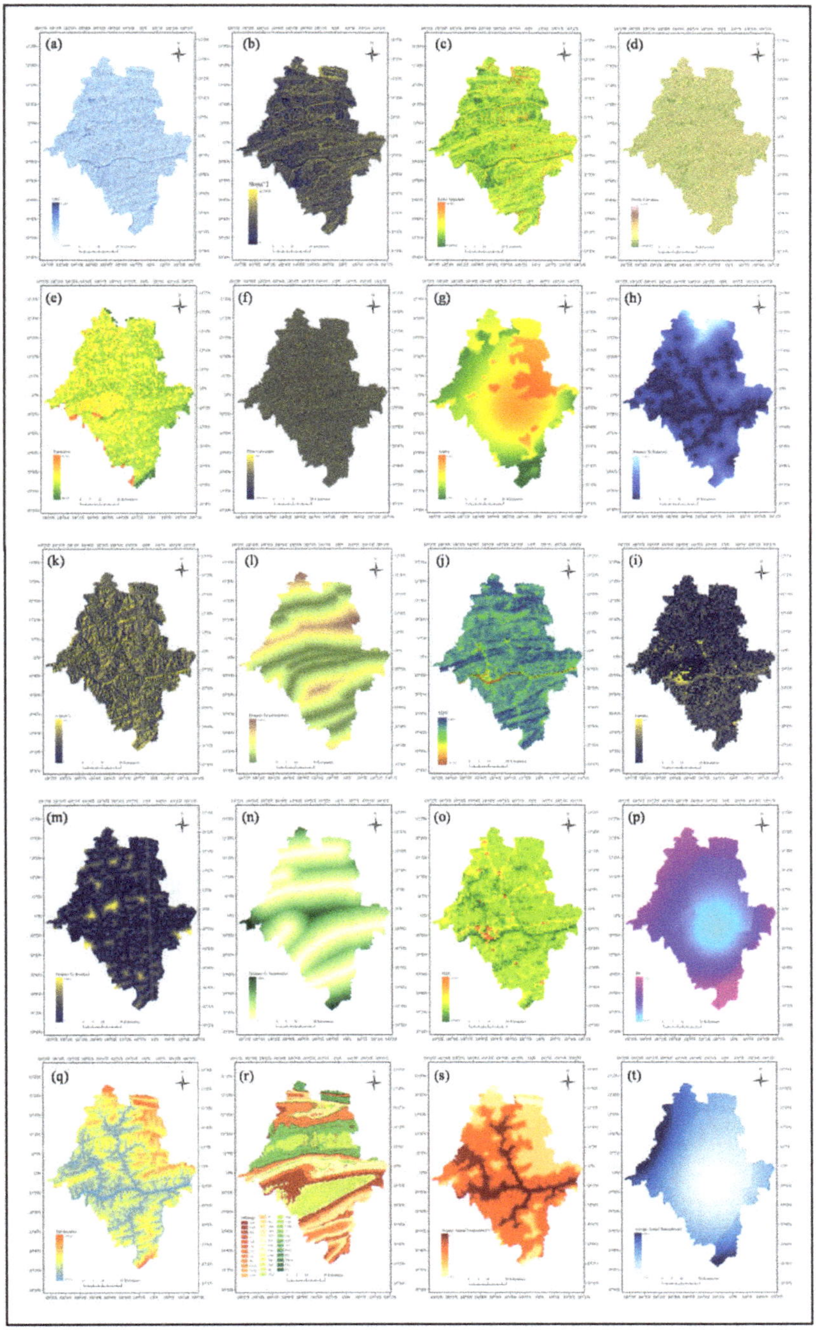

Figure 7. Conditioning factors of landslides susceptibility: (**a**) TWI; (**b**) Slope; (**c**) Relief amplitude; (**d**) Profile curvature; (**e**) Population; (**f**) Plane curvature; (**g**) Aridity; (**h**) Distance from rivers; (**i**) Land use; (**j**) NDVI; (**k**) Aspect; (**l**) Distance from anticline axis; (**m**) Distance from road; (**n**) Distance from syncline axis; (**o**) HAILS; (**p**) IM; (**q**) Elevation; (**r**) Lithology; (**s**) Average annual temperature; (**t**) Average annual rainfall.

After obtaining the 827 historical landslide cells in the study area as positive samples (1), the same number of non-landslide cells were randomly extracted from the landslide-free areas as negative samples (0) [35,36]. To make sure that the sub-models can be constructed with the same data as the parent model, the number of negative samples selected from each sub-zone was determined by the number of the positive samples in the sub-zone (i.e., 89 for sub-zone I, 286 for sub-zone II, 44 for sub-zone III, and 408 for sub-zone IV). Thus the five sample datasets were formed (Table 3). Based on previous research [14,37], 80% of each dataset was randomly selected to train the corresponding statistical model, and the remaining 20% was used for validation purposes.

Table 3. The number of data points in different zonation and datasets.

Sub-Zone	# of Possitive Samples	# of Negative Samples	Total	Percentage of Trainning Set	# of Training Samples
I	89	89	178	80%	142
II	286	286	572	80%	457
III	44	44	88	80%	70
IV	408	408	816	80%	652
Whole Region	827	827	1654	80%	1323

4.2. Model Development and Application

Five independent random forest-based models were established and validated based on the sample datasets constructed before. The hyper-parameters of each model were tuned by the grid search method. Being considered as the most straightforward optimization method [38], the grid search method needs to iterate over the entire interval of each hyper-parameter. Therefore, the number and the iterating interval of the tuned hyper-parameters have an obvious impact on its search efficiency. The number of estimators determines the stability of a random forest model, and adding more estimators will lower its mean squared prediction delta (MSPD) and hence improve model stability [39], but will increase computational cost [40]. The maximum depth of a tree controls the stability of a random forest model in a different way. The stability will decrease as depth increases since increasing the depth will make the model tend to just memorize the training data, but if the forest is too shallow, the model will underfit, resulting in low AUC [39]. In this case the number of estimators and the maximum depth of a tree are the two hyper-parameters to be tuned; the results are shown in Table 4.

Table 4. Hyper-parameters of models.

Model	Number of Estimators	Maximum Depth of a Tree
Parent Model	175	21
Sub-Model One	10	6
Sub-Model Two	95	12
Sub-Model Three	10	4
Sub-Model Four	100	12

After obtaining well-trained models, we used all 5,831,382 cells prepared in Section 4.1 as input to generate the landslide susceptibility maps for the study area.

5. Results

The landslide susceptibility maps generated by the parent model and sub-models are displayed in Figure 8, where Figure 8a represents the map outputted by the parent model, and Figure 8b is the map produced by sub-models, Figure 8c,d are the detailed scopes for the part of both maps. The entire region is divided and classified into five zones of susceptibility to different levels of landslides (very low, low, moderate, high, and very high) by the method of natural breaks method. From the result of the parent model in Table 5, 14.6% of the whole area is classified as a very low landslide-prone zone, 22.8% as a low landslide-prone zone, 25.23% as a moderate landslide-prone zone, 21.47% as a high landslide-prone zone, and 15.90% as a very high landslide-prone zone. From the result of the sub-models, their ratios are 16.42%, 20.81%, 22.56%, 21.54%, and 18.67%, respectively.

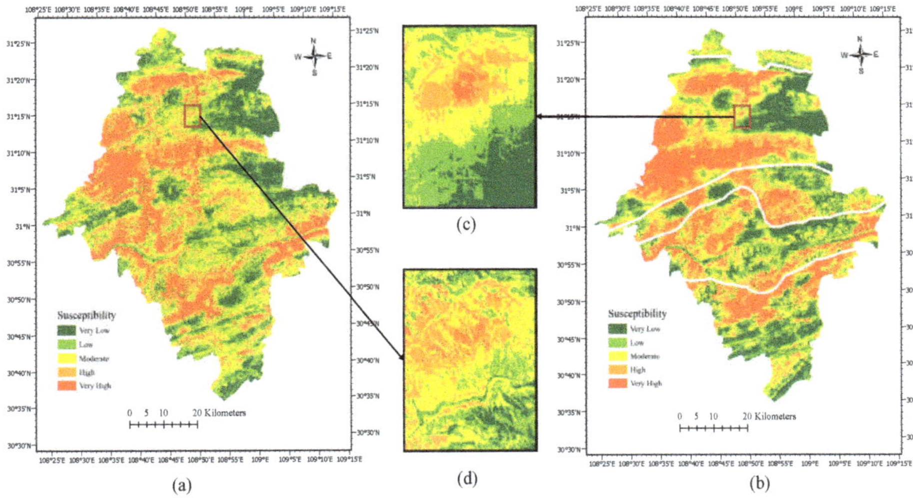

Figure 8. Comparison between the landslide susceptibility mapping of the parent model and sub-models: (**a**) Landslide susceptibility map generated by the parent model, (**b**) Landslide susceptibility map generated by the sub-models, (**c**) Detailed view of landslide susceptibility map generated by the sub-models, (**d**) Detailed view of landslide susceptibility map generated by the parent model.

Table 5. Statistic results of landslide susceptibility in different levels of different models.

Model	Landslides Susceptibility Level	Landslides Susceptibility Threshold	Landslides Ratio	Area Ratio	Landslides/Area Ratio
Parent Model	Very Low	0.000–0.239	0.24%	14.60%	0.016
	Low	0.239–0.408	2.18%	22.80%	0.096
	Moderate	0.408–0.573	6.05%	25.23%	0.248
	High	0.573–0.741	19.22%	21.47%	0.895
	Very High	0.741–1.000	72.31%	15.90%	4.548
Sub-Models	Very Low	-	0.12%	16.42%	0.007
	Low	-	1.21%	20.81%	0.058
	Moderate	-	4.48%	22.56%	0.199
	High	-	18.55%	21.54%	0.861
	Very High	-	75.64%	18.67%	4.051
Sub-Model One	Very Low	0.000–0.235	0.00%	20.22%	0.000
	Low	0.235–0.435	0.00%	23.88%	0.000
	Moderate	0.435–0.631	8.99%	22.36%	0.402
	High	0.631–0.831	25.84%	18.64%	1.386
	Very High	0.831–1.000	65.17%	14.90%	4.374
Sub-Model Two	Very Low	0.000–0.278	0.35%	15.14%	0.023
	Low	0.278–0.435	2..46%	22.03%	0.112
	Moderate	0.435–0.588	4.56%	24.99%	0.182
	High	0.588–0.749	13.33%	22.45%	0.594
	Very High	0.749–1.000	79.30%	15.39%	5.153

Table 5. *Cont.*

Model	Landslides Susceptibility Level	Landslides Susceptibility Threshold	Landslides Ratio	Area Ratio	Landslides/Area Ratio
Sub-Model Three	Very Low	0.000–0.303	0.00%	13.68%	0.000
	Low	0.303–0.421	0.00%	24.81%	0.000
	Moderate	0.421–0.536	9.10%	25.78%	0.353
	High	0.536–0.662	31.81%	22.06%	1.442
	Very High	0.662–1.000	59.09%	13.67%	4.323
Sub-Model Four	Very Low	0.000–0.227	0.00%	16.59%	0.000
	Low	0.227–0.427	0.74%	16.20%	0.0457
	Moderate	0.427–0.612	2.94%	19.38%	0.152
	High	0.612–0.792	19.12%	22.26%	0.859
	Very High	0.792–1.000	77.20%	25.57%	3.020

Logically, the landslides/area ratio should increase from a very low landslide-prone zone to a very high landslide-prone zone, which is exactly what our models indicate. According to the results of the parent model, the landslides/area ratio increases from 0.015 to 4.548, from a very low landslide-prone zone to a very high landslide-prone zone, and that also increases from 0.007 to 4.051 from the results of the sub-models. Figure 9 displays such a tendency, and it can be seen that the outputs of the sub-models have more obvious gaps between the very low landslide-prone zone and the very high landslide prone-zone, which reflects another merit of the sub-models compared with the parent model.

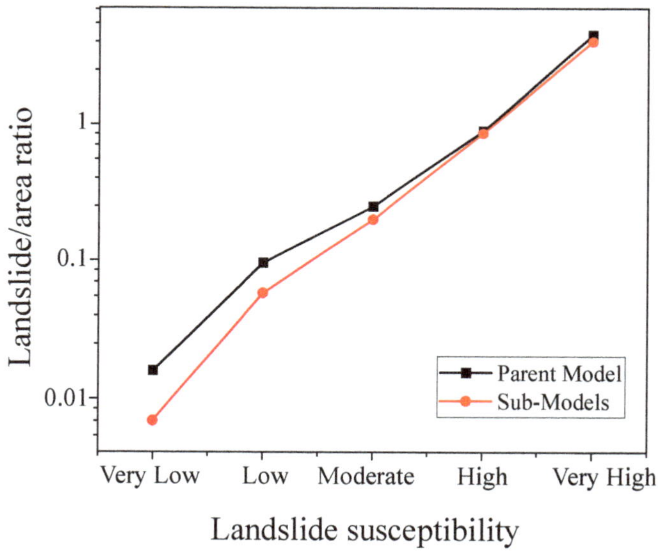

Figure 9. Landslides/area ratio.

The validation AUC values of the five models are shown in Figure 10; the AUC value of the parent model is 0.872, while that of sub-model one to sub-model four are 0.949, 0.892, 0.889, and 0.951, respectively. All of the sub-models outperformed the parent model, which proved the aforementioned hypothesis. However, the increases are not at the same level; sub-model four achieved the highest improvement (9.1%), while the lowest improvement is 1.9%, which was obtained by sub-model two.

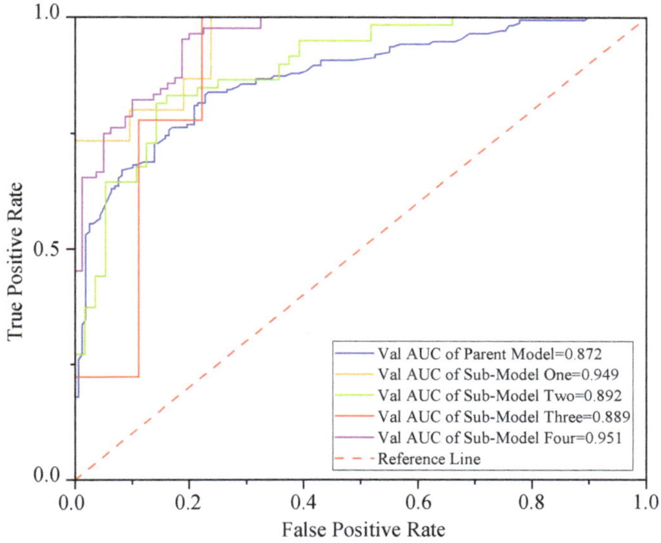

Figure 10. Validation AUC values of different models.

The importance of features generally represents how much a specific feature contributes to the decision-making process of a model. In this case, the most important features can be the key factors in identifying landslide/non-landslide points. As shown in Table 6 and Figure 11, the distance from syncline axis, aridity, elevation, distance from rivers, and average annual temperature are of the highest importance for the parent model. NDVI, average annual rainfall, distance from road, and elevation are the main features that are associated with the formation of landslides in sub-zone I. For sub-model two, elevation is the most important feature, which is followed by average annual rainfall, distance from rivers, distance from anticline axis, and average annual temperature. Distance from road, HAILS, elevation, plane curvature, and average annual temperature are the top features for sub-model three. Last but not least, distance from syncline axis, average annual temperature, elevation, aridity, and average annual rainfall play important roles during the predicting process of sub-model four.

Table 6. Top five important factors in different sub-zones.

Zone	Feature Importance Rank				
	1st	2nd	3rd	4th	5th
Whole Region	Distance from syncline axis (m)	Aridity	Elevation (m)	Distance from rivers (m)	Average Annual Temperature (°C)
Sub-Zone I	Distance from syncline axis (m)	NDVI	Average Annual Rainfall (mm)	Distance from Road (m)	Elevation (m)
Sub-Zone II	Elevation (m)	Average Annual Rainfall (mm)	Distance from rivers (m)	Distance from anticline axis (m)	Average Annual Temperature (°C)
Sub-Zone III	Distance From Road (m)	HAILS	Elevation (m)	Plane Curvature	Average Annual Temperature (°C)
Sub-Zone IV	Distance from syncline axis (m)	Average Annual Temperature (°C)	Elevation (m)	Aridity	Average Annual Rainfall (mm)

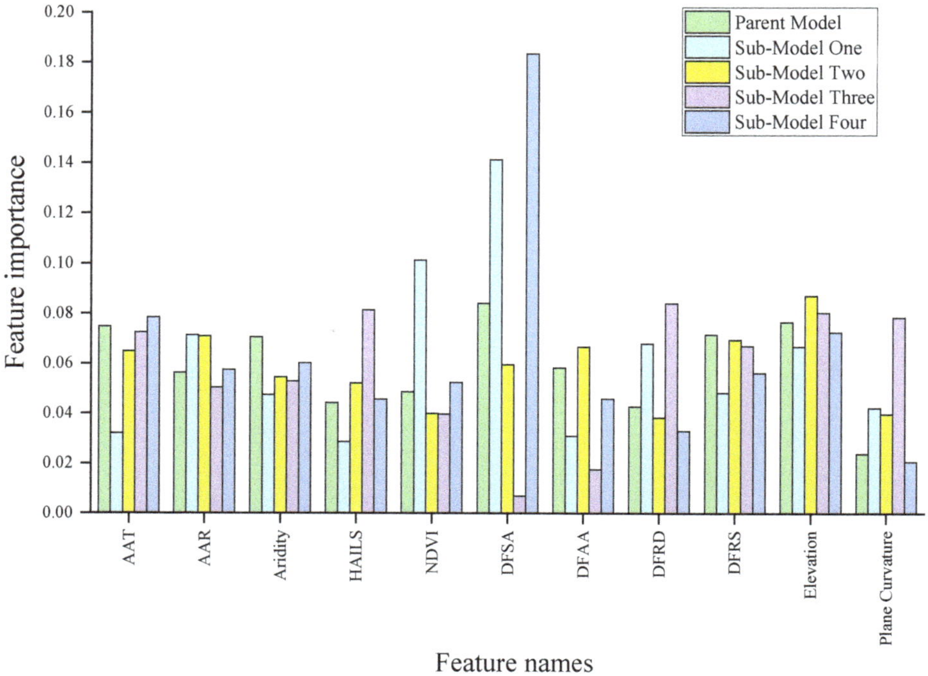

Figure 11. Top five important factors of different models. AAT: Average annual temperature; AAR: Average annual rainfall; DFSA: Distance from syncline axis; DFAA: Distance from anticline axis; DFRD: Distance from road; DFRS: Distance from rivers.

6. Discussion

6.1. Feature Importance Analysis

As indicated in Table 6, the distance from syncline axis is at the dominant place among the twenty factors for the parent model, sub-model one, and sub-model four. Factors associated with water, such as average annual rainfall and the distance from rivers, are also important places. The triggering factors of shallow landslides are highly dependent on the rainfall water infiltration and its further redistribution [41]. Due to its softening effect and long-term erosion effect, the distance from rivers has a significant influence on the development of landslides [42]. Temperature has a remarkable effect on landslide formation; experimental results indicated that the shear strength of slip surface soils reduces with decreasing temperature, which will negatively affect slope instability [43]. Therefore, obtained by the accumulated temperature and rainfall, aridity is also of great importance. Although different plants generally have a different contribution of rainfall to soil water [44], which will affect slope stability differently, on average, the effect of water uptake from the plant cover makes the vegetated slopes averagely 12.84% drier and matric suctions three times higher than the fallow slope, which contributes to slope stability [45]; so NDVI is taken as one of the main considerations. Unlike vegetation cover, human-made land covers usually have negative influences on slope stability. The development of new building areas can potentially increase susceptibility to landslides [46]. Road networks not only directly destabilize existing slopes by disturbing their original structures during construction processes, but transport activities also have negative effects on slope stability. Similarly, HAILS is another influential factor. For topological factors, the two most important factors, in this case, are elevation and plane curvature.

6.2. Model Comparison

6.2.1. Parent Model

When dealing with the whole region, the model tends to capture information that is more general, and thus its feature importance has a higher universality. In our case, six major synclines are lying in the whole domain, and their generally dominant effects on identifying landslide/non-landslide points are successfully captured by the parent model. Aridity, distance from rivers, and average annual temperature affect the formation of landslides in meteorological and hydrological aspects. Although ranked fourth as the most important one, elevation is the most important topological factor in the research area.

6.2.2. Sub-Models

After dividing the whole region into different sub-zones, the models will have the chance to learn the knowledge that is specifically right for each sub-zone. Thus their performances will be improved.

Sub-model One. Located in the south of Yunyang County, sub-zone I occupies 741.7 square kilometers, which is only 20% of the entire research area. Nevertheless, it is crossed by two of the six major synclines, which makes the syncline effect more obvious, and, therefore, the distance from the syncline axis is still the most important factor for sub-zone I. Similarly, the landslide susceptibility in sub-zone I is also affected by both hydrological and topological factors, namely average annual rainfall and elevation. However, specifically, as it is crossed by the major highway S305, land cover (NDVI) and human activities (distance from road) are the other two main influential factors.

Sub-model Two. As we discussed before, sub-zone II is the most special sub-zone since it is crossed by the Three Gorges Reservoir area, most of the landslides in Yunyang County were impacted by water occurred here. Therefore, theoretically, the landslide formation in sub-zone II is sensitive to hydrological factors. In this case, the average annual rainfall and the distance from rivers are ranked as the second and the third-most important factors. The reason why elevation is of the greatest importance is that lower places are usually prone to the influence of the periodic variation of reservoir water level. Statistically, 76% of the landslides impacted by water in sub-zone II occurred below the elevation of 249 m. The tectonic action and natural factors also contribute a lot to the landslide formation in sub-zone II; the distance from anticline axis and the average annual temperature are ranked in fourth and fifth place, respectively.

Sub-model Three. For sub-zone III, specifically, its data points are relatively far away from the major synclines, which mitigates the importance of the feature. Instead, as an area crossed by major highways and dense road networks, human activities, including distance from road and HAILS, are the two most important factors that have the main contribution to the landslide formation of the sub-zone. Topographical factors, namely elevation and plane curvature, are ranked in third and fourth place, respectively, and the average annual rainfall is evaluated as the fifth-most important factor.

Sub-model Four. Crossed by three major synclines, the landslide identification in sub-zone IV is also significantly dependent on the distance from the syncline axis. Furthermore, since it is part of the low hills section crossing Yunyang, Fengjie, and Kaizhou, the formation of landslides in this area is primarily influenced by the factors associated with the stability of mountain bodies, such as the average annual temperature, elevation, aridity, and the average annual rainfall.

7. Conclusions

In this study, Yunyang County is manually zoned into four parts based on the qualitative analysis of geological hazards exploration in Chongqing City, including the mechanism of landslide formation and sliding failure and geomorphic unit characteristics. Based on the qualitative analysis result, five random forest landslide susceptibility models are constructed using historical landslides data points and twenty relating factors for the following quantity analysis. These models, including a parent model and four sub-models, are optimized by the grid search method individually. A comparison between the parent model and the combination of the sub-models is conducted. The following conclusions are drawn:

The AUC value of the parent model achieves 0.872, which shows that the traditional RF with the hyper-parameters tuned by the grid search method has a reliable performance on landslide susceptibility mapping. In this study, synclines have the most important effects on the formation of landslides in Yunyang County, followed by aridity, elevation, distance from rivers, and average annual temperature.

However, more general information extracted from "mainstream" landslides would usually cover that of the "minority" landslides when treating a large region equally, resulting in low information utility and the inability to identify potential landslides under special geological conditions. With enough data points, experience-based zoning before modeling is proved to be an effective solution to the issue; the qualitative analysis serves the purpose of pre-classification based on the information from geological hazards exploration, which groups the landslides that occurred under similar geological conditions, and thus enables the models to obtain the specific knowledge under each condition. Therefore, in our case, while the traditional RF obtained the general prediction skill for the entire region of Yunyang County, all the sub-models have become "experts" in their respective sub-areas. The test AUC values of sub-model one to four are 8.8%, 2.3%, 1.9%, and 9.1% higher than those of the parent model. Furthermore, the proposed method also contributes to further revealing the key factors that include local landslide instability under specific geological conditions, which can be used by planners and policymakers for a more specific and accurate landslide control in certain areas, thus further improving the safety of life and public property.

For sub-zone I, the top five conditioning factors are distance from syncline axis, NDVI, average annual rainfall, distance from road, and elevation. For sub-zone III, without the influence of major synclines, its top factors are distance from road, HAILS, elevation, plane curvature, and average annual temperature. For sub-zone IV, the distance from syncline axis becomes the most important factor again, and it is followed by average annual temperature, elevation, aridity, and average annual rainfall.

Sub-zone II is crossed by the Three Gorges Reservoir area. Suffered by periodic variation in reservoir water level and the impacts of other factors related to the reservoir band, the modified method based on general conditioning factors has relatively less effect on improving the accuracy of the mapping. The effect of more specific factors on the formation of landslides on the banks of the reservoir will be analyzed in further research. In the case of this paper, the results of sub-model two point out that elevation, average annual rainfall, distance from rivers, distance from anticline axis, and average annual temperature are the top five conditioning factors among the existing twenty factors for sub-zone II.

Author Contributions: Methodology, S.L.; validation, L.W.; software, Y.H.; resources, P.S.; formal analysis, M.C.; supervision, W.Z. All authors have read and agreed to the published version of the manuscript.

Funding: This research was funded by the National Key R&D Program of China (2019YFC1509605), the China Postdoctoral Science Foundation funded project (2021M700608), the Natural Science Foundation of Chongqing, China (cstc2021jcyj-bsh0047), High-end Foreign Expert Introduction program (G20200022005 and DL2021165001L), and Science and Technology Research Program of Chongqing Municipal Education Commission (HZ2021001).

Data Availability Statement: Restrictions apply to the availability of these data. Data was obtained from National Earth System Science Data Center, Resource and Environment Science and Data Center and Chongqing Geological monitoring station and are available at http://www.geodata.cn/;jsessionid=4FDC4730785878003B1ED71B62212634 (accessed on 1 May 2022), and https://www.resdc.cn/Default.aspx (accessed on 1 May 2022) with the permission of National Earth System Science Data Center and Resource and Environment Science and Data Center.

Conflicts of Interest: The authors declare that they have no known competing financial interests or personal relationships that could have appeared to influence the work reported in this paper.

References

1. Huang, F.; Huang, J.; Jiang, S.; Zhou, C. Landslide displacement prediction based on multivariable chaotic model and extreme learning machine. *Eng. Geol.* **2017**, *218*, 173–186. [CrossRef]
2. Petley, D. Global patterns of loss of life from landslides. *Geology* **2012**, *40*, 927–930. [CrossRef]
3. Lin, L.; Lin, Q.; Wang, Y. Landslide susceptibility mapping on a global scale using the method of logistic regression. *Nat. Hazards Earth Syst. Sci.* **2017**, *17*, 1411–1424. [CrossRef]
4. Sheng, L.; Wang, W.; Zhu, W. *China Statistical Yearbook 2016*; China Statistics Press: Beijing, China, 2016. (In Chinese)
5. Sun, D.; Xu, J.; Wen, H.; Wang, D. Assessment of landslide susceptibility mapping based on Bayesian hyperparameter optimization: A comparison between logistic regression and random forest. *Eng. Geol.* **2021**, *281*, 105972. [CrossRef]
6. Brabb, E. Innovative approaches to landslide hazard mapping. In Proceedings of the 4th International Symposium on Landslides, Toronto, ON, Canada, 16–21 September 1984; Volume 1, pp. 307–324.
7. Wu, C.; Hong, L.; Wang, L.; Zhang, R.; Pijush, S.; Zhang, W. Prediction of wall deflection induced by braced excavation in spatially variable soils via convolutional neural network. *Gondwana Res.* **2022**. [CrossRef]
8. Wang, Y.; Tang, H.; Huang, J.; Wen, T.; Ma, J.; Zhang, J. A comparative study of different machine learning methods for reservoir landslide displacement prediction. *Eng. Geol.* **2022**, *298*, 106544. [CrossRef]
9. Huang, F.; Chen, J.; Liu, W.; Huang, J.; Hong, H.; Chen, W. Regional rainfall-induced landslide hazard warning based on landslide susceptibility mapping and a critical rainfall threshold. *Geomorphology* **2022**, *408*, 108236. [CrossRef]
10. He, Q.; Wang, M.; Liu, K. Rapidly assessing earthquake-induced landslide susceptibility on a global scale using random forest. *Geomorphology* **2021**, *391*, 107889. [CrossRef]
11. Sun, D.; Wen, H.; Wang, D.; Xu, J. A random forest model of landslide susceptibility mapping based on hyperparameter optimization using Bayes algorithm. *Geomorphology* **2020**, *362*, 107201. [CrossRef]
12. Smith, H.G.; Spiekermann, R.; Betts, H.; Neverman, A.J. Comparing methods of landslide data acquisition and susceptibility modelling: Examples from New Zealand. *Geomorphology* **2021**, *381*, 107660. [CrossRef]
13. Lim, C.H.; Kim, H.J. Can Forest-Related Adaptive Capacity Reduce Landslide Risk Attributable to Climate Change?—Case of Republic of Korea. *Forests* **2022**, *13*, 49. [CrossRef]
14. Nhu, V.-H.; Shirzadi, A.; Shahabi, H.; Singh, S.K.; Al-Ansari, N.; Clague, J.J.; Jaafari, A.; Chen, W.; Miraki, S.; Dou, J.; et al. Shallow landslide susceptibility mapping: A comparison between logistic model tree, logistic regression, naïve Bayes tree, artificial neural network, and support vector machine algorithms. *Int. J. Environ. Res. Public Health* **2020**, *17*, 2749. [CrossRef]
15. Zhang, W.; Li, H.; Han, L.; Chen, L.; Wang, L. Slope stability prediction using ensemble learning techniques: A case study in Yunyang County, Chongqing, China. *J. Rock Mech. Geotech. Eng.* 2022, *in press*. [CrossRef]
16. Hu, Q.; Zhou, Y.; Wang, S.; Wang, F. Machine learning and fractal theory models for landslide susceptibility mapping: Case study from the Jinsha River Basin. *Geomorphology* **2020**, *351*, 106975. [CrossRef]
17. Zhou, X.; Wen, H.; Zhang, Y.; Xu, J.; Zhang, W. Landslide susceptibility mapping using hybrid random forest with GeoDetector and RFE for factor optimization. *Geosci. Front.* **2021**, *12*, 101211. [CrossRef]
18. Sun, D.; Gu, Q.; Wen, H.; Shi, S.; Mi, C.; Zhang, F. A Hybrid Landslide Warning Model Coupling Susceptibility Zoning and Precipitation. *Forests* **2022**, *13*, 827. [CrossRef]
19. Zhou, X.; Wen, H.; Li, Z.; Zhang, H.; Zhang, W. An interpretable model for the susceptibility of rainfall-induced shallow landslides based on SHAP and XGBoost. *Geocarto Int. TGEI* **2022**, 1–27, *(online)*. [CrossRef]
20. Akar, Ö.; Güngör, O. Integrating multiple texture methods and NDVI to the Random Forest classification algorithm to detect tea and hazelnut plantation areas in northeast Turkey. *Int. J. Remote Sens.* **2015**, *36*, 442–464. [CrossRef]
21. Hengl, T.; Heuvelink, G.B.M.; Kempen, B.; Leenaars, J.G.B.; Walsh, M.G.; Shepherd, K.D.; Sila, A.; MacMillan, R.A.; Mendes De Jesus, J.; Tamene, L.; et al. Mapping soil properties of Africa at 250 m resolution: Random forests significantly improve current predictions. *PLoS ONE* **2015**, *10*, e0125814. [CrossRef]
22. Feng, Q.; Liu, J.; Gong, J. Urban flood mapping based on unmanned aerial vehicle remote sensing and random forest classifier—A case of Yuyao, China. *Water* **2015**, *7*, 1437–1455. [CrossRef]
23. Taalab, K.; Cheng, T.; Zhang, Y. Mapping landslide susceptibility and types using Random Forest. *Big Earth Data* **2018**, *2*, 159–178. [CrossRef]
24. Yin, Y.; Wang, L.; Zhang, W.; Dai, Z. Research on the collapse process of a thick-layer dangerous rock on the reservoir bank. *Bull. Eng. Geol. Environ.* **2022**, *81*, 109. [CrossRef]
25. Wang, L.; Zhang, Z.; Huang, B.; Hu, M.; Zhang, C. Triggering mechanism and possible evolution process of the ancient Qingshi landslide in the Three Gorges Reservoir. *Geomat. Nat. Hazards Risk* **2021**, *12*, 3160–3174. [CrossRef]
26. Nakileza, B.R.; Nedala, S. Topographic influence on landslides characteristics and implication for risk management in upper Manafwa catchment, Mt Elgon Uganda. *Geoenviron. Disasters* **2020**, *7*, 27. [CrossRef]
27. Nefeslioglu, H.A.; Gokceoglu, C.; Sonmez, H.; Gorum, T. Medium-scale hazard mapping for shallow landslide initiation: The Buyukkoy catchment area (Cayeli, Rize, Turkey). *Landslides* **2011**, *8*, 459–483. [CrossRef]
28. Breiman, L. Bagging predictors. *Mach. Learn.* **1996**, *24*, 123–140. [CrossRef]

29. Cutler, A.; Stevens, J.R. Random forests for microarrays. *Methods Enzymol.* **2006**, *411*, 422–432. [CrossRef]
30. Zhang, W.; Wu, C.; Zhong, H.; Li, Y.; Wang, L. Prediction of undrained shear strength using extreme gradient boosting and random forest based on Bayesian optimization. *Geosci. Front.* **2020**, *12*, 469–477. [CrossRef]
31. Lee, S.; Pradhan, B. Landslide hazard mapping at Selangor, Malaysia using frequency ratio and logistic regression models. *Landslides* **2007**, *4*, 33–41. [CrossRef]
32. Raghuvanshi, T.K.; Negassa, L.; Kala, P.M. GIS based grid overlay method versus modeling approach—A comparative study for landslide hazard zonation (LHZ) in Meta Robi District of west showa zone in Ethiopia. *Egypt. J. Remote Sens. Space Sci.* **2015**, *18*, 235–250. [CrossRef]
33. Girma, F.; Raghuvanshi, T.K.; Ayenew, T.; Hailemariam, T. Landslide hazard zonation in Ada Berga district, central Ethiopia- A GIS based statistical approach. *J. Geomat.* **2015**, *9*, 25–38.
34. Reichenbach, P.; Rossi, M.; Malamud, B.D.; Mihir, M.; Guzzetti, F. A review of statistically-based landslide susceptibility models. *Earth-Sci. Rev.* **2018**, *180*, 60–91. [CrossRef]
35. Yesilnacar, E.; Topal, T. Landslide susceptibility mapping: A comparison of logistic regression and neural networks methods in a medium scale study, Hendek Region (Turkey). *Eng. Geol.* **2005**, *79*, 251–266. [CrossRef]
36. Mathew, J.; Jha, V.K.; Rawat, G.S. Landslide susceptibility zonation mapping and its validation in part of Garhwal Lesser Himalaya, India, using binary logistic regression analysis and receiver operating characteristic curve method. *Landslides* **2008**, *6*, 17–26. [CrossRef]
37. Wang, X.; Li, S.; Liu, H.; Liu, L.; Liu, Y.; Zeng, S.; Tang, Q. Landslide susceptibility assessment in Wenchuan County after the 5.12 magnitude earthquake. *Bull. Eng. Geol. Environ.* **2021**, *80*, 5369–5390. [CrossRef]
38. LaValle, S.M.; Branicky, M.S. *On the Relationship between Classical Grid Search and Probabilistic Roadmaps*; Springer Tracts in Advanced Robotics: Berlin/Heidelberg, Germany, 2004; pp. 59–75. [CrossRef]
39. Liu, C.H.; Chamberlain, B.P.; Little, D.A.; Cardoso, Â. Generalising random forest parameter optimisation to include stability and cost. *Mach. Learn. Knowl. Discov. Databases* **2017**, *10536*, 102–113. [CrossRef]
40. Lujan-Moreno, G.A.; Howard, P.R.; Rojas, O.G.; Montgomery, D.C. Design of experiments and response surface methodology to tune machine learning hyperparameters, with a random forest case-study. *Expert Syst. Appl.* **2018**, *109*, 195–205. [CrossRef]
41. Ivanov, V.; Arosio, D.; Tresoldi, G.; Hojat, A.; Zanzi, L.; Papini, M.; Longoni, L. Investigation on the role of water for the stability of shallow landslides—insights from experimental tests. *Water* **2020**, *12*, 1203. [CrossRef]
42. Wang, F.; Xu, P.; Wang, C.; Wang, N.; Jiang, N. Application of a GIS-based slope unit method for landslide susceptibility mapping along the Longzi River, southeastern Tibetan Plateau, China. *ISPRS Int. J. Geo-Inf.* **2017**, *6*, 172. [CrossRef]
43. Shibasaki, T.; Matsuura, S.; Okamoto, T. Experimental evidence for shallow, slow-moving landslides activated by a decrease in ground temperature. *Geophys. Res. Lett.* **2016**, *43*, 6975–6984. [CrossRef]
44. Zhang, Y.; Zhang, B.; Xu, Q.; Gao, D.; Xu, W.; Ren, R.; Jiang, J.; Wang, S. The Effects of Plant and Soil Characteristics on Partitioning Different Rainfalls to Soil in a Subtropical Chinese Fir Forest Ecosystem. *Forests* **2022**, *13*, 123. [CrossRef]
45. Gonzalez-Ollauri, A.; Mickovski, S.B. Hydrological effect of vegetation against rainfall-induced landslides. *J. Hydrol.* **2017**, *549*, 374–387. [CrossRef]
46. Promper, C.; Puissant, A.; Malet, J.-P.; Glade, T. Analysis of land cover changes in the past and the future as contribution to landslide risk scenarios. *Appl. Geogr.* **2014**, *53*, 11–19. [CrossRef]

forests

MDPI

Article

Analysis of the Influence of Forests on Landslides in the Bijie Area of Guizhou

Yu Zhang [1], Chaoyong Shen [1,2,3,*], Shaoqi Zhou [1,*] and Xuling Luo [3]

1 College of Resources and Environmental Engineering, Guizhou University, Guiyang 550025, China; zhangyu@gzchsy.cn
2 Guizhou Academy of Sciences, Guiyang 550009, China
3 The Third Surveying and Mapping Institute of Guizhou Province, Guiyang 550004, China; luoxuling@mail.gyig.ac.cn
* Correspondence: shency@gzchsy.cn (C.S.); comradeshaoqi@163.com (S.Z.); Tel.: +86-139-8553-6169 (C.S.); +86-182-8613-3628 (S.Z.)

Abstract: Forests are an important part of the ecological environment, and changes in forests not only affect the ecological environment of the region but are also an important factor causing landslide disasters. In order to correctly evaluate the impact of forest cover on landslide susceptibility, in this paper, we build an evaluation model for the contribution of forests to the landslide susceptibility of different grades based on survey data for forest land change in Bijie City and landslide susceptibility data, and discuss the effects of forest land type, origin, age group, and dominant tree species on landslide susceptibility. We find that forests play a certain role in regulating landslide susceptibility: compared with woodland, the landslide protection ability of shrubland is stronger. Furthermore, natural forests have a greater inhibitory effect on landslides than artificial forests, and compared with young forest, mature forest and over-mature forest, middle-aged forest and near-mature forest have stronger landslide protection abilities. In addition, the dominant tree species in different regions have different impacts on landslides. Coniferous forests such as Chinese fir and Cryptomeria fortunei in Qixingguan and Dafang County have a low ability to prevent landslides. Moreover, the soft broad tree species found in Qianxi County, Zhijin County, Nayong County and Jinsha County are likely to cause landslides and deserve further research attention. Additionally, a greater focus should be placed on the landslide protection of walnut economic forests in Hezhang County and Weining County. Simultaneously, greater attention should be paid to the Cyclobalanopsis glauca tree species in Weining County because the area where this tree species is located is prone to landslides. Aiming at addressing the landslide susceptibility existing in different forests, we propose forest management strategies for the ecological prevention and control of landslides in Bijie City, which can be used as a reference for landslide susceptibility prevention and control.

Keywords: landslide; susceptibility; tree species; age group; woodland type; forest origin

Citation: Zhang, Y.; Shen, C.; Zhou, S.; Luo, X. Analysis of the Influence of Forests on Landslides in the Bijie Area of Guizhou. *Forests* **2022**, *13*, 1136. https://doi.org/10.3390/f13071136

Academic Editors: Chong Xu, Haijia Wen, Weile Li and Hiromu Daimaru

Received: 17 May 2022
Accepted: 18 July 2022
Published: 19 July 2022

1. Introduction

Geological disasters are widely encountered by human society and cause significant suffering [1]. They are among the most prevalent natural disasters, causing thousands of deaths and billions of dollars of property damage each year [2,3]. China is one of the countries that is most severely affected by geological disasters in the world, and the intensity and frequency of disasters are increasing year by year [4]. A total of 4772 geological disasters have occurred in China, resulting in 80 deaths and 11 missing people, as well as direct economic losses of 3.2 billion yuan. Among these, landslide disasters occurred most frequently and had the widest impact in 2021. A total of 2335 landslides have occurred in total nationwide and landslides have accounted for 49% of all geological disasters this year [5].

Forests can play a useful role with respect to water retention, which is extremely important for the stability of slopes [6–8] and can effectively inhibit the occurrence of geological disasters such as debris flows and landslides [9,10], reducing their degree of harm [11]. In recent years, increasing attention has been paid to the role of forests in landslides, and relevant studies have shown that forests have a huge inhibitory effect on the occurrence of landslides. Hwang et al. found that characteristics of forests such as composition and spatial distribution can greatly affect slope stability [12]. Nelson et al. analyzed landslide events in Colombia and pointed out that the landslide susceptibility in non-forest areas is much higher than that in forest areas [8]. Installation and maintenance costs for implementing technical measures for landslide risk prevention tend to be lower in locations with forested areas [13]. Huang et al. studied the effect of plant roots on landslides and found that plant roots can be anchored to deeper soil layers to enhance their integrity with the topsoil, thereby generating additional shear forces and reducing the possibility of landslides [14,15]. Peduzzi et al. found that plants can reduce soil moisture to maintain high shear, thereby increasing the threshold for triggering landslides [16].

However, different types of vegetation have different root depths and hydrophilic properties [17,18], resulting in different types of forests with different inhibitory effects on landslides [19,20]. The root system of an arbor has a good effect on the shear strength of the surface soil [21,22], but it is not very useful for the control of deep landslides [23]. The root system of shrubland can improve soil physical properties and inhibit shallow landslides [8,24]. Similarly, different forest origins have different effects on landslides [25]. Natural forests have stronger soil fixation effects than planted forests, and higher soil porosity allows better penetration of rainfall and affects other landslide factors [26]. For the same forest, different age groups of trees have different impacts on landslides. For example, in a single shallow-rooted coniferous forest, the slope stability increases first and then decreases after planting from young forest to mature forest [27]. Therefore, reasonable forest management can enhance the geotechnical stability and reduce the susceptibility to landslides to a certain extent [28–30], thereby realizing the ecological control of landslides.

In recent years, with the implementation of the Chinese government's policy of "Grain for Green", the forest coverage rate has increased significantly, but Grain for Green is dominated by a single tree species [31], ignoring the impact on landslides. It is therefore necessary to conduct research on the degree of landslide impact of different forest stand structures to reduce landslide susceptibility while planting trees. In this study, we took Bijie City, Guizhou Province, a high-susceptibility area with respect to geological disasters, as the research area. Based on the contribution model, the impact of different forest types on landslide susceptibility was analyzed in order to provide a reference for local governments to formulate disaster-prevention strategies and regional forest-management systems.

2. Materials and Methods

2.1. Study Area

Bijie City (103°36′–106°43′ E, 26°21′–27°46′ N) is located in southwest China, northwest of Guizhou Province. The terrain fluctuates greatly and is high in the west and low in the east [32]. Simultaneously, it is the highest elevation area in Guizhou Province. The mountains are high and steep, the peaks are densely overlapped, the ravines are vertical and horizontal, the river valleys are deep, and the land is damaged in these locations due to mining activities and other human factors. At the same time, plateaus, mountains, basins, valleys, flat dams, peak clusters, troughs, depressions, karst lakes, and other landforms are intertwined [33]. Karst landforms are extremely developed and diverse in shape. Most of the western part of the territory is characterized by plateaus, karsts, gentle hills, and basins. The middle part comprises mostly peaks, valleys, hills, and depressions. The east and south comprise mostly valleys, peak clusters, gentle hills, and depressions [34]. A total of 93.3% of its area comprises plateaus and mountains, and there is abundant rainfall, with an average annual rainfall of 1200 mm in Bijie. At the same time, the spatial and temporal distribution of precipitation is uneven. About 75% of the precipitation is concentrated in the period from April to September, and most precipitation comprises heavy rains and rainstorms, which can easily cause landslide disasters. Bijie is one of the five high-incidence areas of geological disasters in Guizhou Province [35,36]. According to the material division of landslide proposed by Varnes et al. [37,38]. The landslide types in the study area can be divided into bedrock, debris and earth. The types of landslides are mainly earth, followed by debris; they mainly occur in the shallow surface.

2.2. Data Sources

2.2.1. Geological Condition

The tectonic structure of Bijie City is located at the junction of the two global megatectonic domains, the Tethys-Himalaya and the marginal-Pacific tectonic system, and belongs to the southwestern margin of the Upper Yangtze Block within the first-level tectonic unit of the Yangtze Block. There are NE-trending, NW-trending, near-EW-trending faults, folds and other secondary faults (Figure 1). Its geomorphological manifestations are mainly linear boundaries of intermountain basins, fault triangles, linear valleys, etc. They are clearly visible in fault-controlled valleys, fault triangles and fault cliffs. The peak acceleration of ground motion in Bijie generally shows a decreasing trend from southwest to northeast, with 0.10 g on the southwest side and 0.05 g on the northeast side. In recent years, earthquakes of different magnitudes have occurred in Weining County, Hezhang County, Nayong County, Qixingguan and other places, causing a series of landslides.

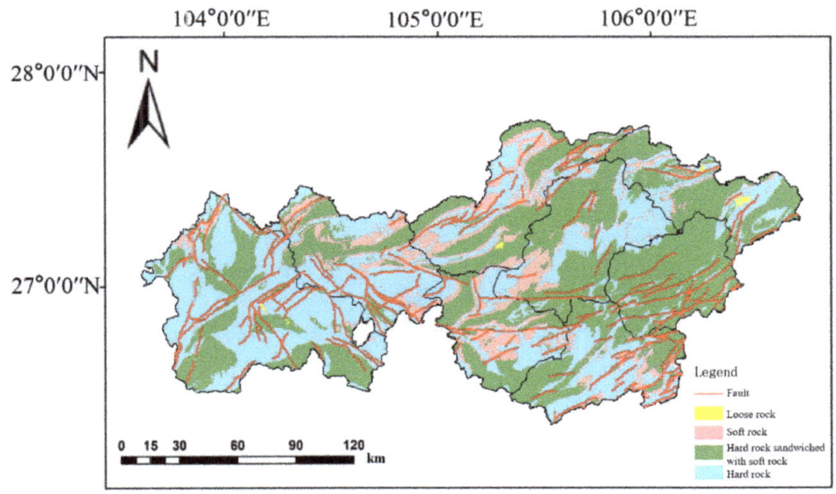

Figure 1. Geological condition map in Bijie City.

According to field reconnaissance and comprehensive analysis of regional geological data, as well as rock and soil hardness, structure, and physical and mechanical properties, the rock formations in the study area are divided into four types: hard rock, hard rock sandwiched with soft rock, soft rock and loose rock (Figure 1). Hard rocks mainly include medium-thickness limestone, dolomitic limestone, dolomite, sandstone, etc. These rock formations are distributed in all counties and districts of Bijie City. These rocks are compact and hard in structure, with high mechanical strength of intact rocks, good engineering mechanical properties, and relatively low landslide susceptibility. Hard rock sandwiched with soft rock mainly includes carbonate rock sandwiched with mudstone, sandstone, shale, sandstone sandwiched with mudstone, etc. This type of rock formation is distributed in the entire area of Bijie City. The mechanical strength of this type of rock mass varies greatly, and most of them have weak interlayers. Under the condition of poor combination of strata, structure and terrain, it is easy to produce geological disasters such as landslides. Mudstones and shales, including sandstones, conglomerates, marls, etc., are mainly distributed in the central region. This rock group has weak weathering resistance, and the shallow parts of the surface are severely weathered, which can provide a large amount of material sources for landslides. Loose rocks are mainly clay, loam, sandy soil, loamy soil, and gravel soil. This rock group has a complex structure, loose, large porosity, strong water permeability and low strength. In areas such as inter-mountain valleys and riverbeds where the terrain is gentle, there are few geological disasters; in slope areas, under the influence of groundwater, rainfall, and human engineering activities, slope instability can easily occur, resulting in landslides and other geological disasters. Due to the complex geological conditions, rich mineral resources and intense human activities in the study area, landslides and other geological disasters frequently occur.

2.2.2. Forestry Change Data

The forestry change data provided by the Forestry Bureau of Bijie City show that in recent years, the tree-planting area in Bijie City has continued to expand, and the forest coverage rate has increased from 49.02% in Guizhou Province in 2016 to 54.19% in 2018. As shown in Figure 2, the forest structure comprises mostly woodland and shrubland, and the woodland stand structure is single, with an area of 770,782.36 hm^2, accounting for 55.79% of the total forest area. The areas of other types of forest land, such as sparse forest land, undeveloped forest land, suitable forest land, nursery land, and forest land without standing trees, are small and scattered. The arbor forests in Bijie City are mainly plantations, which are artificially planted coniferous pure forests with simple hierarchies, poor biodiversity, unstable ecosystems, and a higher occurrence of natural disasters such as landslides [39].

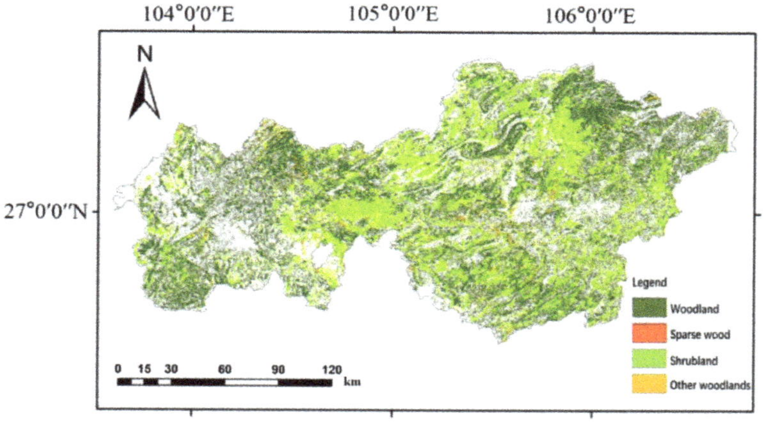

Figure 2. Spatial distribution map of forest land types in Bijie City.

2.2.3. Landslide Susceptibility Mapping

The landslide susceptibility data for Bijie City were provided by the Guizhou Provincial Mountain Geological Hazard Prevention Engineering Technology Research Center. These data originate from a database of hidden landslide danger points surveyed on the ground. According to a grid of 100 m × 100 m, the Bijie area can be divided into 2,394,053 cells, and the various influencing factors in the geographic information database can be rasterized and corresponding influencing factors can be assigned to each surface unit. In order to analyze the comprehensive superposition effect of geological conditions, rainfall, river distribution and other influencing factors on landslide susceptibility, influencing factors are selected based on the comprehensive consideration of natural and human factors. Human engineering activities include roads, mining areas, and land use, while geological environmental conditions include elevation, slope, plane curvature, slope aspect, profile curvature, lithology, rivers, rainfall, and fault zones. Then, the ArcGIS spatial overlay function is used to calculate the distribution density of hidden danger points of landslides with different influencing factors, and the information model is employed to calculate the information value of each influencing factor, and the information values are normalized to form a regional distribution map of various types of influencing factors. Two out of three sets of landslide hidden danger points identified in the field are selected as the training dataset, and the remaining set of landslide hidden danger points is selected as the verification dataset. The support vector machine (SVM) model in Python's scikit-learn library is then called upon to calculate the weights of the 12 influencing factors. Using the obtained weights, grid algebra calculations are performed on the information about all influencing factors, the landslide susceptibility index of each surface unit in the study area is calculated, the results are obtained and drawn in ArcGIS, and an initial susceptibility evaluation of the study area is generated. The higher the value of the grid, the higher the susceptibility of the unit to landslides. Using the natural breakpoint method commonly used in statistics, the evaluation results were divided into four categories: low-susceptibility areas, medium-susceptibility areas, high-susceptibility areas, and extremely high-susceptibility areas. The middle-grade landslide-prone areas in Bijie City account for 30.21% of the city's total area and the high-grade areas account for 41.30% of the city's total area.

According to relevant research, among the 12 factors selected, the one with the greatest impact on landslide susceptibility is mining activities, and the other factors with a greater impact are the slope, rainfall, and distance to a river [40,41]. Among them, from Figure 3a,d, it can be seen that the distribution of slopes is consistent with the landslide-prone areas, and the slopes with low landslide-prone areas are generally not high. The distribution of rivers in Bijie is relatively uniform, and there is no obvious difference between east and west, but the annual rainfall in the east is greater (Figure 3b), which makes the east more susceptible to landslides. At the same time, coal mines in the eastern region are more widely distributed and mining activities are more intensive (Figure 3c). Under the combined effect of these factors, geological landslide disasters are more intensive in the eastern region. During the plant growth period, in the stratum with developed fissures, the growth of root system can lead to fissure expansion and root splitting. In clastic rock regions, root growth can mechanically break up the rock surrounding the roots, providing soil and nutrients. In the windy and rainy season, the tall trees have a loosening effect on the rock and soil at the root due to the dynamic action of the wind, which easily increases the susceptibility to landslides. At the same time, the root system of vegetation has the effect of strengthening the stability of the soil and slope. Therefore, whether vegetation factors can inhibit or promote landslides in the area remains to be studied. Geological conditions are the main factors affecting landslide susceptibility, while vegetation has less of an effect on landslides. However, the geological conditions are difficult to change in the natural state and can only be transformed through engineering activities. In contrast, vegetation is more plastic, and China is implementing afforestation activities. Quantifying the contribution of vegetation to landslides and selecting reasonable tree species and methods for afforestation can improve landslide susceptibility while restoring forests.

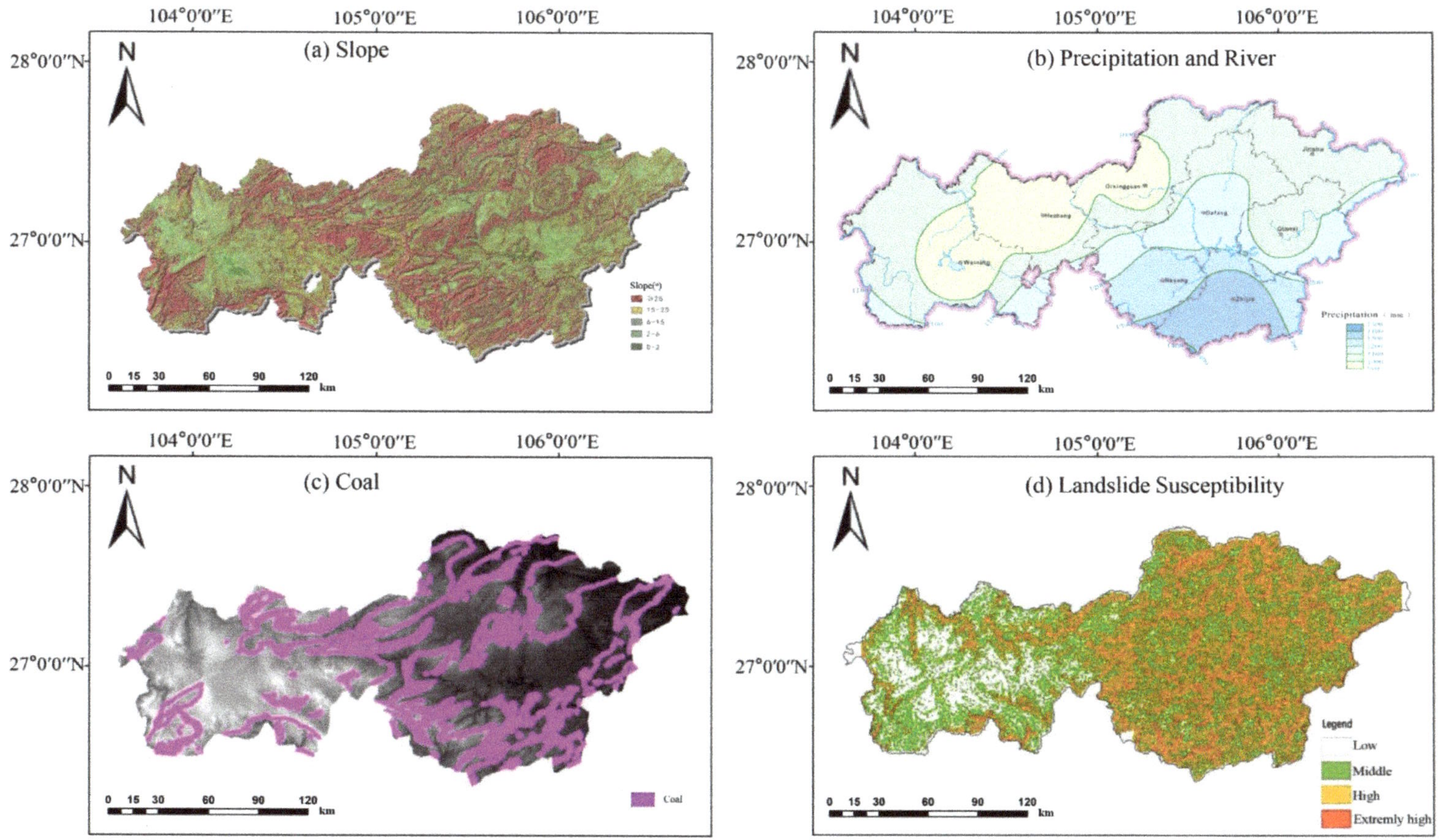

Figure 3. (**a**) The slope distribution of Bijie; (**b**) the precipitation and river distribution of Bijie; (**c**) the coal mine distribution of Bijie; (**d**) landslide susceptibility.

2.3. Methods

2.3.1. Contribution Evaluation Model

Plant roots are closely related to soil shock resistance and shear strength [42,43]. At the same time, different forests have different interception effects on atmospheric precipitation. In order to eliminate the impact of differences in forest areas, we quantitatively evaluated the relationship between forests and landslide susceptibility considering the aspects of forest land type, forest origin, age group, and dominant tree species. This paper is based on previous research results [44,45]. A normalized contribution evaluation model is constructed:

$$P(i, y) = 1/(1 + e^{(1 - (ALSM(i, y)/TALSM(i))/(ASEA(y)/TAREA))}) \quad (1)$$

In the formula, i is the landslide susceptibility grade; y is the forest type; P(i, y) is the contribution of the i grade susceptibility to the y-type forest in the region, and the value varies from 0 to 1; ALSM(i, y) is the area of the y-type forest with the i-level vulnerability; TALSM(i) is the sum of the i-level vulnerability area in all forests in the region; ASEA(y) is the area of the y-type forest, and TAREA is the sum of the areas of all types of forests in the region.

When P(i, y) is closer to 1, the y-type forest contributes more to the i-level susceptibility; that is, the protection ability of this type of forest with respect to landslides is weaker. When P(i, y) is closer to 0.5, this indicates that the regional y-type forest does not contribute much to the i-level susceptibility. When P(i, y) is closer to 0, this indicates that the regional y-type forest contributes little to the i-level susceptibility; that is, this type of forest has a protective effect with respect to landslides.

2.3.2. Spatial Pattern Analysis

We collected the forestry change survey database information for eight districts and counties in Bijie City in 2019 from the Forestry Bureau of Bijie City and collected the landslide susceptibility evaluation results for Bijie City from the Guizhou Provincial Geological Disaster Emergency Technical Guidance Center. The ArcGIS spatial analysis module was used to extract the forest land type, forest origin, age group, and spatial location information of dominant tree species, and a spatial overlay analysis was performed using the landslide susceptibility data. Formula (1) was used to calculate the contribution of different types of forests to landslide susceptibility of different grades We quantitatively evaluated the relationship between forests and landslide susceptibility. Landslide susceptibility was divided into four levels: low, medium, high, and extremely high. Considering that landslides are less likely to occur for the low-susceptibility level, it is impossible to compare the contribution of different forest types. Therefore, we only analyzed the relationship between different forest types and medium and high levels, and extremely high-grade landslide susceptibility.

3. Results

3.1. Influence of Woodland Types on Landslide Susceptibility

Woodland and shrub forest land account for 98.49% of the total forest land area, and other types of forest land are small and scattered in Bijie City. Therefore, the contribution of woodland and shrubland to the landslide susceptibility of different grades is mainly discussed. The distribution of the landslide susceptibility of each grade in different forest land types is shown in Table 1 and Figure 4.

Table 1. Relationship between forest land types and landslide susceptibility in Bijie City.

Forest land Use	Landslide Susceptibility						This Type of Forest		Contribution		
	Medium		High		Extremely High						
	Area (hm^2)	Ratio (%)	Area (hm^2)	Ratio (%)	Area (hm^2)	Ratio (%)	Area (hm^2)	Ratio (%)	Medium	High	Extremely High
Woodland	195,589	52.94	297,205	56.42	221,121	60.12	770,782	55.79	0.49	0.50	0.52
Sparse forest	122.74	0.03	129.44	0.02	87.77	0.02	372.26	0.03	0.57	0.49	0.48
Shrubland	148,250	40.13	219,637	41.69	144,049	39.17	582,091	42.7	0.50	0.51	0.49

Note: "Medium" refers to the medium-susceptibility area, "High" refers to the high-susceptibility area, and "Extremely High" refers to the extremely high-susceptibility area.

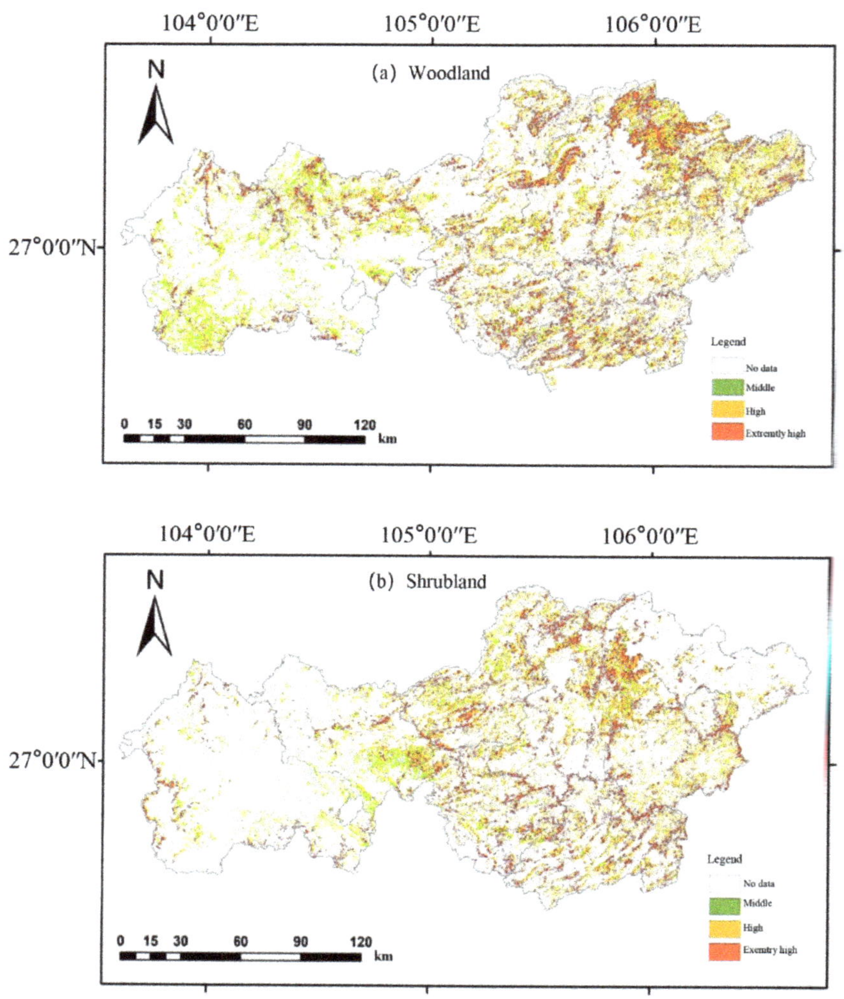

Figure 4. Spatial distribution map of landslide susceptibility in shrubland in Bijie City.

More than 50% of the forest land in Bijie City is woodland, with an area of 770,782 hm^2. The landslide-prone areas of medium susceptibility, high susceptibility, and extremely high-susceptibility grades are 195,589 hm^2, 297,205 hm^2, and 221,121 hm^2, and account for 52.94%, 56.42%, and 60.12%, of area of the corresponding landslide susceptibility grades, respectively (Table 1). The extremely high-grade landslide-prone areas are mainly distributed in the eastern part of Bijie, the high-grade landslide-prone areas are mainly in the central and eastern parts, and the medium-grade landslide-prone areas are most widely distributed in the western part (Figure 4a). In addition, the woodland has the greatest impact on the susceptibility of extremely high-grade landslide areas, with a contribution of 0.52. The contribution of the susceptibility of high-grade landslide areas is 0.50, and the impact on medium-grade landslide susceptibility is the smallest, with a contribution of 0.49 (Table 1). On the whole, the impact of woodland on the susceptibility to landslides of high susceptibility and extremely high-susceptibility grade areas is up to 0.5, which indicates an area prone to landslides.

The landslide-prone areas of high susceptibility and extremely high-susceptibility grades within the shrubland range were 219,637 hm^2 and 144,049 hm^2, accounting for 42% and 39% of area of the corresponding landslide susceptibility grades, respectively, mainly in the eastern part of Bijie City (Figure 4b); the landslide-prone areas of medium susceptibility grades were 148,250 hm^2, mainly distributed in the central area. The contribution of shrubland to extremely high-grade susceptibility is 0.49, which is lower than that of woodland, indicating that the landslide susceptibility of shrubland is smaller than that of woodland; however, the contribution to high-grade landslide susceptibility is 0.51, which is higher than 0.5, indicating that the landslide protection ability is still not high.

3.2. Forest Origin and Landslide Susceptibility

It can be seen from Table 2 and Figure 5 that the area ratio of natural forest to plantation forest in Bijie City is 3:2, and the natural forests with high and extremely high landslide susceptibility are mainly distributed in the central and eastern parts of Bijie City (Figure 5a). Plantation forests are mainly in medium-susceptibility areas and are less distributed in the west (Figure 5b). The landslide-prone area of high and extremely high grades within the natural forest land comprises 538,509 hm^2, accounting for 62.28% of all natural forest land. The sum of the landslide-prone areas of high and extremely high grades within the plantation forest is 384,263 hm^2, accounting for 66.62% of the total plantation area. The contributions of high- and extremely high-grade landslide susceptibility in the plantation forest are 0.52 and 0.51, respectively; the contributions for natural forest are 0.49 and 0.51, respectively, which are closer to 0.5. This shows that the plantation forest has weaker protection ability against landslides, while the landslide protection ability of natural forest is better than that of plantation forest.

Table 2. Relationship between forest origin and landslide susceptibility in Bijie City.

Forest Origin	Landslide Susceptibility						This Type of Forest		Contribution		
	Medium		High		Extremely High		Area (hm^2)	Ratio (%)	Medium	High	Extremely High
	Area (hm^2)	Ratio (%)	Area (hm^2)	Ratio (%)	Area (hm^2)	Ratio (%)					
Natural	221,323	61.9	323,311	57.77	215,198	59.27	864,556	59.92	0.50	0.49	0.50
Artificial	136,210	38.1	236,364	42.23	147,899	40.73	576,821	40.08	0.49	0.52	0.51

Note: "Medium" refers to the medium-susceptibility area, "High" refers to the high-susceptibility area, and "Extremely High" refers to the extremely high-susceptibility area.

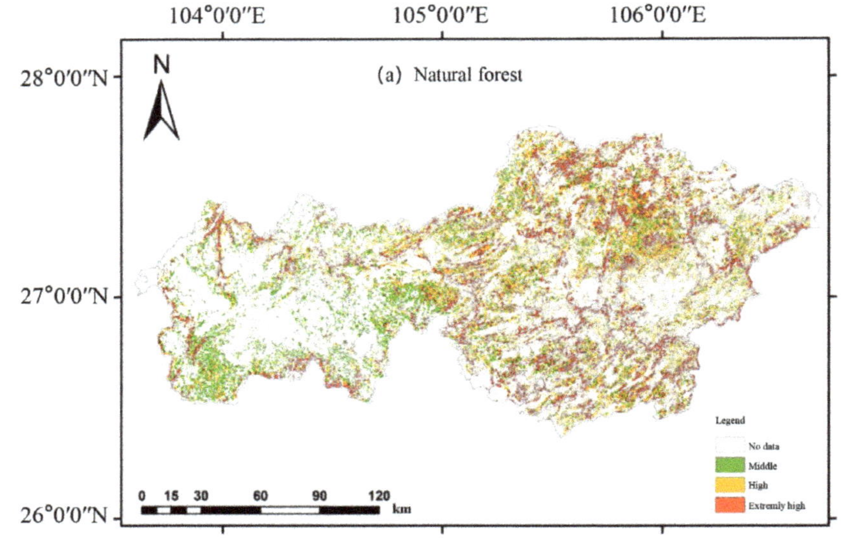

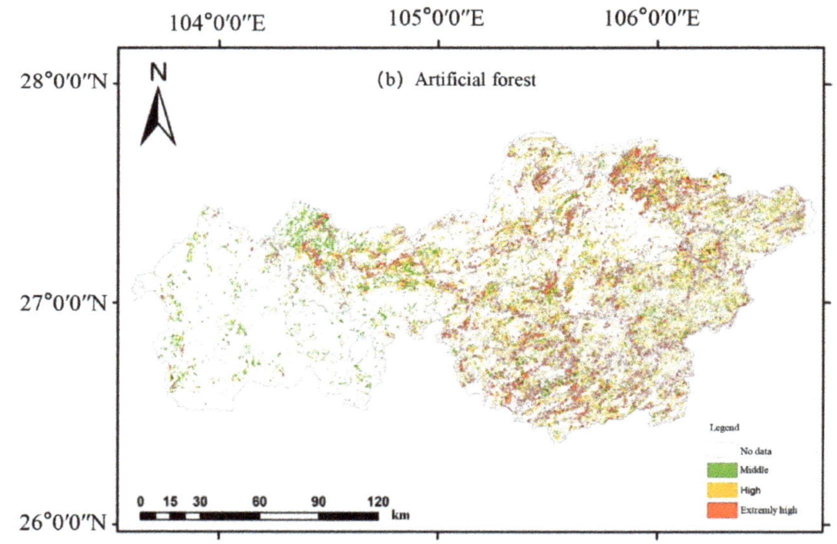

Figure 5. Spatial distribution of forest origin and landslide susceptibility in Bijie City.

3.3. Forest Age Group and Landslide Susceptibility

It can be seen from Table 3 and Figure 6 that the areas of young- and medium-aged forests are the largest, accounting for 43.36% and 33.53%, respectively. Near-mature forests and mature forests account for 16.11% and 6.54%, respectively, and over-mature forests account for only 0.57%. The distribution of young-aged forests (Figure 6a) and medium-aged forests (Figure 6b) is similar, with high- and extremely high-landslide-susceptibility areas mainly distributed in the central and eastern areas, and the western areas being mainly medium-susceptibility areas. The medium-susceptibility areas of near-mature forests (Figure 6c) and mature forests (Figure 6d) are mainly concentrated in the western region, while high- and extremely high-susceptibility areas are scattered sporadically. Over-mature forests (Figure 6e) are small in area and are scattered in distribution. The contributions of young forest to high and extremely high landslide susceptibility are both 0.53, and the protection ability with respect to landslides is the weakest.

Table 3. Relationship between forest age groups and landslide susceptibility in Bijie City.

Age Group	landslide Susceptibility						This Type of Forest		Contribution		
	Medium		High		Extremely High						
	Area (hm^2)	Ratio (%)	Area (hm^2)	Ratio (%)	Area (hm^2)	Ratio (%)	Area (hm^2)	Ratio (%)	Medium	High	Extremely High
Young forest	73,380	37.52	142,055	47.8	89,794	47.82	334,187	43.36	0.47	0.53	0.53
Medium aged forest	71,280	36.45	91,323	30.73	60,353	32.14	258,411	33.53	0.52	0.48	0.49
Near mature forest	36,896	18.86	42,438	14.28	25,115	13.38	124,144	16.11	0.54	0.47	0.46
Mature forest	13,037	6.67	19,290	6.49	11,533	6.14	49,612	6.44	0.51	0.50	0.49
Over mature forest	992	0.51	2098	0.71	966	0.51	4429	0.57	0.47	0.56	0.47

Note: "Medium" refers to the medium-susceptibility area, "High" refers to the high-susceptibility area, and "Extremely High" refers to the extremely high-susceptibility area.

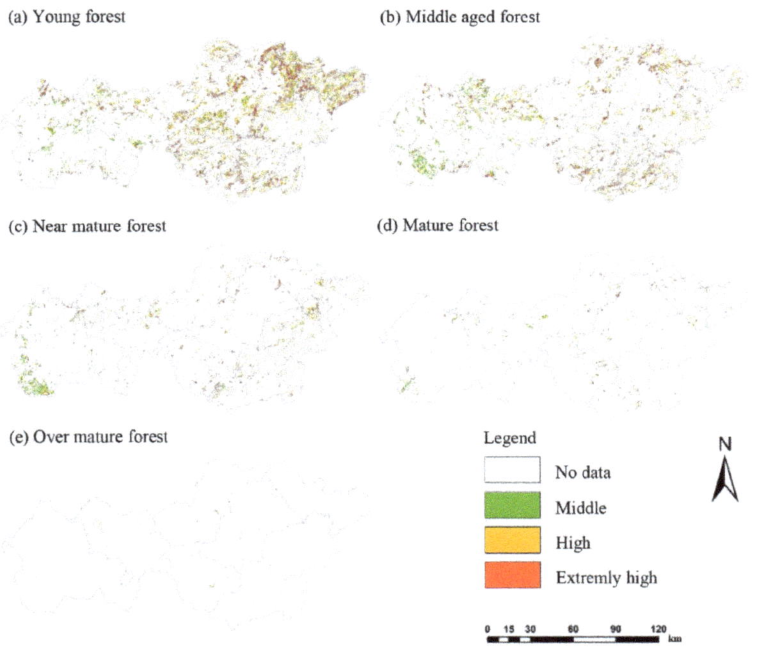

Figure 6. Spatial distribution map of forest age landslide susceptibility in Bijie City.

3.4. Dominant Tree Species and Landslide Susceptibility

Considering that Bijie City is located in the slope transition zone from the Yunnan Plateau to the Guizhou Plateau, the terrain is high in the west and low in the east. The natural geographical environment, such as the altitude, varies greatly among districts and counties [42]. Large differences and diverse geographical environments lead to different dominant tree species constituting arbor forest land in most districts and counties, and their forestry management strategies are also different. In order to improve the pertinence of the analysis of the impact of dominant tree species in arbor forest land on landslides and to guide the operability of forest management, in this paper, we used forestry change survey data to calculate the top-five dominant tree species in each county. We took a county as a unit (Table 4) to analyze the relationship between dominant tree species and landslide susceptibility, and to provide a reference for the region to formulate precise afforestation plans.

Table 4. Relationship between dominant tree species and landslide susceptibility in various districts and counties in Bijie City.

District	Dominant Tree Species	Landslide Susceptibility						This type of Forest		Contribution		
		Medium		High		Extremely High						
		Area (hm^2)	Ratio (%)	Area (hm^2)	Ratio (%)	Area (hm^2)	Ratio (%)	Area (hm^2)	Ratio (%)	Medium	High	Extremely High
Qixingguan	Birch	2611	7.52	7805	9.05	3646	7.16	14,342	8.06	0.48	0.53	0.47
	Walnut	2095	6.03	6114	7.09	3592	7.06	12,145	6.83	0.47	0.51	0.51
	Chinese fir	2140	6.16	5519	6.4	4171	8.2	12,140	6.83	0.48	0.49	0.55
	Cryptomeria fortunei	1762	5.07	4093	4.74	1810	3.56	7878	4.43	0.54	0.52	0.45
	Pinus yunnanensis	1003	2.89	3176	3.68	1842	3.62	6194	3.48	0.46	0.52	0.51
Dafang	Birch	3555	10.72	10,638	10.73	5101	8.26	19,689	9.94	0.52	0.52	0.46
	Walnut	2608	7.87	7494	7.56	3527	5.71	13,890	7.02	0.53	0.52	0.45
	Cryptomeria fortunei	2199	6.63	6039	6.09	3549	5.75	12,073	6.1	0.52	0.50	0.49
	Chinese fir	1715	5.17	5123	5.17	3059	4.95	10,029	5.07	0.51	0.51	0.49
	Poplar	955	2.88	2772	2.8	2820	4.57	6710	3.39	0.46	0.46	0.59
Qianxi	Osmanthus	8245	31.67	20,993	31.1	14,409	41.13	44,568	33.89	0.48	0.48	0.55
	Masson pine	3841	14.75	9147	13.55	4435	12.66	17,851	13.57	0.52	0.50	0.48
	Other soft broad	2421	9.3	6583	9.75	3240	9.25	12,528	9.53	0.50	0.51	0.49
	Cedarwood	1892	7.27	4608	6.83	2163	6.18	8888	6.76	0.52	0.50	0.48
	Broad-leaved mixed forest	1827	7.02	4742	7.03	1885	5.38	8634	6.57	0.52	0.52	0.46
Jinsha	Other soft broad	3305	14.53	11,153	16.45	9488	20.49	24,385	17.53	0.46	0.49	0.54
	Masson pine	3287	14.45	9273	13.68	5333	11.51	18,128	13.03	0.53	0.51	0.47
	Quercus fabri	2887	12.69	7949	11.72	4359	9.41	15,522	11.16	0.54	0.51	0.46
	Cedarwood	1623	7.14	6045	8.92	4143	8.94	11,916	8.57	0.46	0.51	0.51
	Chinese fir	1788	7.86	5501	8.11	3733	8.06	11,295	8.12	0.49	0.50	0.49
Zhijin	Cryptomeria fortunei	3707	13.63	9778	13.6	7291	12.06	21,150	12.98	0.51	0.51	0.48
	Chinese fir	2947	10.84	7506	10.44	6751	11.17	17,575	10.79	0.50	0.49	0.51
	Birch	2457	9.04	6732	9.36	5130	8.48	14,636	8.98	0.50	0.51	0.49
	Other soft broad	1785	6.56	4962	6.9	4267	7.06	11,188	6.87	0.49	0.50	0.51
	Walnut	1137	4.18	3168	4.4	3186	5.27	7551	4.63	0.48	0.49	0.54
Nayong	Birch	2864	11.43	8991	15.79	6538	13.74	18,875	13.94	0.46	0.53	0.49
	Cryptomeria	3347	13.36	6230	10.94	5762	12.11	16,206	11.97	0.53	0.48	0.51
	Chinese fir	2014	8.04	5798	10.18	4479	9.41	12,681	9.37	0.47	0.52	0.50
	Peach	1046	4.17	2910	5.11	2236	4.7	6243	4.61	0.48	0.53	0.51
	Other soft broad	1099	4.39	2460	4.32	2243	4.71	6142	4.54	0.49	0.49	0.51

Table 4. *Cont.*

District	Dominant Tree Species	Landslide Susceptibility						This type of Forest		Contribution		
		Medium		High		Extremely High						
		Area (hm^2)	Ratio (%)	Area (hm^2)	Ratio (%)	Area (hm^2)	Ratio (%)	Area (hm^2)	Ratio (%)	Medium	High	Extremely High
Hezhang	Pinus armandi	19,016	25.89	10,475	19.6	4311	18.7	45,373	22.87	0.53	0.47	0.46
	Walnut	5755	7.83	8250	15.43	4172	18.1	20,247	10.2	0.44	0.63	0.68
	Pinus yunnanensis	3869	5.27	5721	10.7	2138	9.28	14,167	7.14	0.44	0.62	0.57
	Quercus fabri	4626	6.3	2835	5.3	1328	5.76	12,007	6.05	0.51	0.47	0.49
	Chinese fir	1473	2.01	1605	3	618	2.68	4276	2.16	0.48	0.60	0.56
Weining	Pinus yunnanensis	45,751	43.48	16,011	38.91	8142	29.14	99,533	38.55	0.53	0.50	0.44
	Pinus armandi	19,326	18.37	4694	11.41	1682	6.02	48,384	18.74	0.50	0.40	0.34
	Sawtooth oak	3739	3.55	1418	3.45	776	2.78	10,468	4.05	0.47	0.46	0.42
	Cyclobalanopsis glauca	849	0.81	1258	3.06	1977	7.07	4381	1.7	0.37	0.69	0.96
	Other soft broad	1485	1.41	722	1.75	509	1.82	3687	1.43	0.50	0.56	0.57

Note: "Medium" refers to the medium-susceptibility area, "High" refers to the high-susceptibility area, and "Extremely High" refers to the extremely high-susceptibility area.

3.4.1. Dominant Tree Species and Landslide Susceptibility in Qixingguan District

It can be seen from Table 4 and Figure 7 that the top-five dominant tree species in the forests of Qixingguan District are Birch, Walnut, Chinese fir, Cryptomeria fortunei, and Pinus yunnanensis. The sums of high- and extremely high-grade landslide-prone areas for the various tree species are 11,451 hm^2, 9706 hm^2, 9690 hm^2, 5903 hm^2, and 5018 hm^2, respectively, accounting for 79.8%, 79.9%, 79.8%, 74.9%, and 81% of the total areas of these tree species. Among the dominant tree species, Birch has the largest area, but it is less distributed in the southwest and northeast regions (Figure 7a); the areas with high and extremely high landslide susceptibility for Walnut are mainly in the southern region (Figure 7b); Chinese fir is more concentrated and is mainly distributed in the northern region (Figure 7c); Cryptomeria fortunei is scattered in a sporadic distribution (Figure 7d); Pinus yunnanensis has the smallest area and is distributed in sheets almost only in the south (Figure 7e).

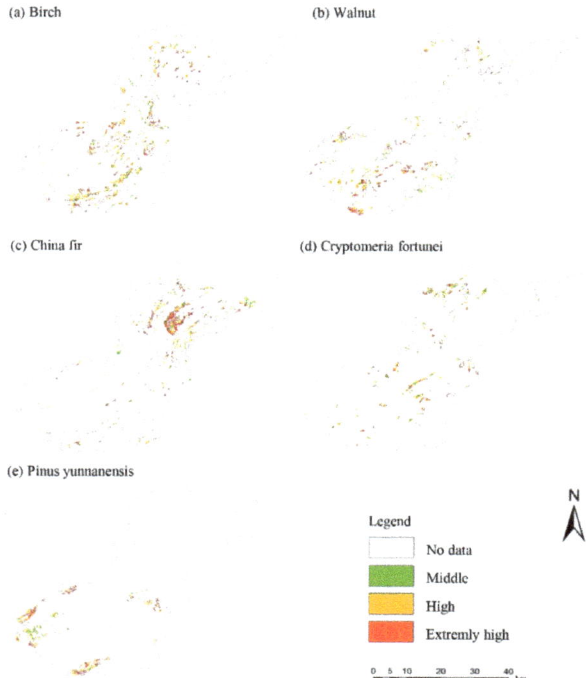

Figure 7. Spatial distribution map of landslide susceptibility of dominant tree species in Qixingguan District.

Chinese fir makes the largest contribution to the susceptibility of extremely high-grade landslides, which is 0.55. The extremely high-susceptibility area is 4171 hm^2, accounting for 34.4% of its own area, and the protection ability with respect to landslides is the lowest. The contributions of Walnut and Pinus yunnanensis to extremely high-grade landslide susceptibility are 0.51 and 0.51, indicating that Pinus yunnanensis and Walnut have low protection ability with respect to landslides. The contributions of Birch to high-grade landslide susceptibility are 0.53 and 0.52, and the protection ability with respect to landslides is not high. Chinese fir, Walnut, and Pinus yunnanensis have an impact greater than 0.5 on the susceptibility to extremely high grades, and the landslide protection ability is not as good as that of Birch and Cryptomeria fortunei.

3.4.2. Dominant Tree Species and Landslide Susceptibility in Dafang County

From Table 4 and Figure 8, it can be seen that the top-five dominant tree species in Dafang County are Birch, Walnut, Cryptomeria fortunei, Chinese fir, and poplar, and the sums of the high- and extremely high-grade landslide-prone areas of various tree species are 15,739 hm^2, 11,024 hm^2, 9588 hm^2, 8182 hm^2, and 5592 hm^2, respectively, accounting for 79.9%, 79.4%, 79.4%, 81.6%, and 83.3% of the areas of these tree species. Among the dominant tree species, Birch is less distributed in the southeast and north (Figure 8a); Walnut is mainly distributed in the central area and the south (Figure 8b); Cryptomeria fortunei (Figure 8c) and Chinese fir (Figure 8d) are mainly concentrated in the west and north; poplar is less distributed in the north, and its extremely high-susceptibility area is mainly in the center (Figure 8e).

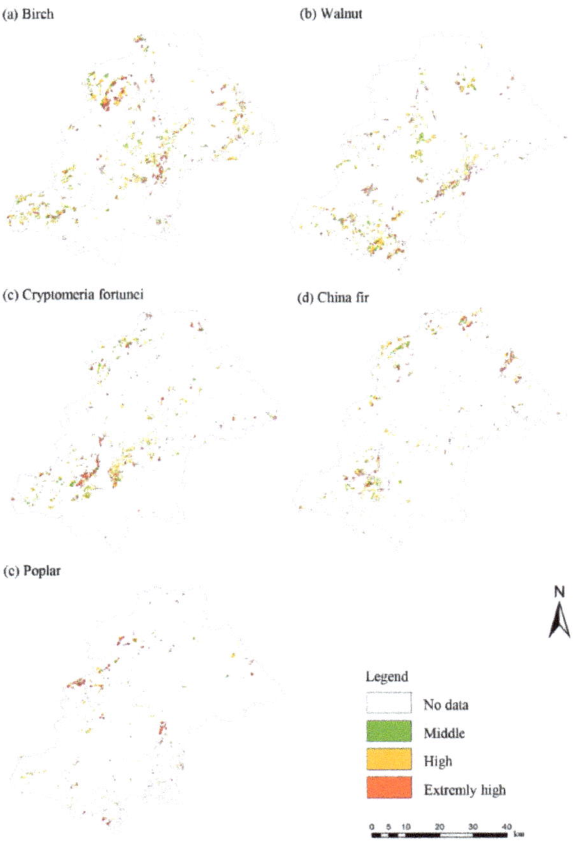

Figure 8. Spatial distribution map of landslide susceptibility of dominant tree species in Dafang County.

Poplar has the smallest area, accounting for 3.39%. Its contribution to extremely high-grade landslide susceptibility reaches 0.59, and it is the only tree species with an extremely high-grade landslide susceptibility of more than 0.5, indicating that poplar has the weakest landslide protection ability compared with other tree species. The areas of Birch and Walnut are the largest, accounting for 8.26% and 6.71%, respectively; their contribution to high-grade landslide susceptibility is the largest (0.52), and their protection ability with respect to landslides is not high. Cryptomeria fortunei and Chinese fir contribute 0.5 and 0.51 to the high-grade landslide susceptibility and are also prone to landslide. Except for

Cryptomeria fortunei, the impact of other tree species on high- and extremely high-grade landslide susceptibility is more than 0.5. Relatively speaking, Cryptomeria fortunei has the lowest possibility of landslide.

3.4.3. Dominant Tree Species and Landslide Susceptibility in Qianxi County

It can be seen from Table 4 and Figure 9 that the top-five dominant tree species in the forests of Qianxi County are Osmanthus trees, Masson pine, other soft broad tree species, Cedarwood, and Broad-leaved mixed tree species. The sums of high- and extremely high-grade landslide-prone areas are 35,402 hm^2, 13,582 hm^2, 9823 hm^2, 8182 hm^2, and 6771 hm^2, accounting for 79.4%, 76.1%, 78.4%, 81.6%, and 76.2% of the areas of these tree species, respectively. The area of Osmanthus is much larger than that of other tree species, and the extremely high-grade landslide susceptibility area is more obvious in the marginal zone (Figure 9a); Masson pine is mainly concentrated in the central and northeastern parts (Figure 9b); other soft and broad tree species (Figure 9c), Cedarwood (Figure 9d), and broad-leaved mixed tree species (Figure 9e) are relatively small in area and are sporadically distributed.

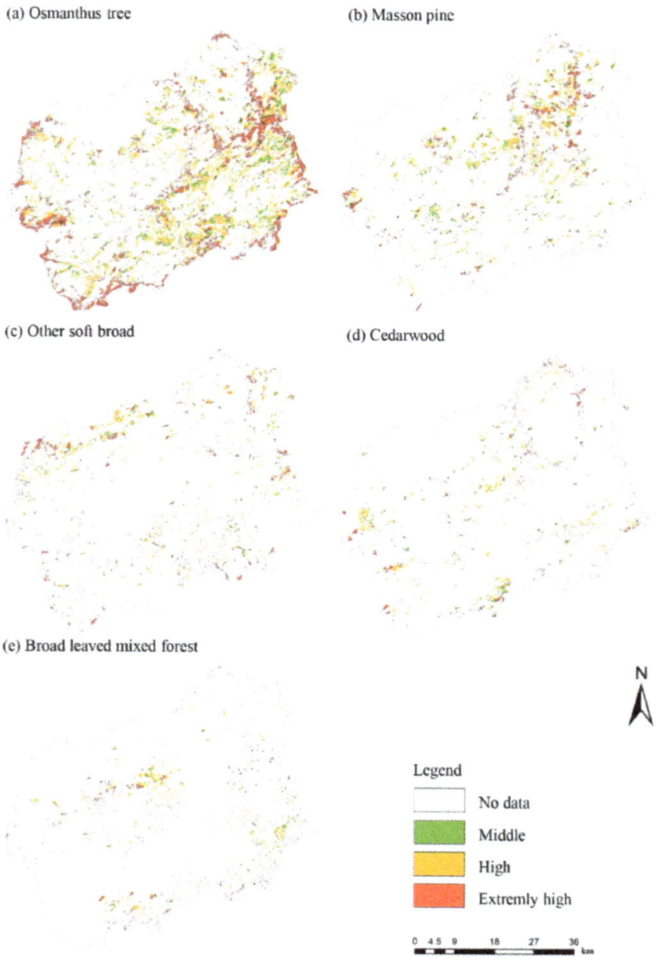

Figure 9. Spatial distribution map of landslide susceptibility of dominant tree species in Qianxi County.

Among the dominant tree species, the contribution of Osmanthus trees to the extremely high-grade landslide susceptibility is 0.55, and it is the only dominant tree species with an extremely high landslide susceptibility value greater than 0.5. Its extremely high-grade landslide-prone area comprises 14,409 hm^2, accounting for 55.1% of the total extremely high-grade landslide-prone area, and its proportion is high, indicating that the Osmanthus has the lowest ability to protect against landslides. The contribution of other soft broad tree species and broad-leaved mixed tree species to high-grade landslide susceptibility is greater than 0.5, indicating that these tree species have poor protection ability with respect to landslides. Masson pine and Cedarwood species account for the lowest proportion of landslide-prone areas at high and extremely high grades, and their contribution to high and extremely high landslide susceptibility is relatively low, both less than 0.5, indicating that they provide relatively good protection against landslides.

3.4.4. Dominant Tree Species and Landslide Susceptibility in Jinsha County

It can be seen from Table 4 and Figure 10 that the top-five dominant tree species in Jinsha County's forests are other soft broad tree species, Masson pine, Quercus fabri, Cedarwood, and Chinese fir. The sums of the landslide prone areas of various tree species with high and extremely high grades are 20,641 hm^2, 14,606 hm^2, 12,308 hm^2, 10,188 hm^2, and 9234 hm^2, respectively, accounting for 84.6%, 80.6%, 79.3%, 85.4%, and 81.6% of the areas of these tree species. Other soft and broad tree species with the largest area have a large distribution in the whole county (Figure 10a); Masson pine is rare in the south, while the high-susceptibility area is most obvious in the west (Figure 10b); Quercus fabri is less distributed in the east (Figure 10c); Cedarwood is mainly found in the west and east (Figure 10d); fir is more abundant in the center, and the extremely high-susceptibility areas are also mostly in the center (Figure 10e).

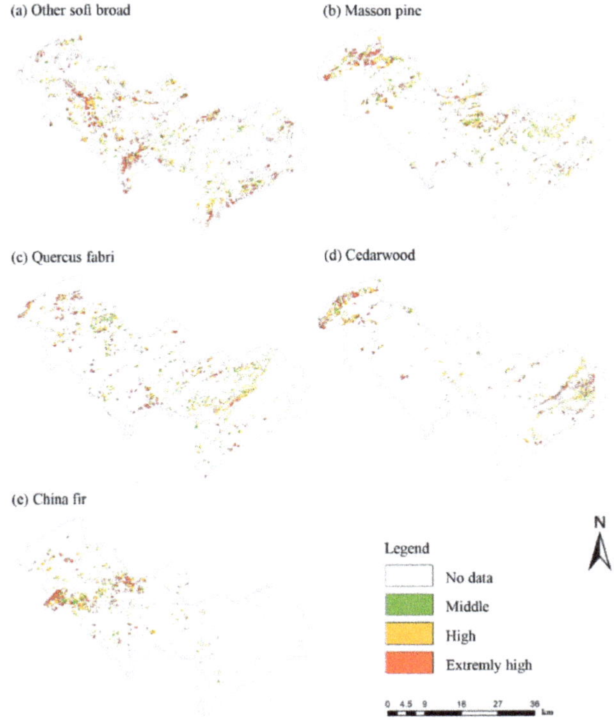

Figure 10. Spatial distribution map of landslide susceptibility of dominant tree species in Jinsha County.

Among the dominant tree species, the extremely high-susceptibility area of other soft and broad tree species makes the largest contribution of 0.54, and its extremely high-grade landslide-prone area is 9488 hm^2, accounting for 35.1% of the total extremely highly prone landslide area; the landslide protection ability is the lowest. The contribution of Cedarwood to the landslide susceptibility of high and extremely high grades is 0.51, accounting for 8.92% and 8.94% of the total landslide susceptibility area of high and extremely high grades, respectively, and the landslide susceptibility is second only to other soft and broad tree species. The contribution of Masson pine and Quercus fabri to the high-grade landslide susceptibility is greater than 0.5, indicating that these tree species have low protection ability with respect to landslides. The area proportion of Chinese fir tree species in the landslide-prone area of high and extremely high grades is relatively low, and the contribution to the landslide susceptibility of high and extremely high grades is relatively low; it is not greater than 0.5. This shows that compared with other tree species, Chinese fir has relatively better protection ability against landslides.

3.4.5. Dominant Tree Species and Landslide Susceptibility in Zhijin County

From Table 4 and Figure 11, it can be seen that the top-five dominant tree species in Zhijin County are Cryptomeria fortunei, Chinese fir, Birch, other soft broad tree species, and Walnut. The sums of the landslide prone areas of various tree species with high and extremely high grades are 20,641 hm^2, 14,606 hm^2, 12,308 hm^2, 10,188 hm^2, and 9234 hm^2, accounting for 80.7%, 81.1%, 81%, 82.5%, and 84.1% of the areas of these tree species respectively. The distribution of Cryptomeria fortunei is relatively concentrated, with fewer areas in the north, and the high- and extremely high-susceptibility areas are mainly in the center (Figure 11a); Chinese fir (Figure 11b) and Birch (Figure 11c) are less distributed in parts of the north and northeast; other soft and broad tree species are more concentrated in the central aera of south (Figure 11d); Walnut has the smallest area and a scattered distribution (Figure 11e).

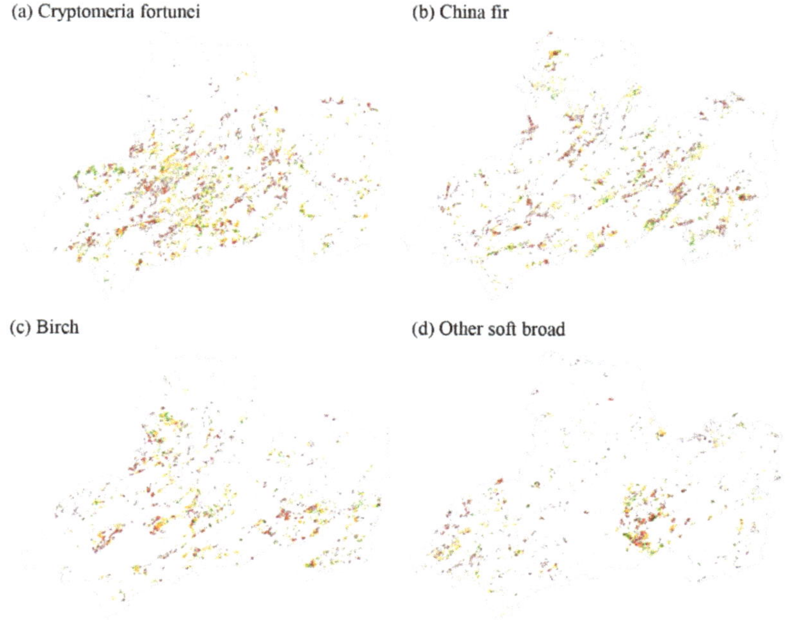

Figure 11. *Cont.*

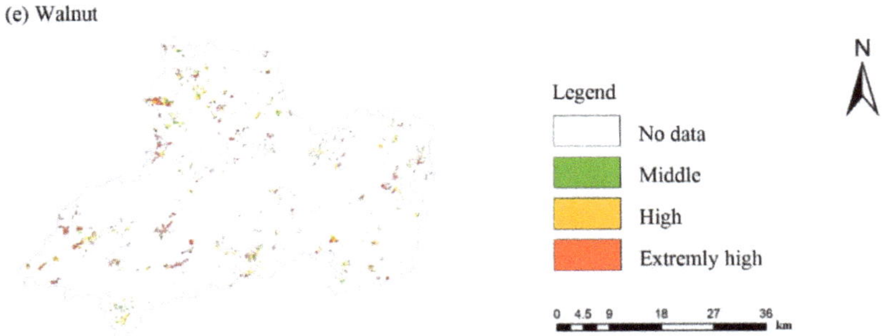

Figure 11. Spatial distribution map of landslide susceptibility of dominant tree species in Zhijin County.

Among the dominant tree species, the extremely high-grade landslide-prone area of Walnut tree species comprises 3186 hm^2, accounting for 42.2% of the area of this tree species. The contribution of Walnut to extremely high-grade landslide susceptibility is the largest, with a maximum contribution of 0.54 and the lowest landslide protection ability. Chinese fir and other soft broad tree species contribute 0.51 to extremely high-grade landslide susceptibility, with areas of 17,575 hm^2 and 11,188 hm^2, respectively. These tree species have low landslide protection ability. The influence of Cryptomeria fortunei and Birch on the high-grade landslide susceptibility is 0.51, which is higher than 0.5, indicating that the landslide protection ability of these tree species is not high, but they are more effective than the other three tree species.

3.4.6. Dominant Tree Species and Landslide Susceptibility in Nayong County

It can be seen from Table 4 and Figure 12 that the top-five dominant tree species in the forests of Nayong County are Birch, Cryptomeria fortunei, Chinese fir, Peach, and other soft and broad tree species. The sums of the landslide-prone areas for various tree species with high and extremely high grades are 15,529 hm^2, 11,992 hm^2, 10,277 hm^2, 5146 hm^2, and 4703 hm^2, respectively, accounting for 82.3%, 74%, 81%, 82.4%, and 76.6% of the areas of these tree species. Birch is widely distributed in the whole county (Figure 12a); Cryptomeria fortunei is mainly distributed in the central and southern areas (Figure 12b); Chinese fir is not as dense as in other areas in the west (Figure 12c); Peach (Figure 12d) and other soft broad tree species (Figure 12e) have the lowest area and a more sporadic distribution.

Among the dominant tree species in Nayong County, Peach makes the highest contribution to the landslide susceptibility of high and extremely high grades, being 0.53 and 0.51, respectively. The landslide protection ability is the worst, but its area is not large, accounting for only 4.61%. In addition, Chinese fir contributes 0.5 and 0.51 to the landslide susceptibility of high and extremely high grades, respectively, indicating that it is prone to landslides, and its ability to stop landslides is only better than that of Peach. The contribution of Cryptomeria fortunei and other soft and broad tree species to the extremely high-grade landslide susceptibility is 0.51, and the landslide protection ability is poor. The largest area of Birch species is 18,875 hm^2, but its contribution to high landslide susceptibility is 0.53, and the landslide protection ability is not high; however, it is better than other tree species.

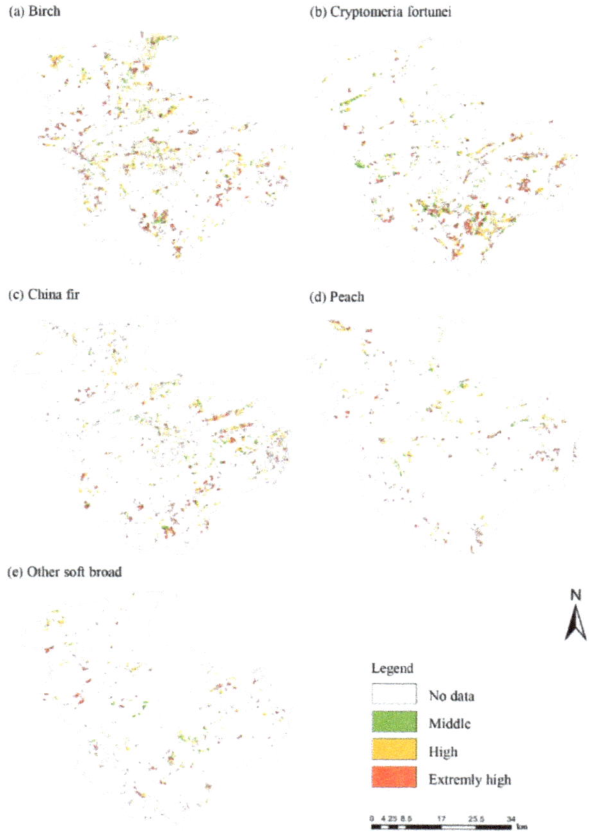

Figure 12. Spatial distribution map of landslide susceptibility of dominant tree species in Nayong County.

3.4.7. Dominant Tree Species and Landslide Susceptibility in Hezhang County

It can be seen from Table 4 and Figure 13 that the top-five dominant tree species in Hezhang County are Pinus armandi, Walnut, Pinus yunnanensis, Quercus fabri, and Chinese fir. The sums of the landslide-prone areas of the various tree species with high and extremely high grades are 14,786 hm^2, 12,422 hm^2, 7859 hm^2, 4163 hm^2, and 2223 hm^2, respectively, accounting for 32.6%, 61.4%, 55.5%, 34.7%, and 60% of the areas of these tree species. The dominant tree species, Pinus armandi, has the largest area, is evenly distributed, and is mainly found in the medium-susceptibility area (Figure 13a). Walnut is mainly distributed in the central area (Figure 13b); Pinus yunnanensis is mainly distributed in the northwest (Figure 13c); Quercus fabri is small in area but concentrated in the south (Figure 13d); Chinese fir is scattered (Figure 13e).

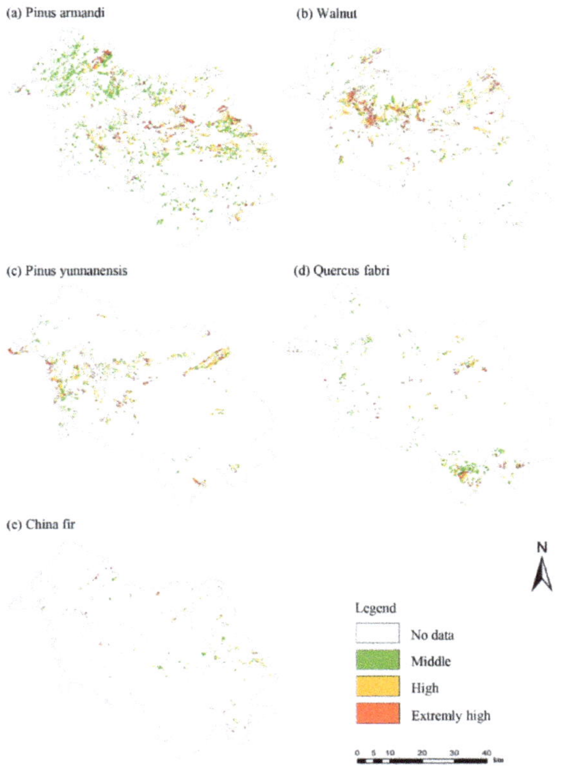

Figure 13. Spatial distribution map of landslide susceptibility of dominant tree species in Hezhang County.

The area of Walnut comes second, accounting for 10.2%, and the highest contributions to the susceptibility to landslides of high and extremely high grades are 0.63 and 0.68, respectively, indicating that its landslide susceptibility is the greatest. Pinus yunnanensis and Chinese fir contribute 0.57 and 0.56 to extremely high-grade landslide susceptibility, respectively, and contribute 0.62 and 0.60 to high-grade landslide susceptibility, indicating that their landslide protection capabilities are also poor. The contributions of Pinus armandi and Quercus fabri to the susceptibility to landslides of high and extremely high grades are lower than 0.5, indicating that their landslide susceptibility is low; at the same time, the area of Pinus armandi is 45,373 hm^2, and the proportion reaches 22.87%. In general, the influence of Walnut, Pinus yunnanensis, and Chinese fir on the susceptibility to landslides of high and extremely high grades is much greater than 0.5, and the possibility of landslide events is high.

3.4.8. Dominant Tree Species and Landslide Susceptibility in Weining County

It can be seen from Table 4 and Figure 14 that the top-five dominant tree species in Weining County's forests are Pinus yunnanensis, Pinus armandi, Sawtooth oak, Cyclobalanopsis glauca, and other soft broad tree species. The sums of the landslide-prone areas for the various tree species with high and extremely high grades are 24,153 hm^2, 6376 hm^2, 2194 hm^2, 3235 hm^2, and 1231 hm^2, accounting for 24.3%, 13.2%, 21%, 73.8%, and 33.4% of the areas of these tree species, respectively. Pinus yunnanensis is concentrated in the southwest region (Figure 14a); Pinus armandi is less distributed in the west (Figure 14b); Sawtooth oak (Figure 14c) and Cyclobalanopsis glauca (Figure 14d) are small in area, and both are concentrated in parts of the north; the distribution of other soft broad species is sporadic (Figure 14e).

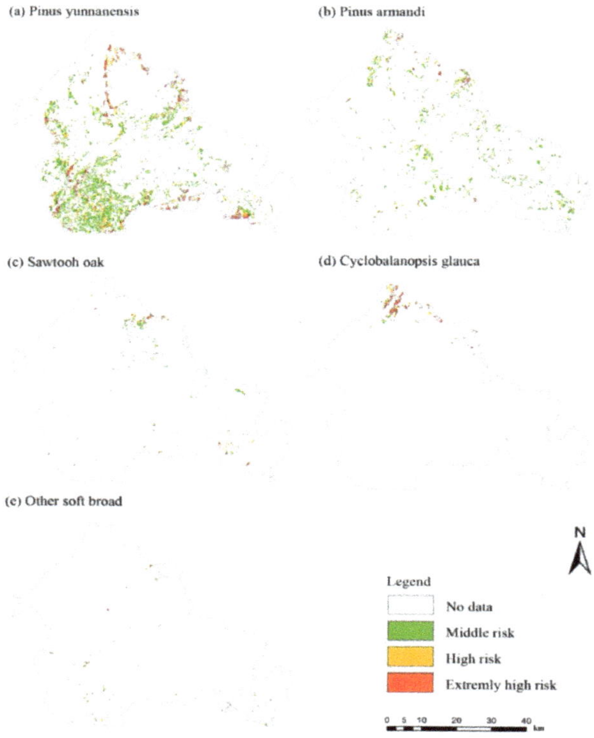

Figure 14. Spatial distribution map of landslide susceptibility of dominant tree species in Weining County.

Among the dominant tree species, the area of Cyclobalanopsis glauca is 4381 hm^2, accounting for 1.7%. Its contribution to extremely high-grade landslide susceptibility is 0.96, and it is the only tree species with a contribution close to 1 in all regions. It has the lowest protection ability against landslides and may even promote landslides. The contributions of other soft and broad tree species to the landslide susceptibility of high and extremely high grades are 0.56 and 0.57, indicating that the landslide protection ability is poor, but the tree species has the smallest area, 509 hm^2, and the lowest proportion is 1.43%. Pinus yunnanensis has a large area with a proportion of 38.55%, which is larger than the sum of the other four tree species, and contributes 0.5 to high-grade landslide susceptibility. Its landslide protection ability is not high but is better than Cyclobalanopsis glauca and Pinus yunnanensis. The contributions of Pinus armandi and Sawtooth oak to the landslide susceptibility of medium, high, and extremely high grades is low, and neither exceed 0.5 indicating that their landslide protection ability is the best and is worthy of promotion.

4. Discussion

At present, research on the impact of forests on landslide susceptibility mostly focuses on the forest land type, origin, and forest root system, and concentrates on a single influencing factor. Few systematic studies have been conducted on forests, and the specific impacts of forests have not been quantified. By constructing a contribution model and combining the landslide susceptibility data in Bijie, we have quantified the contributions of forest type, origin, age, and dominant tree species to landslide susceptibility for the first time and deeply analyzed the impact of forests on landslides. We found that different forest types, origins, ages, and dominant tree species have different landslide susceptibilities.

4.1. Forest Types

Woodland has a higher landslide susceptibility than shrubland. Bijie City has a large proportion of coniferous forests represented by Pinus armandi and Pinus yunnanensis in the woodland. These kinds of trees have short root systems, single-tree stand structures, unstable ecosystems, thick soil layers for the growth of the forest, rich material sources for landslides, high centers of gravity, and easy lodging under the influence of rainfall and wind. These factors make the woodland more prone to landslides, which is contrary to the findings of Sun et al. [46]. This is because the soil layer in the karst area is thin and shrubland sites have a thin soil layer, less landslide material sources, and a low shrub height. Compared with tall trees, shrubland has a lower center of gravity, is less affected by rainfall and wind, and is less prone to lodging. Moreover, according to Li et al. [47], the soil moisture content can affect the runoff and slope stability. The distribution of water content in soil is also affected by plant species and root systems. Shrub roots can increase the water content and porosity in the rhizosphere soil; fine and medium roots provide the main channels for the formation of soil runoff, enhancing soil porosity and infiltration. Most of the fine roots of shrubs are on the soil surface, which can provide more water channels, promote infiltration, reduce surface soil moisture, and make the soil shear strength higher, which is conducive to slope stability. Therefore, the landslide protection ability of shrubland is relatively strong.

4.2. Forest Origin

The landslide susceptibility for artificial forests is greater than that of natural forests. According to forestry survey data, artificial forests include shallow-root arbor forests such as Pinus armandi and Pinus yunnanensis, and economic forests such as Walnut. Compared with natural forests, the stand structure of artificial forests is relatively simple, more disturbed by human activity, and less stable. The economic benefits are higher, but the ecological benefits of artificial forests are lower than those of natural forests. Under severe meteorological conditions such as heavy rain, plantations face a greater susceptibility of landslides; natural forests have a stronger soil fixation effect, and higher soil porosity enables them to better absorb rainfall and cope with other landslide factors; this is confirmed by Wang et al. [48]. This is because the root biomass per unit of soil in natural forests is much larger than that in artificial forests, and a large number of roots will be lost during transplanting, with some roots dying after transplanting. This will cause rainwater to fail to infiltrate in time when it rains, increase the soil moisture, soften the soil, reduce the adhesion and friction of soil particles, reduce the shear strength of the soil, and increase the susceptibility of landslides. At the same time, the soil will be turned over during the transplanting process, breaking the natural soil crust, making the surface moisture increase faster during rainfall, and reducing the adhesion of the soil. According to research results for soil samples collected from plantations and natural forests by relevant scholars, the proportions of soil aggregates in natural forests and plantations were 43.22% and 23.01%, respectively, indicating that the influence of natural vegetation on the aggregation of the soil structure is greater than that of plantations [28]. Therefore, the landslide protection ability of a plantation forest is reduced. As a result, in the establishment of plantations, the stand structure should be optimized, mixed forests should be planted, the combination of arbor and shrub should be reasonable, and plantation and ecological restoration should be carried out on the principle of imitating natural plant communities.

4.3. Forest Age

The landslide protection ability of forests in different age groups showed a trend of increasing first and then decreasing. Young forests have lower comprehensive efficiency, lower plant canopy density, fewer root systems, and fewer soil and water conservation functions. Under the influence of heavy rainfall, landslides are likely to occur and the protection ability with respect to landslides is weak. The protection ability of medium-aged forests against landslides is strong. This is because with the change from young forests to medium-aged forests, the stability of the forest land increases, the root system begins to develop gradually, the soil adhesion is stronger, and landslides cannot easily occur. The contribution was less than 0.5, which is consistent with the research results of Šilhán et al. and Sati et al. [49,50]. Near-mature forests provide the strongest protection against landslides. At this time, plant roots are developed, leaves are dense, respiration and transpiration are strong, roots absorb water from soil quickly, and infiltration of soil surface water is also faster, all of which can effectively reduce soil surface moisture, increase the soil shear strength, and reduce the landslide susceptibility. When the forest grows into mature forest and over-mature forest, the center of gravity of tall trees with dense canopies is higher than for the ground. Against the background of heavy rainfall and other extreme weather conditions, mature and over-mature forests are prone to lodging, thereby uprooting, and the soil stability declines. The protection ability with respect to landslides is weaker than that of near-mature forests, so the contribution of mature and over-mature forests is greater than 0.5, and their ability to prevent landslides is poor.

4.4. Dominant Tree Species

In terms of the impact of dominant tree species, artificial forests such as economic forests, other soft broad tree species, and shallow-rooted coniferous forests have limited inhibitory effects on landslides. Compared with near-mature forests, the slope stability of mature forests and over-mature forests in shallow-rooted arbor forests such as Chinese fir is reduced, and the probability of landslides is high. The dominant tree species in Qixingguan and Dafang are Chinese fir, Cryptomeria fortunei, and other coniferous species. Thus, the forest structure should be optimized and deep-rooted shrubs should be planted in steep-slope and high landslide-prone areas to increase the slope stability. Walnut, as a characteristic economic forest planted on a large scale in the process of poverty alleviation, has high economic benefits. However, due to frequent human activity and loose soil created in the process of economic forest management, it is easily affected by heavy rainfall and is prone to landslides. The dominant tree species in Hezhang and Weining counties is mainly Walnut. When planning and developing Walnut economic forests, attention should be paid to avoiding areas that are prone to landslides. Other soft broad tree species are widely planted in Qianxi County, Zhijin County, Nayong County, and Jinsha County. Compared with coniferous forests such as Chinese fir and Cedarwood, other soft broad tree species contribute more in extremely high landslide-prone areas and have weaker landslide protection capabilities. Consistent with the results of Bordoni et al. [23], this is because the average tensile strength of the roots of broad-leaved tree species is weak, and the enhancement effect on the soil shear strength is relatively poor. In particular, the area of soft broad tree species in Jinsha County is 24,385 hm^2, a large proportion, covering 17.53% of the area. The extremely high-grade landslide susceptibility contributes the most, and landslides occur frequently, which should arouse the attention of the forestry department. It is notable that Cyclobalanopsis glauca trees are widely planted in Weining County, and most of them are distributed in mountainous areas with more gravel. The soil erosion is serious and, coupled with the steep terrain and abundant sources of weathered materials and other factors, the susceptibility of landslides is extremely high, which should arouse great attention from the forestry department.

4.5. Forest Management Strategy

In forestry planning and design and afforestation, landslide factors should be comprehensively considered, and the potential for landslide protection in the process of forest management should be brought into play. It is necessary to strengthen the protection and restoration of natural forests, gradually restore natural forests with high slopes and broken soils in areas prone to landslides, enhance the ability of natural forests to store water and soil, control soil erosion, reduce vegetation damage, protect the stability of the original ecosystem, and improve the landslide protection ability. This can be carried out by optimizing the plantation stand structure and following the principle of imitating natural plant communities in the process of afforestation, planting mixed forests, paying attention to the combination of arbor and shrub, and improving the diversity of plantation stands. For existing pure plantations, this should be based on the dominant tree species in different regions, combined with the distribution of landslide susceptibility. The forest stand structure should be transformed, plants that can stabilize the slope and prevent landslides should be added, the soil's shear resistance should be improved, and ecosystem stability should be enhanced. When planting economic forests, it is necessary to avoid areas with high landslide susceptibility and pay attention to soil and water conservation.

5. Conclusions

We investigated the contributions of the woodland type, forest origin, forest age group, and dominant tree species to landslide susceptibility, and the results show that forests play a role in regulating landslide susceptibility. In general, natural forests have significantly stronger landslide protection capabilities than artificial forests, shrubland has stronger landslide protection capabilities than woodland, near-mature forests have stronger landslide protection capabilities than young forests and mature forests, and the dominant tree species in different regions have different effects on landslides. Among the dominant tree species in each county, the shallow-rooted coniferous forests represented by Chinese fir and Pinus yunnanensis, as well as the economic forests represented by Walnut, Peach, and Osmanthus, have a single stand structure, and most of them are plantations and young forests, which are highly susceptible to landslides. In forest management, it is necessary to optimize the stand structure according to the distribution of landslide susceptibility and take into account the organic unity of economic, social, and ecological benefits.

Author Contributions: All authors contributed to the study conception and design. Material preparation, data collection and analysis were performed by Y.Z., C.S., S.Z. and X.L. The first draft of the manuscript was written by Y.Z. and all authors commented on previous versions of the manuscript. All authors have read and agreed to the published version of the manuscript.

Funding: This research work was supported by the Outstanding Youth Science and Technology program of Guizhou Province of China ([2021]5615).

Institutional Review Board Statement: Not applicable.

Informed Consent Statement: Not applicable.

Data Availability Statement: Not applicable.

Acknowledgments: Thanks to Guizhou Provincial Mountain Geological Hazard Prevention Engineering Technology Research Center and Forestry Bureau of Bijie City for providing us with the data, which are important to our research.

Conflicts of Interest: We declare that this manuscript entitled "Analysis of the Influence of Forests on Landslides in the Bijie Area of Guizhou" is original, has not been published before and is not currently being considered for publication elsewhere. We confirm that the manuscript has been read and approved by all named authors and that there are no other persons who satisfied the criteria for authorship but are not listed. We further confirm that the order of authors listed in the manuscript has been approved by all of us. We understand that the Corresponding Author is the sole contact for the Editorial process. He is responsible for communicating with the other authors about progress, submissions of revisions and final approval of proofs.

References

1. Li, Z.; Zhang, F.; Gu, W.; Dong, M.L. The Niushou landslide in Nanjing City, Jiangsu Province of China: A slow-moving landslide triggered by rainfall. *Landslides* **2020**, *17*, 2603–2617. [CrossRef]
2. Stanley, T.; Kirschbaum, D.B. A heuristic approach to global landslide susceptibility mapping. *Nat. Hazards* **2017**, *87*, 145–164. [CrossRef] [PubMed]

3. Salvati, P.; Petrucci, O.; Rossi, M.; Bianchi, C.; Pasqua, A.A.; Guzzetti, F. Gender, age and circumstances analysis of flood and landslide fatalities in Italy. *Total Environ.* **2018**, *610–611*, 867–879. [CrossRef] [PubMed]
4. Hu, X.S.; Brierley, G.; Zhu, H.L.; Li, G.R.; Fu, J.T.; Mao, X.Q.; Yu, Q.Q.; Qiao, N. An exploratory analysis of vegetation strategies to reduce shallow landslide activity on loess hillslopes, Northeast Qinghai-Tibet Plateau, China. *J. Mt. Sci.* **2013**, *10*, 668–686. [CrossRef]
5. China Government Network. National Geological Disasters in 2021 [EB/OL](2022-01-12) (2022-01-13). Available online: http://www.mnr.gov.cn//dt/ywbb/202201/t20220113_2717375.html (accessed on 13 January 2022).
6. Ghestem, M.; Sidle, R.C.; Stokes, A. The influence of plant root systems on subsurface flow. Implications for slope stability. *Bioscience* **2011**, *61*, 86961. [CrossRef]
7. Depicker, A.; Jacobs, L.; Mboga, N.; Smets, B.; Rompaey, A.V.; Lennert, M.; Wolff, E.; Kervyn, F.; Michellier, C.; Dewitte, O.; et al. Historical dynamics of landslide risk from population and forest-cover changes in the Kivu Rift. *Nat. Sustain.* **2021**, *4*, 965–974. [CrossRef]
8. Grima, N.; Edwards, D.; Edwards, F.; Petley, D.; Fisher, B. Landslides in the Andes: Forests can provide cost-effective landslide regulation services. *Sci. Total Environ.* **2020**, *745*, 141128. [CrossRef]
9. Kim, H.G.; Park, C.Y. Landslide susceptibility analysis of photovoltaic power stations in Gangwon-do, Republic of Korea. *Geomat. Nat. Hazards Risk* **2021**, *12*, 2328–2351. [CrossRef]
10. Schmaltz, E.M.; Steger, S.; Glade, T. The influence of forest cover on landslide occurrence explored with spatio-temporal information. *Geomorphology* **2017**, *290*, 250–264. [CrossRef]
11. Griffiths, J.W.; Lukens, C.E.; May, R. Increased forest cover and limits on clear-felling could substantially reduce landslide occurrence in Tasman, New Zealand. *N. Z. J. For. Sci.* **2020**, *50*. [CrossRef]
12. Hwang, T.; Kangw, S.; Kim, J.; Kim, Y.; Lee, D.; Band, L. Evaluating drought effect on MODIS Gross Primary Production (GPP) with an eco-hydrological model in the mountainous forest, East Asia. *Glob. Chang. Biol.* **2008**, *14*, 1037–1056. [CrossRef]
13. Dorren, L.; Schwarz, M. Quantifying the Stabilizing Effect of Forests on Shallow Landslide-Prone Slopes. In *Ecosystem-Based Disaster Risk Reduction and Adaptation in Practice*; Springer: Berlin/Heidelberg, Germany, 2016; pp. 255–270.
14. Huang, G.; Zheng, M.; Peng, J. Effect of Vegetation Roots on the Threshold of Slope Instability Induced by Rainfall and Runoff. *Geofluids* **2021**, *2021*, 6682113. [CrossRef]
15. Li, P.; Xiao, X.; Wu, L.; Li, X.; Zhang, H.; Zhou, J. Study on the Shear Strength of Root-Soil Composite and Root Reinforcement Mechanism. *Forests* **2022**, *13*, 898. [CrossRef]
16. Peduzzi, P. Landslides and vegetation cover in the 2005 North Pakistan earthquake: A GIS and statistical quantitative approach. *Nat. Hazards Earth Syst. Sci.* **2010**, *10*, 623–640. [CrossRef]
17. Jamal, M.; Mandal, S. Monitoring forest dynamics and landslide susceptibility in Mechi–Balason interfluves of Darjiling Himalaya, West Bengal using forest canopy density model (FCDM) and Landslide Susceptibility Index model (LSIM). *Model. Earth Syst. Environ.* **2016**, *2*, 1–17. [CrossRef]
18. Davoudi, M.H.; Aghda, S.M.F.; Pour, G.R.S.A. Landslide stabilization by tree root reinforcement. *WIT Trans. Ecol. Environ.* **2004**, *75*, 39–45.
19. Dias, A.S.; Pirone, M.; Urciuoli, G. Review on the methods for evaluation of root reinforcement in shallow landslides. In *Advancing Culture of Living with Landslides, Vol 2: Advances in Landslide Science*; Springer: Berlin/Heidelberg, Germany, 2017; pp. 641–648.
20. Moos, C.; Bebi, P.; Graf, F.; Mattli, J.; Rickli, C.; Schwarz, M. How does forest structure affect root reinforcement and susceptibility to shallow landslides? *Earth Surf. Process. Landf.* **2016**, *41*, 951–960. [CrossRef]
21. Kobayashi, Y.; Mori, A.S. The potential role of tree diversity in reducing shallow landslide risk. *Environ. Manag.* **2017**, *59*, 807–815. [CrossRef]
22. Fan, C.C.; Su, C.F. Role of roots in the shear strength of root-reinforced soils with high moisture content. *Ecol. Eng.* **2008**, *33*, 157–166. [CrossRef]
23. Tan, H.; Chen, F.; Chen, J.; Gao, Y.F. Direct shear tests of shear strength of soils reinforced by geomats and plant roots. *Geotext. Geomembr.* **2019**, *47*, 780–791. [CrossRef]
24. Bordoni, M.; Cislaghi, A.; Vercesi, A.; Bischetti, G.B.; Meisina, C. Effects of plant roots on soil shear strength and shallow landslide proneness in an area of northern Italian Apennines. *Bull. Eng. Geol. Environ.* **2020**, *79*, 3361–3381. [CrossRef]
25. Montgomery, D.; Schmidt, K.; Greenberg, H.; Dietrich, W. Forest clearing and regional landsliding. *Geology* **2000**, *28*, 311–314. [CrossRef]
26. Tosi, M. Root tensile strength relationships and their slope stability implications of three shrub species in the Northern Apennines (Italy). *Geomorphology* **2007**, *87*, 268–283. [CrossRef]
27. Facelli, J.M.; Temby, A.M. Multiple effects of shrubs on annual plant communities in arid lands of South Australia. *Austral. Ecol.* **2002**, *27*, 422–432. [CrossRef]
28. Wang, X.; Huang, Z.; Hong, M.M.M.; Zhao, Y.F.; Ou, Y.S.; Zhang, J. A comparison of the effects of natural vegetation regrowth with a plantation scheme on soil structure in a geological hazard-prone region. *Eur. J. Soil Sci.* **2019**, *70*, 674–685. [CrossRef]
29. Zhang, G.P.; Xu, J.; Bi, B.G. Relationship between landslide and debris flow hazards and environmental factors. *Chin. J. Appl. Ecol.* **2009**, *20*, 653–658.
30. Iqbal, J.; Tu, X.; Xu, L. Landslide hazards in reservoir areas: Case study of Xiangjiaba reservoir, Southwest China. *Nat. Hazards Rev.* **2017**, *18*, 04017009. [CrossRef]

31. Yang, D.; Qiu, H.; Zhu, Y.; Liu, Z.; Pei, Y.; Ma, S.; Du, C.; Sun, H.; Liu, Y.; Cao, M. Landslide Characteristics and Evolution: What We Can Learn from Three Adjacent Landslides. *Remote Sens.* **2021**, *13*, 4579. [CrossRef]
32. Hua, F.Y.; Wang, L.; Fisher, B.; Zheng, X.L.; Wang, X.Y.; Yu, D.W.; Tang, Y.; Zhu, J.G.; Wilcove, D.S. Tree plantations displacing native forests: The nature and drivers of apparent forest recovery on former croplands in Southwestern China from 2000 to 2015. *Biol. Conserv.* **2018**, *222*, 113–124. [CrossRef]
33. Wang, Q.; Yang, R. Study on REEs as tracers for late permian coal measures in Bijie City, Guizhou Province, China. *J. Rare Earths* **2008**, *26*, 121–126. [CrossRef]
34. Wang, M.; Zhang, L.; He, Y.; Liu, L.; Chen, D.; Shan, A.; Feng, Y.; Yang, X. Soil fungal communities affect the chemical quality of flue-cured tobacco leaves in Bijie, Southwest China. *Sci. Rep.* **2022**, *12*, 2815. [CrossRef] [PubMed]
35. Li, T. *Research on Earthquake Disaster Risk Ssment Model and Its Application in Bijie City, Guizhou Province*; Kunming University of Science and Technology: Kunming, China, 2015.
36. Li, L.; Huang, M.; Liu, Z.J.; Xu, Q.Y.; Tong, X.J.; Lu, P.L. Landslide hazard risk assessment in Bijie area based on RS and GIS. *J. Nat. Disasters* **2011**, *20*, 177–182.
37. Varnes, D.J. Slope movement types and processes. *Spec. Rep.* **1978**, *176*, 11–33.
38. Cruden, D.M.; Varnes, D.J. Landslide types and processes. In *Landslides-Investigation and Mitigation*; Turner, K.A., Schuster, R.L., Eds.; National Research Council Transportation Research Board, Special Report; National Research Council: Washington, DC, USA, 1996; Volume 247, pp. 36–75.
39. Xie, X.N.; Li, Y.C.; Liu, Y. *Summary Report on Professional Investigation of Hidden Hidden Dangers of High-Level Hidden Geological Hazards in Guizhou Province*; Guizhou Provincial Institute of Geological Environment Monitoring: Guiyang, China, 2019.
40. Yao, K.; Yang, S.; Wu, S.; Tong, B. Landslide Susceptibility Assessment Considering Spatial Agglomeration and Dispersion Characteristics: A Case Study of Bijie City in Guizhou Province, China. *ISPRS Int. J. Geo-Inf.* **2022**, *11*, 269. [CrossRef]
41. Shen, C.Y.; Feng, Z.K.; Xie, C.; Fang, H.R.; Zhao, B.B.; Ou, W.H.; Zhu, Y.; Wang, K.; Li, H.W.; Bai, H.L.; et al. Refinement of Landslide Susceptibility Map Using Persistent Scatterer Interferometry in Areas of Intense Mining Activities in the Karst Region of Southwest China. *Remote Sens.* **2019**, *11*, 2821. [CrossRef]
42. De Baets, S.; Poesen, J.; Reubens, B.; Baerdemaeker, J.D.; Muys, B. Root tensile strength and root distribution of typical Mediterranean plant species and their contribution to soil shear strength. *Plant Soil* **2008**, *305*, 207–226. [CrossRef]
43. Lee, J.T.; Chu, M.Y.; Lin, Y.S.; Kung, K.N.; Lin, W.C.; Lee, M.J. Root traits and biomechanical properties of three tropical pioneer tree species for forest restoration in landslide areas. *Forests* **2020**, *11*, 179. [CrossRef]
44. Yu, K.Y.; Liu, J.; Yang, P.; Yao, X. Research on the relationship between forest management process and regional landslides. *J. Cent. South Univ. For. Technol.* **2016**, *36*, 5–10, 16.
45. Kim, D.; Im, S.; Lee, C.; Woo, C. Modeling the contribution of trees to shallow landslide development in a steep, forested watershed. *Ecol. Eng.* **2013**, *61*, 658–668. [CrossRef]
46. Fusun, S.; Jinniu, W.; Tao, L.; Yan, W.; Haixia, G.; Ning, W. Effects of different types of vegetation recovery on runoff and soil erosion on a Wenchuan earthquake-triggered landslide, China. *J. Soil Water Conserv.* **2013**, *68*, 138–145. [CrossRef]
47. Li, J.; Wang, X.; Jia, H.X.; Liu, Y.; Zhao, Y.F.; Shi, C.M. Assessing the soil moisture effects of planted vegetation on slope stability in shallow landslide-prone areas. *Soils Sediments* **2021**, *21*, 2551–2565. [CrossRef]
48. Phillips, C.; Hales, T.; Smith, H.; Basher, L. Shallow landslides and vegetation at the catchment scale: A perspective. *Ecol. Eng.* **2021**, *173*, 106436. [CrossRef]
49. Šilhán, K. A new tree-ring-based index for the expression of spatial landslide activity and the assessment of landslide hazards. *Geomat. Nat. Hazards Risk* **2021**, *12*, 3409–3428. [CrossRef]
50. Sati, S.P.; Sundiyal, Y.P. Role of some tree species in slope instability. *Himal. Geol.* **2007**, *28*, 75–78.

forests

MDPI

Article

Research on a Regional Landslide Early-Warning Model Based on Machine Learning—A Case Study of Fujian Province, China

Yanhui Liu [1,*], Junbao Huang [2], Ruihua Xiao [1], Shiwei Ma [3] and Pinggen Zhou [1]

1 China Institute of Geo-Environment Monitoring (Technical Guidance Center for Geo-Hazards Prevention of MNR), Beijing 100081, China
2 Fujian Monitoring Center of Geological Environment, Fuzhou 350002, China
3 Institute of Geology and Geophysics, Chinese Academy of Sciences, Beijing 100029, China
* Correspondence: lyanhui@mail.cgs.gov.cn

Abstract: China's landslide disasters are serious, and regional landslide disaster early-warning is one of the important means of disaster prevention and mitigation. The traditional regional landslide disaster early-warning model, however, is limited by the complex landslide induction mechanism, limited data accumulation, and insufficient big data analysis methods, and has problems such as limited early-warning accuracy and insufficient refinement. In this paper, a machine learning method was introduced into the field of regional landslide disaster warning. From the model construction process of training sample-set construction, sample learning and training, model parameter optimization, model preservation, warning output, and so on, a method for constructing a regional landslide early-warning model based on machine learning was systematically proposed. In the sample learning and training, 80% of the training sample-set was used as the training set, and 20% was used as the test set for five-fold cross validation. The Bayesian Optimization algorithm was used to optimize the model parameters, and the accuracy, ROC curve, and AUC value were used to verify the model accuracy and model generalization ability. With China's Fujian province as an example, based on nine years of geological and meteorological data (2010–2018), geological environment factors, factors of hazard-affected bodies and historical disaster situations, and rainfall-induced factors in four categories, a total of 26 indicators were used as input characteristic parameters. Six machine learning algorithms were adopted to improve model training; the results showed that the Random Forest algorithm performed the best, giving an accuracy of 92.3%, and was the model with the best generalization ability (AUC was 0.955). The second best was the Artificial Neural Network model, with an accuracy of 0.937 and an AUC of 0.935. Next were the Nearest Neighbor model, the Logistic Regression model, and the Support Vector Machine; the poorest results were from the Decision Tree model. Finally, the typical rainfall-type landslide disaster process in Fujian Province was selected as an example to verify the Random Forest algorithm model. The results showed that compared with the early-warning results of the original explicit statistical model, the hit rate of the new model was 6 times, or equal to that of the original model, and the landslide density in the early-warning area of the new model was 1.6–1.7 times that of the original model. Preliminary verification showed that the new model based on the Random Forest method has obvious advantages, a higher hit rate and a smaller warning area, and can achieve more accurate warnings. The follow-up will continue to track the new landslide disaster situation in the study area and carry out model verification and correction.

Keywords: landslide; early-warning model; machine learning; Random Forest; model study

Citation: Liu, Y.; Huang, J.; Xiao, R.; Ma, S.; Zhou, P. Research on a Regional Landslide Early-Warning Model Based on Machine Learning—A Case Study of Fujian Province, China. *Forests* **2022**, *13*, 2182. https://doi.org/10.3390/f13122182

Academic Editors: Haijia Wen, Weile Li, Chong Xu, Hiromu Daimaru and Víctor Resco de Dios

Received: 15 October 2022
Accepted: 30 November 2022
Published: 19 December 2022

1. Introduction

China is one of the countries with the most widespread and serious geological disasters in the world. Geological disasters spread throughout the country's mountainous and highland areas, with nearly one million known locations, causing hundreds of deaths and billions of yuan of direct economic losses every year (according to the National Geological

Disaster Bulletin issued by the Ministry of Natural Resources). More than 20 countries or regions in the world, including Hong Kong, China, the United States, Italy, Brazil, Japan, etc., have also performed or are carrying out regional geological disaster early-warning and mitigation services to varying degrees [1]. Since 2003, the Chinese mainland has carried out a meteorological early-warning for regional geological disasters and achieved good results in disaster prevention and mitigation [2–10]. Additionally, in the monitoring and early-warning demonstration area [11,12], the Three Gorges Reservoir Area [2], Wenchuan earthquake disaster area [13], and other key regions, extensive research and practice have also been carried out. According to statistics, since 2003, owing to the efforts of various parties, the number of deaths and missing caused by geological disasters has decreased from about 1000 per year during the Tenth Five-Year Plan period to about 500 since the Twelfth Five-Year Plan period, which indicates that important contributions have been made to meteorological early-warning and prediction of geological disasters [3,4].

The model study is the basic scientific problem of regional landslide early-warning. A large number of scholars have carried out long-term and unremitting research on it. The first and most widely used model is the statistics-based critical rainfall threshold model, which was systematically studied in Hong Kong, China, and the United States [14,15], and has been widely used as a reference in other countries or regions due to its advantages of simplicity and ease of generalization [16–19]. Based on the statistical principle, the explicit statistical early-warning model proposed by Liu Chuanzheng et al. [2] has been deeply explored and applied at all levels of early-warning, key research areas, and monitoring and early-warning areas in mainland China [4–7,9,11,12]. A regional dynamic early-warning model based on the mechanism process analysis of rainfall–seepage and disaster occurrence has been continuously studied. By coupling slope stability analysis with a hydrogeological model, the critical rainfall index for landslide initiation has been determined [20–23], and the physical significance of the dynamic early-warning model is clear, but due to the complex parameter input and uncertainty in the model, this method is mostly used in the small-scale research process, and its practical operation is also limited.

In recent years, with the rapid development of artificial intelligence technology, machine learning and deep learning algorithms based on artificial intelligence have become increasingly mature and widely used in various industries. In the field of geological disaster prevention and control, a variety of machine learning algorithms have been widely used in landslide spatial evaluation and prediction in recent years, such as Artificial Neural Networks, Decision Trees, Support Vector Machines, Random Forests, etc. [24–29]. Most of the above-mentioned studies introduce machine learning algorithms into landslide spatial evaluation and prediction to evaluate regional landslide sensitivity or susceptibility. After the spatial evaluation is completed, the critical precipitation threshold is determined by traditional statistical methods [30]. However, there are few related achievements of realizing spatial and temporal warning for regional landslide disasters based directly on a machine learning algorithm.

Aiming at the problems that exist in the traditional regional landslide early-warning models, such as complex landslide-inducing mechanisms, limited data accumulation, and insufficient data analysis methods, past results lead to limited warning accuracy and a lack of indicator precision. Through the training sample-set construction, comparative analysis of various machine learning methods could improve the precision of early-warning models and other aspects of the research to solve these problems. This paper systematically expounded on the construction method for a regional landslide disaster early-warning model based on crucial components of machine learning algorithms: training sample-set construction, model training, optimization evaluation, and early-warning modeling. Taking Fujian Province of China as an example, the Random Forest algorithm and Nearest Neighbor algorithm were applied, and Support Vector Machine, Logistic Regression, Decision Tree, Artificial Neural Network, and other algorithms, based on the geological and meteorological data from 2010 to 2018, were used to construct a regional landslide disaster early-warning model for Fujian Province. We selected two typical rainfall-induced landslide disasters in Fujian Province in 2019. Taking the process as an example, we carried out the live operation and application verification of the early-warning model.

2. Geological Background of the Study Area

Fujian Province is located in a mountainous highland area along the southeast coast of mainland China. It is one of the provinces with complex terrain, abundant rainfall, and frequent landslide disasters. At the present, the forest coverage rate of Fujian Province rate has reached 66.8 percent, ranking first in China for 43 consecutive years. The highlands and mountainous areas of the study area account for about 80% of the total area, mainly in the central and western regions of Fujian Province. The terrain of Fujian Province is generally high in the northwest and low in the southeast. Under the control of the New Cathaysia tectonic system, the western Min-Dashan belt and the middle Min-Dashan belt are formed in the west and central part of Fujian Province, and there are disconnecting valleys and basins between the two belts. There are many river systems in Fujian Province, and their flow direction is mostly from west to east. The rivers are mostly mountainous rivers with abundant water volume, great seasonal changes, and rapid flow. The province has a subtropical humid monsoon climate, with an average annual rainfall of 1000~1900 mm, abundant rainfall, strong monsoon circulation, and a remarkable monsoon climate. Therefore, Fujian is an area prone to natural disasters, frequently-occurring areas, and worst-hit areas, especially landslide disasters are most widely distributed (Figure 1). The residual slope soil layer in the study area is widely developed. The slope gradient is generally 0~30°, locally greater than 30°. Due to the poor geological environment conditions, most of the villages in the mountainous area were built on the slopes, forming a large number of high and steep slopes in front of or behind the houses. During the flood season, sudden geological disasters such as collapses and landslides occur frequently. Although the scale of landslides is small, most of them occur in front of or behind residents' houses, which can easily cause casualties and property losses. The occurrence of landslide disasters in Fujian Province is closely related to the terrain and induced by heavy rainfall and human engineering activities. The disaster-vulnerable terrain is prone to occur on gradients with a slope of more than 20°; generally, the hazards are liable to happen during the heavy rainfall period from May to June and the typhoon and rainy period from July to September, and they usually take place in front of and behind houses sections such as the cutting slopes, planting economic forests, and spoiling soil along the slopes [31].

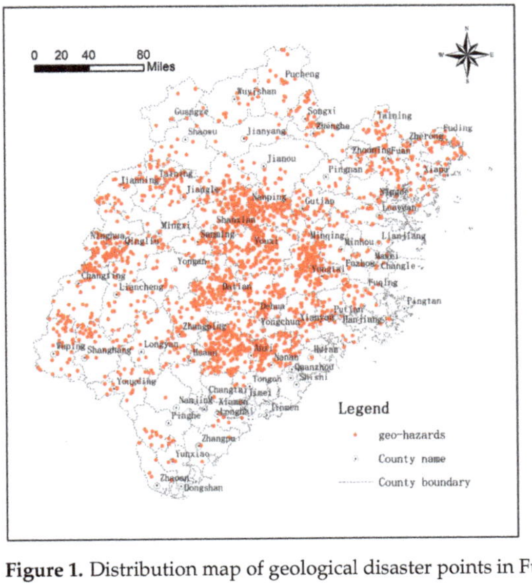

Figure 1. Distribution map of geological disaster points in Fujian Province.

3. Determination of Evaluation Index and Selection of Algorithm

3.1. Extraction of Evaluation Indexes

Landslide disasters are influenced and affected by the combination of topography, geological conditions, environmental conditions, and human engineering. Analyzing influence factors of landslide disasters is the basis for carrying out mechanism research, early warning and forecasting, and disaster prevention and mitigation. According to the existing survey and monitoring data in the study area, as well as the development and distribution of regional landslide disasters and the analysis of influencing factors [31–33], the input features of influencing factors are extracted from four aspects: geological environment factors, hazard-affected body factors, historical disasters, and rainfall-induced factors. Geological environment factors are the geological environment background factors of landslide disasters in the study area, which determine the potential susceptibility degree of landslide disaster in this area. Hazard-affected body factors are the evaluation indexes of the hazard-affected body that may be caused when a landslide occurs. Historical disasters refer to the number of historical landslide disasters in the study area. Rainfall-induced factors are the direct rainfall-inducing factors of landslides in the forecast period. These four categories cover all the geological environment aspects that influence the prediction of the possibility and risk of landslides disaster in this study area.

The extracted geological environment factors mainly include grade, geomorphic type, stratigraphic lithology, annual rainfall, vegetation type, water system influence, etc. Based on the 1:200,000 and 1:500,000 geological environment and geological disaster survey database of Fujian Province, the hourly precipitation data of Fujian Province from 2010 to 2018 (nearly 2000 stations), and the grid precipitation Real-time (QPE) data of Fujian Province in 2021 (grid scale: 5 km $\times$ 5 km), the correlation analysis between these six geo-environmental factors evaluation indexes and the spatial distribution of landslide hazards in the region was carried out. According to the results of correlation analysis, the grade factors were classified into five categories: 0~15°, 15~25°, 25~35°, 35~50°, $\geq$50° (Figure 2a); the geomorphic types were classified into five categories: plains, hills, low mountains, medium mountains, and high mountains (Figure 2b). The lithological factors of the formation were classified into a massive hard granite rock group, hard-harder diorite rock group, massive hard-harder tuff lava rock group, medium-thick layer, and relatively hard sandstone rock group, thin layer soft mudstone, shale rock group, medium-thick hard quartz gneiss rock group, medium-thick hard carbonate rock group, and loose sandy clay soil layer (Figure 2c). The annual average rainfall types were classified into 13 categories such as 1400–1450 mm, 1450–1500 mm, and so on, up to >2000 mm (Figure 2d). Vegetation affects landslides through coverage, density, abundance, height, underground biomass, leaf area index, and aboveground biomass. According to the correlation between the spatial distribution of landslides and the distribution of vegetation types, the vegetation types were classified into six types (Figure 2e); the water system was classified into two categories based on whether the impact distance is greater than 500 meters (Figure 2f). Detailed information is shown in Table 1.

The main indicators of hazard-affected body factors extracted are roads, houses, and population density. The road distribution layer was extracted from 1:250,000 DLG data and the distance from the road 0~500 m and $\geq$500 m were classified into two categories and quantified separately (Figure 3a). The house distribution layer was extracted from 1:250,000 DLG data, and the distances from the house 0~500 m and $\geq$500 m were classified into two categories and quantified separately (Figure 3b). According to the data of the sixth National Population Census, the population density (unit: pieces /km^2) types were classified into seven categories, 50–100, 100–150, 150–300, 300–450, 450–600, 600–750, >750, and quantified, respectively (Figure 3c). The specific content is shown in Table 1.

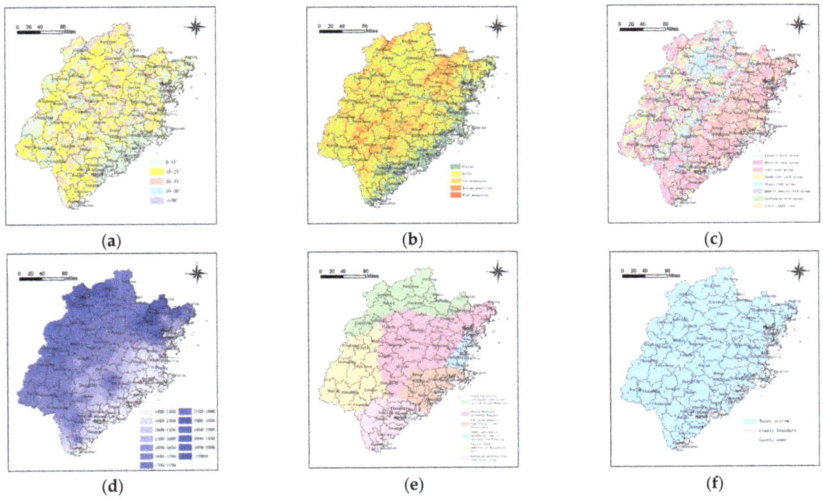

Figure 2. Geological environment factor index of the study area: (**a**) grade, (**b**) geomorphic type, (**c**) formation lithology, (**d**) annual rainfall, (**e**) vegetation type, (**f**) water system.

Table 1. Input features and parameters of the training sample set.

Category	Serial Number	The Input Features	Characteristic Parameters	Data Sources and Processing Methods
Geological environment factors	1	Grade/(°)	① 0~15; ② 15~25; ③ 25~35; ④ 35~50; ⑤ ≥50	1:100,000 grade map of Fujian Province, classification and quantification
	2	Geomorphic type	① plains; ② hills; ③ low mountains; ④ medium mountains; ⑤ high mountains	1:200,000 geomorphic type map of Fujian Province, classification and quantification
	3	Stratigraphic lithology	① massive hard granite rock group; ② hard-harder diorite rock group; ③ massive hard-harder tuff, tuff lava rock group; ④ medium-thick layer and relatively hard sandstone rock group; ⑤ Thin layer soft mudstone, shale rock group; ⑥ medium-thick hard quartz gneiss rock group; ⑦ medium-thick hard carbonate rock group; ⑧ loose sandy clay soil layer	1:200,000 stratigraphic lithology map of Fujian Province, classified and quantified
	4	Annual rainfall/(mm)	① 1400–1450; ② 1450–1500; ③ 1500–1550; ④ 1550–1600; ⑤ 1600–1650; ⑥ 1650–1700; ⑦ 1700–1750; ⑧ 1750–1800; ⑨ 1800–1850; ⑩ 1850–1900; ⑪ 1900–1950; ⑫ 1950–2000; ⑬ >2000	1:500,000 geological disaster survey and zoning reports in Fujian Province, vectorization acquisition, classification, and quantification
	5	Vegetation type	① South subtropical rainforest area in the east of Daiyun Mountain; ② Daiyun Mountain Yijiufeng Mountain Range; ③ Evergreen mulberry-semi-evergreen oak forest area; ④ South subtropical rainforest area in southeastern Pingling; ⑤ Coastal South Subtropical Rainforest Area; ⑥ Evergreen mulberry tree leaf forest area	1:500,000 vegetation type map of Fujian Province, classification and quantification
	6	Water system influence/(m)	① 0~500; ② ≥500	1:500,000 water system distribution map of Fujian Province, calculation buffer classification quantification

Table 1. *Cont.*

Category	Serial Number	The Input Features	Characteristic Parameters	Data Sources and Processing Methods
Hazard-affected body factors	7	The distance from road/(m)	① 0~500; ② ≥500	The road distribution layer was extracted from 1:250,000 DLG data, and the buffer classification quantization was calculated
	8	The distance from house/(m)	① 0~500; ② ≥500	The house distribution layer was extracted from 1:250,000 DLG data, and the buffer classification and quantification were calculated
	9	Density of population/(pieces/km^2)	① 50–100; ② 100–150; ③ 150–300; ④ 300–450; ⑤ 450–600; ⑥ 600–750; ⑦ >750	The sixth national population census data, classification and quantification
Historical disaster factors	10	Historical disaster/(pieces)	the number of historical damage points of each grid cell/10	Potential points of landslides in 1:500,000 geological disaster survey data in the study area, National Geological Disaster Database (2010–2018), with a scaled range
Rainfall-inducing factors	11	Rainfall of that day/(mm)	actual rainfall value/10	The data of meteorological and water conservancy hourly precipitation stations from 2010 to 2018 were interpolated and the range was scaled
	12	Rainfall in the previous day/(mm)	actual rainfall value/10	
	13	Rainfall in the last two days/(mm)	actual rainfall value/10	
	…	…	actual rainfall value/10	
	26	Rainfall in the last 15 days/(mm)	actual rainfall value/10	

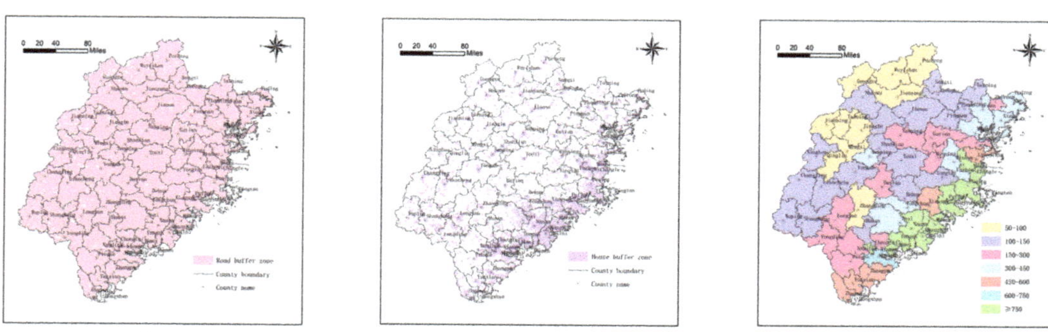

(a) (b) (c)

Figure 3. Hazard-affected body factors: (**a**) distance from the road, (**b**) distance from the house, (**c**) population density.

The historical landslides samples were extracted from the 1:500,000 geological disaster survey data of the study area and the national geological disaster database (2010–2008), and the scope was scaled by ten times of reduction (Figure 1).

The extracted index of rainfall-inducing factors mainly considers the rainfall of the current day and the daily rainfall of the previous 15 days. According to the data of meteorological and water conservancy hourly precipitation stations from 2010 to 2018, the index of rainfall-inducing factors of each grid cell were extracted, and the range was realized by reducing it ten times.

Through the above analysis, the twenty-six input characteristics of four categories of input characteristic factor indexes (geological environment factor, hazard-affected body factor, historical disaster situation, and rainfall-inducing factor) in Fujian Province were obtained. Among them, nine input characteristics of two categories of geological environment factors and hazard-affected body factors were classified and quantified. Data range scaling was carried out for seventeen input features of two categories: historical disaster conditions and precipitation-inducing factors, as shown in Table 1.

3.2. Selection of Machine Learning Algorithms

Commonly used machine learning algorithms mainly solve three major problems of regression, clustering, and classification. The prediction data of the regression algorithm is continuous numerical data, mainly including linear regression, logistic regression, etc., in which logistic regression can also solve the classification problem. The prediction data of the clustering algorithm is categorical, and the category is unknown, which is mainly a K-means clustering algorithm. The prediction data of the classification algorithm is classified data, and the category is known. The main algorithms include the Nearest Neighbor algorithm, Support Vector Machine, Artificial Neural Network, Logistic Regression, Decision Tree, and Random Forest. For regional landslide disaster warning problems, the machine learning classification algorithm is mainly used. In this paper, six algorithms, including the Random Forest algorithm, Nearest Neighbor algorithm, Support Vector Machine, Logistic Regression, Decision Tree, and Artificial Neural Network were selected to establish a landslide early-warning model. The comparative analysis of algorithms is shown in Table 2.

Table 2. Comparison and analysis of commonly used machine learning classification algorithms.

Commonly Algorithm	Principle	Advantages	Disadvantages
Logistic Regression	Based on the existing data, a regression formula (the best-fitting parameter set) was established for the classification boundary.	① The calculation is small and the speed is fast; ② It has a good probability explanation; ③ Can easily update the model.	① It is easy to under-fit and the accuracy is limited; ② It can only deal with two classification problems, and it must be linearly separable.
Nearest Neighbor algorithm	Classification is carried out by measuring the distance between different eigenvalues.	① Suitable for multiple classification problems; ② High accuracy; ③ It is not sensitive to abnormal points.	① Large amount of calculation, poor comprehension; ② When the training data is highly dependent and the samples are unbalanced, the prediction accuracy of rare categories is low.
Decision Tree	The tree structure is used to establish the decision model according to the data attribute.	① Easy to explain and explain, good at dealing with the interaction between features; ② Suitable for analyzing discrete data; ③ Small-scale data sets are effective.	① Poor treatment of continuous variables; ② Online learning is not supported, and the Decision Tree needs to be reconstructed when there are new samples. ③ Easy to overfit.

Table 2. *Cont.*

Commonly Algorithm	Principle	Advantages	Disadvantages
Artificial Neural Network	Simulating biological neural networks, a class of pattern-matching algorithms is a huge branch of machine learning with hundreds of different algorithms.	① High classification accuracy; ② Strong learning ability.	① A large number of parameters are required; ② Unable to observe the learning process, the results are difficult to interpret; ③ The study time is long.
Support Vector Machine	To find the optimal hyperplane, the data can be divided into two parts, each part of the data belongs to the same class.	① It can solve nonlinear classification; ② The idea of classification is simple.	① Large memory consumption, difficult to implement for large-scale training samples; ② It is difficult to solve the multi-classification problem; ③ Difficult to explain, complex to run and optimize.
Random Forest algorithm	A forest is built randomly. The forest is composed of many independent Decision Trees, and finally, the optimal classification result is obtained comprehensively.	① The limited sample can be fully applied; ② It has the advantages of diversity and accuracy.	It will overfit on some noisy problems.

The quality of a machine learning model depends on its evaluation accuracy and model generalization ability. Several common parameters for model evaluation include the following:

(1) Accuracy (ACC), which expresses the model evaluation accuracy. The accuracy of the model is the ratio of the number of samples correctly predicted by the model to the total number of samples. In addition, there are metrics such as precision, recall, and F1 value.

(2) The ROC curve and AUC value express the generalization ability of the model. ROC (Receiver Operating Characteristic) curve refers to the receiver operating characteristic curve, which is a comprehensive index reflecting the continuous variables of sensitivity and specificity. Its main analysis tool is a curve drawn on a two-dimensional plane; AUC (Area Under roc Cure) value is the area under the ROC curve. Usually, the value of AUC is between 0.5 and 1.0, and the larger the AUC value, the better the performance of the model.

(3) Learning curve, which describes the model fitting problem and judges whether the model is over-fitting or under-fitting.

4. Regional Landslide Early-Warning Model Method Based on Machine Learning

4.1. Construction of Training Sample-Set

The accurate construction of the training sample-set is the foundation of the machine-learning regional landslide disaster warning model, and it directly determines the accuracy and generalization ability of the warning model to a certain extent. The occurrence of regional landslide hazards is controlled by the coupling effect of geological environmental conditions and rainfall conditions. From the perspective of input features, machine learning samples include attributes of three aspects: geographic location, geological environmental conditions, and precipitation conditions. Geographical location refers to the spatial geographic coordinates of the point where the sample is located; geological environmental conditions refer to the geological environmental background conditions of the sample; precipitation conditions refer to the induced rainfall factors of the sample. From the perspective of output features, the training sample-set includes positive samples (landslide points, generally denoted as one) and negative samples (non-landslide points, generally denoted as zero). The balance of the number of positive and negative samples should be considered when sampling positive and negative samples.

The construction process of the training sample-set is shown in Figure 4, which mainly includes three steps: geological environment and rainfall factor feature library construction, positive and negative sample sampling, and sample feature attribute extraction.

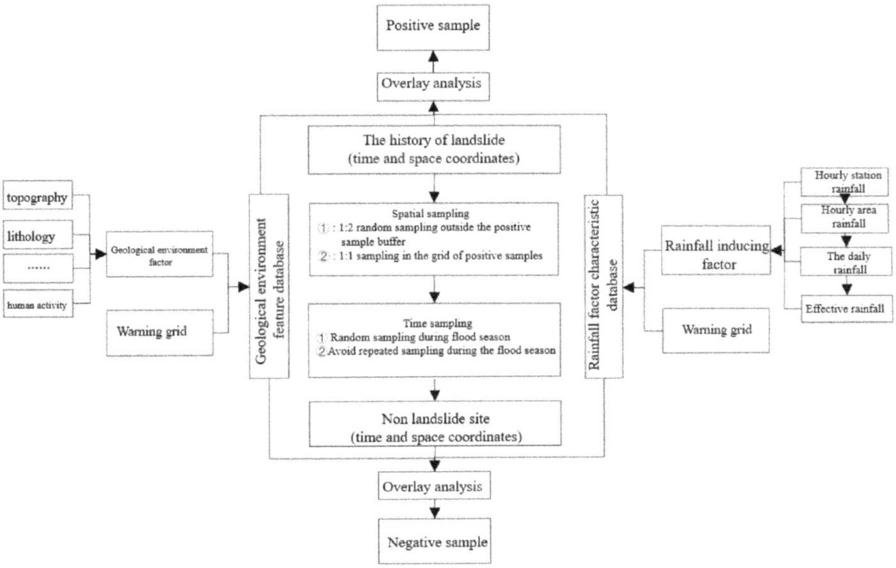

Figure 4. Training sample-set construction flowchart.

4.1.1. Construction of Characteristic Database of Geological Environment and Rainfall Factors

The construction of the geological environment characteristic database and rainfall factor characteristic database is completed based on the analysis of regional landslide disaster distribution patterns and influencing factors in the study area. The geological environmental factors affecting the occurrence of regional landslides generally include topography, strata lithology, human activities, etc.; the rainfall-inducing factors affecting the occurrence of regional landslides generally include daily rainfall, previous rainfall, previous effective rainfall, etc.

The geological environment factors and rainfall-inducing factors are overlaid with the warning grid profiling unit respectively (Figure 5), and the geological environment characteristic library and rainfall factor characteristic library of the early-warning grid unit are obtained. The geological environment characteristic library contains the characteristics attributes of each geological environment influencing factor of each early-warning grid unit; the rainfall factor characteristic library contains the daily rainfall characteristic attribute or effective rainfall characteristic attribute of each early-warning grid unit.

The early-warning grid unit is determined according to the size of the study area and the actual need for early-warning. It is generally a uniform grid unit, which can be determined by referring to the early-warning space accuracy in the relevant early-warning standards (Table 3).

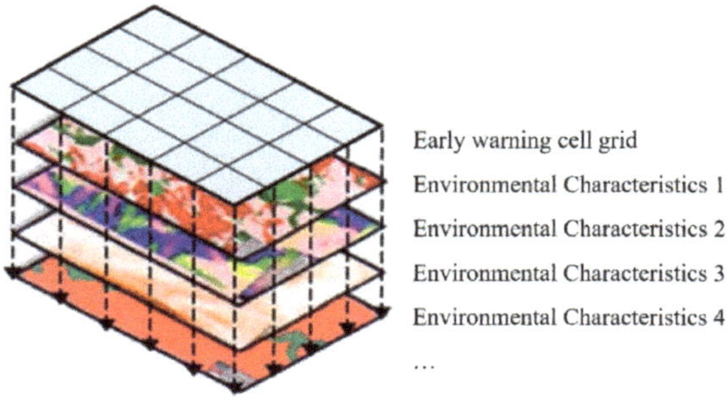

Figure 5. Schematic diagram of spatial superposition analysis.

Table 3. Precision division of early-warning space (reproduced from [34]).

Warning Grading	Recommended Early-Warning Accuracy
At the national level	1:500,000–1:1,000,000
At the provincial level	1:100,000–1:500,000
The municipal	1:50,000–1:100,000
At the county level	1:10,000–1:50,000
Thematic early-warning area	1:2000–1:10,000

4.1.2. Sampling Method of Positive and Negative Samples

Positive samples refer to the points where landslides occur, and the sampling of positive samples is generally based on historical landslide data. The screening requires that landslides should have both definite spatial geographic coordinates (the specific accuracy is determined by the specific conditions of the study area) and time coordinates (generally accurate to the day in the 24-h early-warning). Generally speaking, the sampling accuracy of positive samples is higher.

Negative samples refer to points where landslides do not occur, which cannot be obtained directly. In this paper, the negative samples were sampled in two aspects based on the method of "random sampling under space-time constraints" [35,36]: first, the size of the buffer radius was corrected; secondly, in addition to random sampling outside the positive sample buffer, sampling was supplemented in the grid where the positive sample is located, that is, the negative samples were completed in two parts. The schematic diagram is shown in Figure 6.

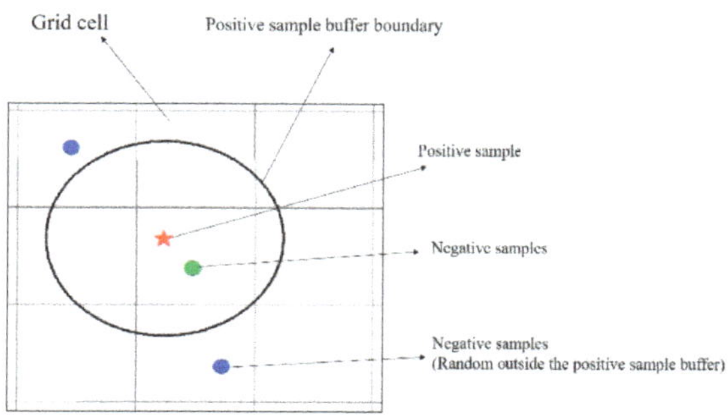

Figure 6. Schematic diagram of negative samples space sampling based on positive samples.

(1) Collect negative samples outside the positive samples buffer

Determination of the spatial location of the negative samples, random sampling in the space outside a certain buffer zone of positive samples, and the determination of the radius of the buffer zone should take into account the minimum early-warning grid cell size in the study area and the distribution of historical landslide points. In this paper, the buffer radius was used as the warning grid unit size to ensure the balance of positive and negative samples; the number of negative samples collected was two times that of the positive samples.

Time attribute assignment of negative samples: under the constraints of a certain period (according to the completeness of rainfall data in the study area, it is generally the whole period of the multi-year flood season), the random function is adopted for sampling. The random function is as follows:

$$T = \mathrm{RAND}\,(T_1,\ T_2) \tag{1}$$

Description:

T, time acquired randomly;

T_1, the lower limit of the time range of random acquisition time;

T_2, the upper limit of the time range of random acquisition time.

(2) Negative samples are collected in the grid where positive samples are located

For the determination of the spatial location of negative samples, random sampling was performed in the grid where the positive samples are located. The number of negative samples collected was about equal to the number of positive samples.

Time attribute assignment of the negative sample also uses the random function shown in Equation (1), with the additional constraint that the negative sample time property sampled should be different from the positive sample.

4.1.3. Feature Attribute Extraction and Data Screening

The positive and negative samples are spatially overlaid with the geological environment feature database to extract the geological environment feature attributes of the positive and negative samples. Based on the temporal attributes of the positive and negative samples, and the rainfall factor feature attributes of the positive and negative samples were extracted by the query. At this point, the construction of the whole training sample-set was completed. In addition, data cleaning is particularly important throughout the training sample-set construction process, and the model evaluation accuracy is higher using the cleaned data set. Data cleaning generally consists of two categories:

(1) Handling data errors: for example, manual errors, data transmission errors, equipment failures, and ambiguity of geological information can affect the original data set, these errors data must be processed and cleaned in advance. In general, this type of data cleaning refers to the imputation or elimination of missing values and the identification of outliers in the data.

(2) Feature attribute preprocessing: considering the dimension difference of the input features of the training samples, it is necessary to perform uniform normalization or feature scaling on the input features of the samples. Different machine learning algorithms differ in their sensitivity to the difference of input feature scales, and the requirements for input feature attribute preprocessing are also different. It is recommended that the input features of the training samples be uniformly normalized or scaled before model training. Generally, the range of input features of the samples should be at least not much different; otherwise, it will directly affect the accuracy of the model.

In summary, the above-mentioned training sample-set construction method was used to complete the positive and negative samples in Fujian Province, and the sample-set covers 15,589 samples in the past nine years (2010–2018). Among them, there are 3562 positive samples and 12,027 negative samples, and the ratio of positive and negative samples is about 1:3.4. The spatial distribution of positive and negative samples is shown in Figure 7.

The positive and negative sample attributes determine the output features of the final training sample-set. The output features of positive samples were taken as one, and the output features of negative samples were taken as zero.

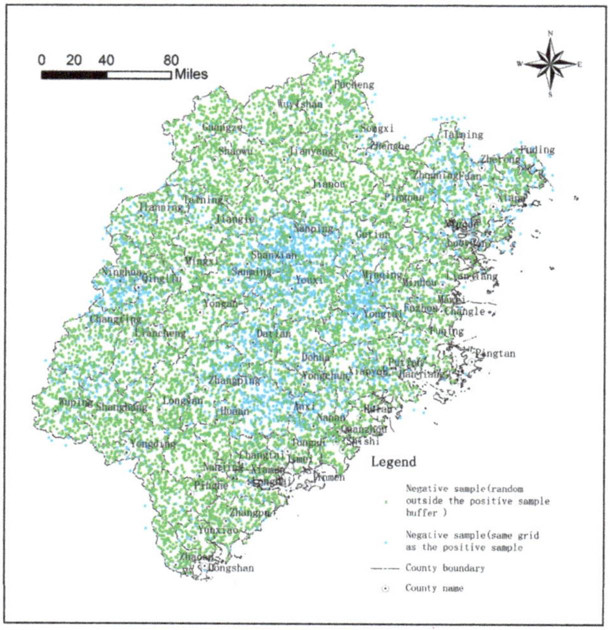

Figure 7. Spatial distribution map of negative samples.

4.2. *Optimization of Model Parameters*

The selection of model parameters has a significant impact on model accuracy, and model optimization is crucial in model construction. The most commonly used model parameter optimization methods include traditional methods and hyperparametric Optimization algorithms. The traditional parameter optimization method is the grid search method, whose optimization accuracy is inversely proportional to its speed. In order to optimize parameters more efficiently, the Bayesian Optimization algorithm has gradually emerged [37,38]. The Bayesian Optimization algorithm adopts the Gaussian process to fit the distribution of the objective function by increasing the number of samples, and the objective function is optimized by cross-verifying the accuracy. Each iteration outputs a hyperparameter, and the hyperparameter is optimized in the process of finding the optimal value.

4.3. *Model Saving and Early-Warning Output*

The model trained by machine learning can be saved by calling the DUMP function in the python environment. The model is generally saved as a .pkl format file.

When the actual early-warning is running, the trained early-warning model is directly called by the LOAD function to output the probability of possible landslide disasters. According to the probability, the warning level is determined by classification. The setting of graded breakpoints can refer to the classification table of geological disaster meteorological risk early-warning grades (according to the geological disaster regional meteorological risk early-warning standard (trial) (T/CAGHP 039-2018)), and can also be fine-tuned according to the specific conditions of the study area. Considering that the output threshold was set to 0.5 in the machine learning algorithm, the geohazard meteorological warning probability class classification was adjusted in conjunction with the specific conditions of the study area. That is, when the output probability $P \geq 50\%$ and $P < 60\%$, the yellow warning of landslide disaster is issued; when the output probability P ≥ 60% and P < 80%, the orange warning of landslide disaster is issued; when the output probability P ≥ 80%, a red warning of landslide disasters is issued, as shown in Table 4.

Table 4. Division of early-warning levels.

The Warning Level	The Risk of Landslides	Output Probability/P
Red alert	highest risk	P ≥ 80%
Orange alert	higher risk	60% ≤ P < 80%
Yellow warning	high risk	50% ≤ P < 60%

5. Result and Verification

5.1. Model Parameter Optimization Training and Effect Evaluation

The training sample-set of regional landslide early-warning in Fujian Province was divided into the training test sets in the ratio of 4:1. The parameters were optimized by a Bayesian Optimization algorithm and five-fold cross-validation. The six commonly used machine learning classification models were compared and evaluated to compare the accuracy and model generalization ability indexes of each model. The optimization parameters and effect evaluation comparison of the six models is shown in Table 5 and Figure 8.

Table 5. Comparison of partial hyperparameter optimization and model evaluation of six machine learning algorithms.

Machine Learning Model	Accuracy	Model Generalization Ability	Hyperparameter	Hyperparameter Value
Random Forest algorithm	0.923	0.955	n_estimators	118
			max_depths	10
			min_samples_split	3
Nearest Neighbor algorithm	0.932	0.924	n_neighbors	10
Decision Tree	0.937	0.904	max_depths	4
Support Vector Machine	0.932	0.920	C	3
			gamma	0.003
Logistic Regression	0.940	0.922	C	5
Artificial Neural Network	0.937	0.935	hidden_layer_sizes	(6,7)
			max_iter	1680

According to the calculation results of the six machine learning algorithms (Table 5), it can be seen that the Random Forest model had the best performance; its accuracy rate was 0.923, the model generalization ability was the best (AUC = 0.955), and the model had no overfitting phenomenon; the learning and ROC curves are shown in Figure 9. The second is the Artificial Neural Network model, with an accuracy rate of 0.937 and an AUC of 0.935, followed by the Nearest Neighbor model, Logistic Regression model, and Support Vector Machine model, with AUCs of 0.924, 0.922, and 0.920, respectively; the worst was decision-making tree, its AUC value being 0.904 and its accuracy 0.937.

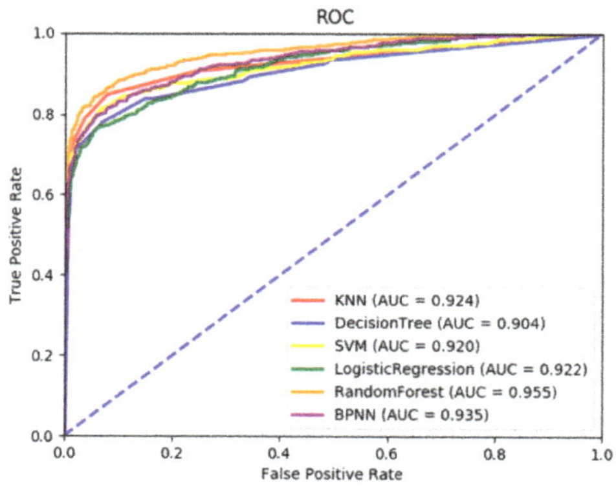

Figure 8. Comparison of ROC and AUC of six machine learning early-warning models.

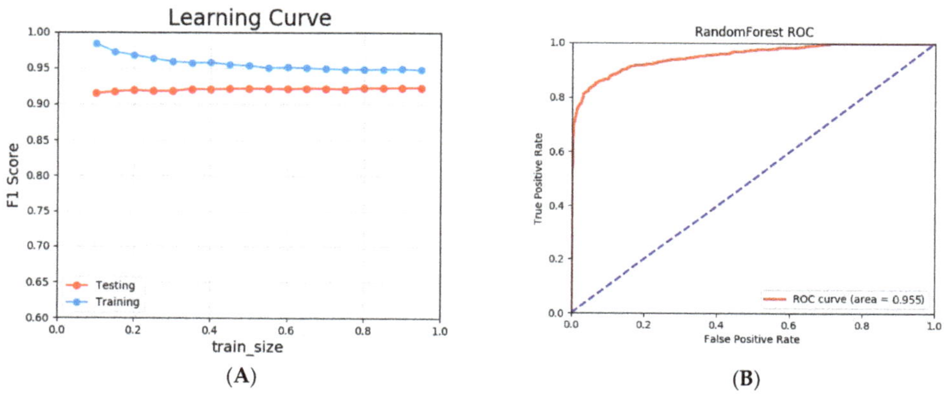

Figure 9. Learning curve (**A**) and ROC curve (**B**) of the Random Forest early-warning model.

5.2. Model Early-Warning Verification—Taking the Random Forest Algorithm as an Example

On two days, 22 June and 28 June 2021, six and four landslide disasters occurred in Youxi County and Shaowu City, Fujian Province, respectively, all of which were small-scale landslide disasters caused by local rainstorms.

Based on the actual precipitation data in Fujian Province (5 km × 5 km QPE data from China Meteorological Administration), the new model based on Random Forest and the original explicit statistical model was used to simulate the early-warning. Then, the objective forecast results of the model on 22 June and 28 were compared with the actual occurrence of landslide disasters, as shown in Table 6 and Figures 10 and 11.

According to the comparison on 22 June 2021 (Table 6, Figure 10), the six actual landslide disasters all fell in the early-warning area of the Random Forest model (one in the yellow early-warning area and six in the orange early-warning area), with a hit rate of 100%; one landslide disaster fell in the early-warning area (yellow early-warning area) of the explicit statistical model, with a hit rate of 16.7%. The landslide densities in the early-warning areas of the two models were 6.2 and 3.8 per 1000 square kilometers, respectively. The actual landslide density in the early-warning area of the Random Forest model was 1.6 times that of the explicit statistical model.

Table 6. Comparison between the Random Forest model and explicit statistical early-warning model.

Actual Landslide (Number)		22 June 2021 Results of Different Models: Explicit Statistical Models	22 June 2021 Results of Different Models: Random Forest Model	28 June 2021 Results of Different Models: Explicit Statistical Models	28 June 2021 Results of Different Models: Random Forest Model
		6		4	
All warning area	Accuracy (%)	16.7	100.0	100.0	100.0
	Number of landslides	1	6	4	4
	Area of warning area (km^2)	262.2	971.9	4119.1	2359.8
	Landslide density (amount of per 1000 km^2)	3.8	6.2	1.0	1.7
Yellow Alert area	Number of landslides	1	1	2	0
	Area of warning area (km^2)	262.2	576.0	3051.3	1631.9
	Landslide density (amount of per 1000 km^2)	3.8	1.7	0.7	0.0
Orange Alert area	Number of landslides	0	5	2	4
	Area of warning area (km^2)	0	396.0	1067.8	728.0
	Landslide density (amount of per 1000 km^2)	/	12.6	1.9	5.5

According to the comparison on 28 June 2021 (Table 6, Figure 11), four landslide disasters all fell in the early-warning area (orange warning area) of the Random Forest model, with a hit rate of 100%; four landslides fell in the warning zone of the explicit statistical model (two yellow warning areas), the hit rate is also 100%. The landslide densities in the early-warning areas of the two models were 1.7 and 1.0 per 1000 square kilometers, respectively. The actual landslide density in the early-warning area of the Random Forest model was 1.7 times that of the explicit statistical model.

By comparing the results of the two models, it can be seen that the hit rate of the new model based on the Random Forest was six times that of the original model (22 June) or equivalent (28 June), and the landslide density in the early-warning area of the new model was 1.6–1.7 times that of the original. Meanwhile, it can be seen from Figure 11 that no landslide occurred in WuYiShanshi area (upper right corner of the figure frame). There is no warning area in the warning results of the Random Forest model, and there is a certain range of yellow warning areas in the warning results of the explicit statistical model warning results. The preliminary verification shows that the new model based on the Random Forest has obvious advantages with a higher hit rate, smaller warning area, and more accurate warning. Since there are few new landslide disasters in the study area, the current model verification work is relatively weak, and we will continue to track the new landslide disasters in the study area and strengthen the model validation and correction.

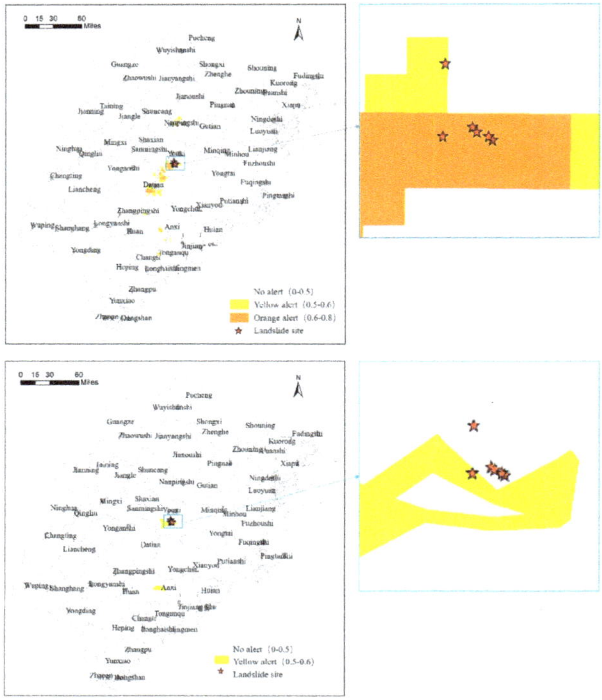

Figure 10. Comparison of early-warning results of different models on 22 June 2021 (above: early-warning results of the Random Forest model; below: explicit statistical model warning results).

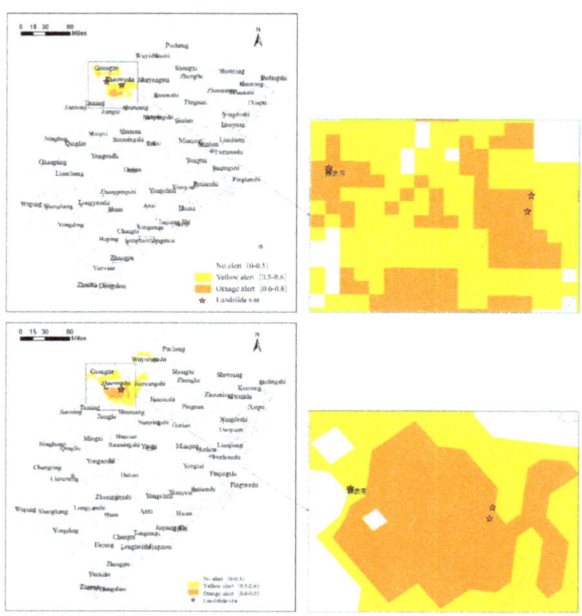

Figure 11. Comparison of early-warning results of different models on 28 June 2021 (above: early-warning results of Random Forest model; below: explicit statistical model warning results).

6. Conclusions

This paper proposed a method and process for building a regional landslide disasters early-warning model based on machine learning. Taking Fujian Province as the geological background, the landslide disaster early-warning based on machine learning in this region was carried out. Six machine learning algorithms were selected, including the Random Forest algorithm, Nearest Neighbor algorithm, Support Vector Machine, Logistic Regression, Decision Tree, Artificial Neural Network, etc. According to the topography, geological conditions, environmental conditions, human engineering activities, and other influencing factors of Fujian Province, evaluation indexes such as grade, geomorphic type, lithology, historical disasters, vegetation type, water system influence, annual rainfall, distance from house, distance from road, density of population, daily rainfall, and daily rainfall from the previous 1 to 15 days were selected. The sample data set of the early-warning model was constructed by including 27 disaster-influencing factors, and the establishment of the geological disaster early-warning model was realized through several key steps such as the construction of the sample-set, the model training and parameter adjustment optimization, and the classification of the model early-warning level. The results of the early-warning model showed that the Random Forest algorithm performed better. This algorithm has the advantages of fully applying limited samples, diversity, and accuracy; and it has the disadvantage of overfitting on certain noisy problems. In this early-warning model, the Random Forest algorithm accuracy rate was the highest (92.3%), and the model generalization ability was the best (AUC is 0.955); the second best was the Artificial Neural Network model, which has the advantages of high classification accuracy and strong learning ability, but because it is the disadvantage of requiring a large number of parameters in this model, the accuracy rate was slightly lower than that of the Random Forest model, being 0.937, and the AUC was 0.935. Contrary, the Nearest Neighbor model, Logistic Regression model, and support vector machine model have problems such as difficulty in solving multi-classification problems and requiring a large amount of training data, their AUC values being 0.924, 0.922, and 0.920, respectively. The worst model was the Decision Tree model, which has problems such as poor handling of continuous variables and easy overfitting, its AUC value being 0.904 and the accuracy 0.937. For the selection of the typical rainfall-type landslide disaster process in Fujian Province from 2019 to 2021 and for the verification of the Random Forest algorithm model, the results showed that the early-warning hit rate of the model was 100%. Compared with the early-warning results of the original explicit statistical model, the hit rate of the new model is higher than that of the original one, since, the landslide density in the early-warning area of the new model was 1.6–1.7 times higher than that of the original model. The preliminary verification showed that the new model based on the Random Forest had obvious advantages with a higher hit rate and smaller warning area, which can achieve more accurate warnings.

Research on the regional geological disasters early-warning model based on machine learning is relatively complex. Through the research in this paper, the problems of insufficient samples, limited methods, and insufficient precision in the traditional regional landslide early-warning model were solved to a certain extent. For the geological disasters early-warning model constructed by machine learning, the larger the sample data set, the higher the accuracy of the trained model, and the follow-up will increase the amount of data to optimize and improve the early-warning models.

Author Contributions: Data resources, J.H., Y.L., P.Z.; research methods, Y.L., P.Z.; writing—original draft preparation, Y.L., S.M.; editing, Y.L., S.M., R.X. All authors have read and agreed to the published version of the manuscript.

Funding: This research was financially supported by the National Natural Science Foundation of China (42077440; 41202217) and the National Key Research and Development Program of China (2018YFC15 05503).

Institutional Review Board Statement: Not applicable.

Informed Consent Statement: Not applicable.

Data Availability Statement: Not applicable.

Conflicts of Interest: There is no conflict of interest in the research methods, routes and data in this study.

References

1. Guzzetti, F.; Gariano, S.L.; Peruccacci, S.; Brunetti, M.T.; Marchesini, I.; Rossi, M.; Melillo, M. Geographical landslide early warning systems. *Earth Sci. Rev.* **2020**, *200*, 102973. [CrossRef]
2. Liu, C.-Z.; Liu, Y.-H. Early warning theory for regional geo-hazards and design of explicit statistical system. *Hydrogeol. Eng. Geol.* **2007**, *34*, 11–18.
3. Liu, C.-Z.; Liu, Y.-H.; Wen, M.-S.; Tang, C.; Xue, Q.-W. *Method and Application of Regional Warning for Geo-Hazards in China*; Geological Publishing: Beijing, China, 2009.
4. Liu, C.-Z.; Liu, Y.-H.; Wen, M.-S.; Tang, C.; Zhao, L.-Q.; Tian, H. Early warning for regional geo-hazards during 2003–2012, China. *Chin. J. Geol. Hazard Control.* **2015**, *26*, 1–8.
5. Liu, Y.-H.; Liu, C.-Z.; Lian, J.-F.; Wen, M.-S.; Tang, C. Preliminary study of geo-hazards regional early warning based on explicit statistical theory. *Geol. China* **2008**, *35*, 344–350.
6. Liu, Y.-H.; Liu, C.-Z.; Wen, M.-S.; Tang, C. Study of Early Warning Models for Regional Geo-Hazards in China. *J. Eng. Geol.* **2015**, *23*, 738–746.
7. Liu, Y.-H.; Zhang, Z.-X.; Su, Y.-C. Case study of vulnerability evaluation for geo-hazards bearing capacity of a region. *J. Eng. Geol.* **2018**, *26*, 1121–1130.
8. Liu, Y.-H.; Su, Y.-C. Early-warning model of regional geological disasters based on meteorological factor in Qingchuan County, Sichuan Province. *J. Eng. Geol.* **2019**, *27*, 134–143.
9. Li, S.-D.; Bai, Y.-H.; Jiang, Y.; Wang, Z.-H.; Wie, W.-H.; Li, X.; Liu, L.-N. Explicit Statistic Meteorological Early-warning Model of Geological Hazards in Xinjiang, China Based on the Genesis Theory of Endogenic and Exogenic Couping. *J. Earth Sci. Environ.* **2017**, *39*, 287–300.
10. Wei, P.-X.; Li, X.-J. The meteorologic early warning research of sudden geo-hazard in Guangdong province. *Chin. J. Geol. Hazard Control.* **2015**, *26*, 138–144.
11. Wen, M.-S.; Liu, Y.-H.; Su, Y.-C.; Fang, Z.-W.; Xiao, R.-H.; Chen, C.-L.; Xu, W.; Chen, Y.-C. *Geo-Hazards Investigation Report of Qingzhujiang River Region, Sichuan*; China Institute of Geo-Environmental Monitoring: Beijing, China, 2016.
12. Hou, S.-S.; Li, A.; Han, B.; Zhou, P.-G.; Ye, H.-J.; Zhu, B.; Ma, W.-F. An approach of geo-hazard warning system in Ya'an, Sichuan and its analysis. *Chin. J. Geol. Hazard Control.* **2014**, *25*, 134–138.
13. Liu, C.-Z.; Wen, M.-S.; Liu, Y.-H.; Liu, Q.-Q.; Gu, X.-X. *Study of Geo-Hazards in Wenchuan Earthquake Zone*; Geology Press: Beijing, China, 2017.
14. Cannon, S.H.; Ellen, S. Rainfall conditions for abundant debris avalanches, San Francisco Bay region, California. *Calif. Geol.* **1985**, *38*, 262–272.
15. Pietro, A. A warning system for rainfall-induced shallow failures. *Eng. Geol.* **2004**, *73*, 247–265.
16. Hong, H.; Pourghasemi, H.R.; Pourtaghi, Z.S. Landslide susceptibility assessment in Lianhua County (China): A comparison between a random forest data mining technique and bivariate and multivariate statistical models. *Geomorphology* **2016**, *259*, 105–118. [CrossRef]
17. Peruccacci, S.; Brunetti, M.T.; Gariano, S.L.; Melillo, M.; Rossi, M.; Guzzetti, F. Rainfall thresholds for possible landslide occurrence in Italy. *Geomorphology* **2017**, *290*, 39–57. [CrossRef]
18. Wei, L.-W.; Huang, C.-M.; Chen, H.; Lee, C.-T.; Chi, C.-C.; Chiu, C.-L. Adopting the I3-R24 rainfall index and landslide susceptibility for the establishment of an early warning model for rainfall-induced shallow landslides. *Nat. Hazards Earth Syst. Sci.* **2018**, *18*, 1717–1733. [CrossRef]
19. Ding, G.-L.; Wang, Y.-H.; Mao, J.; Yao, K.; Liu, H.-H. A study of the rainfall threshold of debris flow forewarning in Beijing based on susceptibility analysis. *Hydrogeol. Eng. Geol.* **2017**, *44*, 136–142.
20. Keefer, D.K.; Wilson, R.C.; Mark, R.K.; Brabb, E.E.; Brown, W.M., III; Ellen, S.D.; Harp, E.L.; Wieczorek, G.F.; Alger, C.S.; Zatkin, R.S. Real-Time Landslide Warning during Heavy Rainfall. *Science* **1987**, *238*, 921–925. [CrossRef]
21. Mulyana, A.R.; Sutanto, S.J.; Hidayat, R.; Ridwan, B.W. Capability of Indonesian Landslide Early Warning System to detect landslide occurrences few days in advance. In Proceedings of the 21st EGU General Assembly, EGU2019, Vienna, Austria, 7–12 April 2019.
22. Pennington, C.V.; Freeborough, K.; Dashwood, C.; Dijkstra, T.; Lawrie, K. The national landslide database of Great Britain: Acquisition, communication and the role of social media. *Geomorphology* **2015**, *249*, 44–51. [CrossRef]
23. Ponziani, F.; Berni, N.; Stelluti, M.; Zauri, R.; Pandolfo, C.; Brocca, L.; Moramarco, T.; Salciarini, D.; Tamagnini, C. LANDWARN: An Operative Early Warning System for Landslides Forecasting Based on Rainfall Thresholds and Soil Moisture. In *Landslide Science and Practice*; Margottini, C., Canuti, P., Sassa, K., Eds.; Springer: Berlin/Heidelberg, Germany, 2013; pp. 627–634.
24. Chen, W.; Pourghasemi, H.R.; Zhao, Z. A GIS-based comparative study of Dempster-Shafer, logistic regression and artificial neural network models for landslide susceptibility mapping. *Geocarto Int.* **2017**, *32*, 367–385. [CrossRef]

25. Bui, D.T.; Tuan, T.A.; Klempe, H.; Pradhan, B.; Revhaug, I. Spatial prediction models for shallow landslide hazards: A comparative assessment of the efficacy of support vector machines, artificial neural networks, kernel logistic regression, and logistic model tree. *Landslides* **2016**, *13*, 361–378.
26. Bui, D.T.; Tuan, T.A.; Hoang, N.-D.; Thanh, N.Q.; Nguyen, D.B.; Van Liem, N.; Pradhan, B. Spatial prediction of rainfall-induced landslides for the Lao Cai area (Vietnam) using a hybrid intelligent approach of least squares support vector machines inference model and artificial bee colony optimization. *Landslides* **2017**, *14*, 447–458.
27. Liu, Y.-L.; Yin, K.-L.; Liu, B. Application of logistic regression and artificial neural networks in spatial assessment of landslide hazards. *Hydrogeol. Eng. Geol.* **2010**, *37*, 92–96.
28. Xu, C.; Xu, X.-W. Logistic regression model and its validation for hazard mapping of landslides triggered by Yushu earthquake. *J. Eng. Geol.* **2012**, *20*, 326–333.
29. Trigila, A.; Iadanza, C.; Esposito, C.; Scarascia-Mugnozza, G. Comparison of logistic regression and random forests techniques for shallow landslide susceptibility assessment in Giampilieri (NE Sicily, Italy). *Geomorphology* **2015**, *249*, 119–136. [CrossRef]
30. Sun, D.; Gu, Q.; Wen, H.; Shi, S.; Mi, C.; Zhang, F. A Hybrid Landslide Warning Model Coupling Susceptibility Zoning and Precipitation. *Forests* **2022**, *13*, 827. [CrossRef]
31. Zhou, T.-Z.; Chen, G.; Zhao, X.-F.; Zhu, Y.-F. Study of distribution laws and genesis of landslides in Fujian province. *J. Chongqing Univ. Sci. Technol. Nat. Sci. Ed.* **2007**, *9*, 17–19.
32. Gao, S.; Liu, A.-M.; Huang, Z.-G.; Zheng, W.-W. The character analysis on heavy rainfall inducing geological hazards in Fujian province. *Geol. Fujian* **2010**, *29*, 64–71.
33. Ye, L.-Z.; Liu, K.; Huang, G.-P.; Zheng, M.-Z. The characteristics and influence factors of catastrophic geological disaster in Fujian province. *Chin. J. Geol. Hazard Control.* **2015**, *26*, 98–102.
34. China Geological Disaster Prevention Engineering Association. *Standard of Geo-Hazards Early-Warning Based on Meteorological Factors (T/CAGHP 039-2018)*; China University of Geosciences Press: Wuhan, China, 2018.
35. Liu, Y.-H.; Fang, R.-K.; Su, Y.-C.; Xiao, R.-H. Machine learning-based model for warning of regional landslide disasters. *J. Eng. Geol.* **2021**, *29*, 116–124.
36. Miao, Y.-M. A New Approach to Generating Absence Samples for Landslide Susceptibility Assessment. Master's Thesis, Nanjing Normal University, Nanjing, China, 2016.
37. Snoek, J.; Larochelle, H.; Adams, R.P. Practical Bayesian Optimization of Machine Learning Algorithms. In Proceedings of the 25th International Conference on Neural Information Processing Systems, Lake Tahoe, NV, USA, 3–6 December 2012.
38. Sun, D.; Wen, H.; Wang, D.; Xu, J. A random forest model of landslide susceptibility mapping based on hyperparameter optimization using Bayes algorithm. *Geomorphology* **2020**, *362*, 107201. [CrossRef]

forests

MDPI

Article

A Data-Driven Model for Spatial Shallow Landslide Probability of Occurrence Due to a Typhoon in Ningguo City, Anhui Province, China

Yulong Cui [1,2,*], Jiale Jin [1], Qiangbing Huang [2,*], Kang Yuan [1] and Chong Xu [3]

1 School of Civil Engineering and Architecture, Anhui University of Science and Technology, Huainan 232001, China; jinjiale1105@163.com (J.J.); yk13739265090@163.com (K.Y.)
2 Key Laboratory of Western China's Mineral Resources and Geological Engineering, Ministry of Education, Chang'an University, Xi'an 710054, China
3 National Institute of Natural Hazards, Ministry of Emergency Management of China, Beijing 100085, China; chongxu@ninhm.ac.cn
* Correspondence: hqb@chd.edu.cn (Q.H.); ylcui@aust.edu.cn (Y.C.)

Abstract: From 9 to 11 August 2019, the southeast coastal areas of China were hit by Typhoon Lekima, which caused a large number of shallow landslides. The typhoon resulted in a maximum rainfall of 402 mm during 3 days in Ningguo City. In this study, satellite images were acquired before and after the rainfall and visual interpretation was used to identify 414 shallow landslides in Ningguo City, and a complete database of shallow landslides caused by the typhoon-induced rainfall in Ningguo City was created. Nine landslide-influencing factors were selected—elevation, slope, aspect, strata, distance to faults, distance to rivers, distance to roads, normalized vegetation difference index, and rainfall—and the relationships between the rainfall-induced landslide distribution and the influencing factors were analyzed. The Bayesian probability method was combined with a logistic regression model to establish a landslide probability map for the study area. The real probabilities of landslide occurrence in the study area under five different rainfall conditions were calculated, and probability maps of landslide occurrence were drawn. The results of this study provide a reference for disaster prevention and reduction of typhoon rainstorm landslides in the southeast coastal areas of China and a future basis for decision making by the Ningguo government departments before a typhoon rainstorm occurs.

Keywords: shallow landslide; probability of occurrence; typhoon; data-driven model; Ningguo City

Citation: Cui, Y.; Jin, J.; Huang, Q.; Yuan, K.; Xu, C. A Data-Driven Model for Spatial Shallow Landslide Probability of Occurrence Due to a Typhoon in Ningguo City, Anhui Province, China. *Forests* **2022**, *13*, 732. https://doi.org/10.3390/f13050732

Academic Editor: Filippo Giadrossich

Received: 8 March 2022
Accepted: 5 May 2022
Published: 8 May 2022

1. Introduction

The southeast coastal areas of China are frequently hit by typhoons. According to the statistics, from 1949 to 2018, a total of 493 typhoons landed in China, an average of 7 typhoons create landfall every year, making China one of the countries with the most frequent typhoon landfalls and suffering the most severe typhoon damage in the world [1]. The more developed the regional economy and the greater the population density, the more serious the casualties and economic losses a typhoon produces [2,3]. The southeast coastal areas of China have a fragile geological environment, hilly mountains, and undergo frequent engineering activities. Under the influence of typhoons and rainstorms, geological disasters are extremely prone to occur, including the largest number of landslides and the most serious damage. For example, in 2004, Typhoon Rananim caused a large number of mudslides and landslides in Yongjia, Yueqing, and other counties in Zhejiang Province, resulting in 28 deaths [4,5]. In 2015, Typhoon Soudelor caused 81 geological disasters, including collapses, landslides, and mudslides in Taishun County, Zhejiang Province, resulting in six casualties and a direct economic loss of 41.18 million yuan [6]. In 2019, Typhoon Lekima caused a landslide in Shanzao Village, Yongjia County, Zhejiang Province.

The landslide blocked the river and quickly formed a barrier lake, resulting in a dam burst. This landslide-barrier-lake disaster chain of events resulted in 32 deaths [7]. Moreover, the typhoon and rainstorm affected 100,000 people and killed 8 people in Ningguo City, Anhui Province, resulting in a direct economic loss of 2594 billion yuan [8]. The heavy casualties and huge economic losses caused by typhoon rainstorms and the geological disasters they induce make it necessary to research the spatial prediction of typhoon rainstorm-induced landslides in the southeast coastal area of China, which can provide a reference for government decision-making and disaster prevention and mitigation of such landslides before typhoons and heavy rainstorms.

At present, there are two methods for the spatial prediction of regional landslides: deterministic and statistical [9–11]. The deterministic method combines a hydrological model, surface runoff model, and slope stability model to evaluate the stability of each grid in the region, after which the spatial prediction of landslides in a region under the action of rainfall is carried out, such as the landslide susceptibility, landslide risk, and landslide probability maps. The statistical methods are mainly divided into two types. One is to determine the statistics of the rainfall factors influencing landslides in specific regions, to determine the rainfall threshold, and to carry out the spatial prediction of regional landslides under the action of rainfall. This method only considers the rainfall factors [12–14]. The other is to use a regression model, machine learning, and other methods to determine the impacts of various geomorphic, geological, road, and rainfall factors on the landslide; these factors are used to evaluate the landslide risk under future rainfall in the region and predict the landslide area [15–17]. The deterministic method requires knowledge of the hydraulic conductivity and the physical-mechanical parameters of the rock and soil mass. Although probabilistic methods such as the Monte Carlo method can be used to deal with the uncertainties of these parameters, using this method in the spatial prediction of landslides in a wide range of areas is still limited [12,18]. Statistical analysis methods based on regression models and machine learning require a large number of landslide cases and a large amount of data on the various landslide impact factors. With the development of computer technology, geographic information system (GISs), and remote sensing technology, these data have become easier to obtain and process. Although this method does not consider the physical and mechanical mechanisms of landslide occurrence, a few studies have shown that statistical analysis methods have high prediction accuracies in terms of the spatial prediction of landslides [19–21].

In this study, the landslides in Ningguo City caused by Typhoon Lekima in the southeast coastal areas of China in 2019 were taken as an example (1) to establish a complete database of the landslides induced by Typhoon Lekima in Ningguo City based on satellite images; (2) to analyze the relationships between the landslide distribution and the elevation, slope, aspect, strata, distance to faults, distance to rivers, distance to roads, normalized vegetation difference index (NDVI), and rainfall and explore the landslide development; (3) to use a logistic regression model to classify the typhoon rainstorm landslide susceptibility in the study area based on the above 9 factors; and (4) to calculate the probability of landslide occurrence under different rainfall conditions based on the Bayesian probability method and a logistic regression model and create a landslide hazard zoning map.

2. Geological Background

The study area (118°38′–119°17′ E, 30°43′–30°42′ N) is located in Ningguo City, Anhui Province, China, with slightly different local boundaries. The total area is 3002 km^2. The geomorphology of the area is mountainous and hilly; the southeastern and southwestern regions are mountainous and the central and northern regions are hilly. The overall terrain is high in the south and low in the north. The highest elevation is 1444 m, and the lowest elevation is 12 m (Figure 1). The strata in the study area are the characteristic strata in the Yangtze region, and the exposed strata include Silurian (S), Cambrian (∈), Ordovician (O), Sinian (Z), Triassic (T), Jurassic (J), Quaternary (Qp), Devonian (D), Permian (P), Cretaceous (K), Carboniferous (C), granite porphyry ($\gamma\pi$), and a small number of unknown strata (using NONE as its code). Among them, the Silurian and Cambrian strata are the most commonly exposed (Figure 2). The study area is located in the Jiangnan uplift belt in the southeastern part of the Yangtze region. The geological structure is relatively complex. The main body of the structure is NW-trending, and the NW-trending faults are also well-developed (Figure 2). The neotectonic movement in the region has been characterized by intermittent slow uplift and local uplift. The seismic activity in the study area is low, the magnitude is small, and there are no active fractures [22]. The study area has a humid subtropical monsoon climate zone in the northern midlatitude region, with four distinct seasons, a mild climate, and abundant rainfall. In the hydrologic regionalization of China, it is located in the wet zone and has abundant water and the hydrogeological conditions are relatively simple. The main rivers in the region are the Shuiyang, Dongjin, Zhongjin, and Xijin rivers. Figure 3 shows the annual average rainfall in Ningguo City. It can be seen from the figure that the average annual rainfall in Ningguo City is more than 1000 mm. In particular, in 2016, the annual average rainfall reached 2267.9 mm.

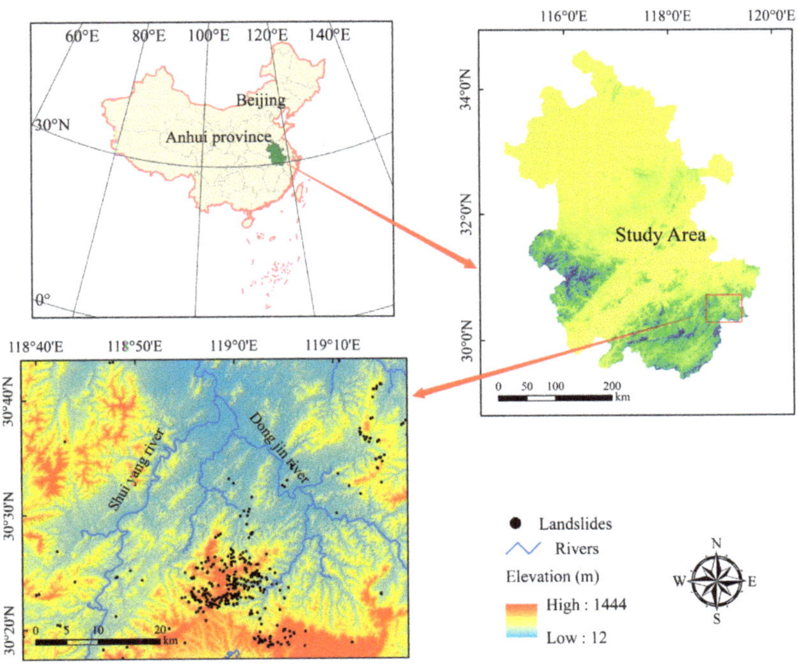

Figure 1. Location of the study area.

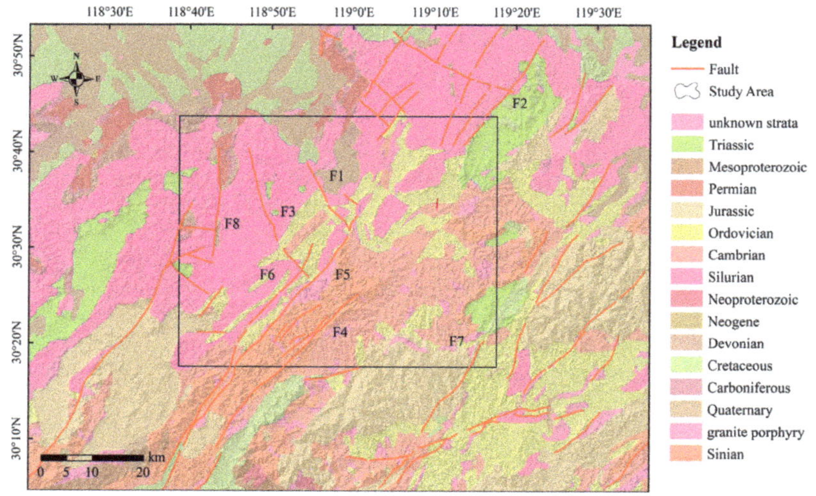

Figure 2. Topography and geomorphology. F1: Zhufengpu fault; F2: Zhangcun fault; F3: Xiyusucun-Pawandian fault; F4: Longzhishan thrust; F5: Jixi-Houkengwu thrust; F6: Jikengkou-Hulesi thrust; F7: Tangjiawan fault; F8: Taohuayuan fault.

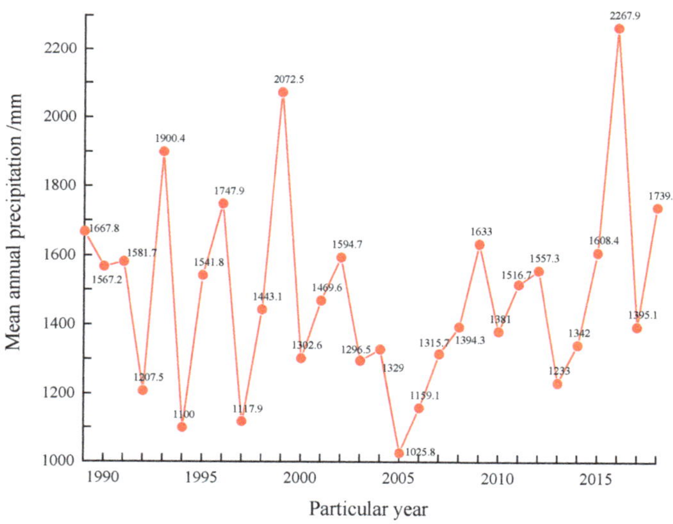

Figure 3. Average rainfall in Ningguo City from 1989 to 2018.

3. Data and Methods

3.1. Data

Artificial visual interpretation of the landslides was carried out based on 3 m resolution planet satellite images. Satellite images acquired on 2 August, 5 August, and 6 August 2019 were selected to represent the area before the landslides occurred in the study area, and satellite images acquired on 17 August and 27 August 2019, were selected to represent the area after the landslides occurred in the study area. The elevation, slope, and aspect were extracted from a 12.5 m resolution digital elevation model (DEM) downloaded from the Advanced Land Observing Satellites (ALOSs) (earth observation satellites of Japan) (https://search.asf.alaska.edu, accessed on 10 January 2022). The faults and rivers are from Deng Qidong's Active Tectonic map of China. The strata were extracted by vectorization of a 1:500,000 geological map in ArcGIS, and the roads were extracted from national roads data (https://malagis.com/, accessed on 10 January 2022). The NDVI data were downloaded from the Geospatial Data Cloud (http://www.gscloud.cn/search, accessed on 10 January 2022); the selected product was a composite of 1–5 August 2019, i.e., corresponding to the date of the landslides. The rainfall data were collected by six meteorological stations in the study area (http://data.cma.cn, accessed on 10 January 2022) and the distances from these meteorological stations to the center of the study area are 56.55 km, 46.85 km, 11.95 km, 59.19 km, 58.07 km, and 72.71 km.

3.2. Methods

Using the latitude and longitude grid in the ArcGIS platform, the study area was divided into several small areas. Then, the satellite images acquired before and after the landslides were compared one by one, mainly including a comparison of the tonal changes of the images, which are caused by the destruction of surface vegetation induced by landslides. The white areas in the images represent vegetation destruction and valley areas, and the green areas in the images represent vegetation coverage areas. If one small area is green before the typhoon and turns white after the typhoon, then this area was affected by a landslide. A complete landslide inventory for the study area was then obtained. Typical landslide examples are shown in Figure 4. Figure 4a is an image acquired before the landslides occurred, and Figure 4b is an image acquired after the landslides occurred.

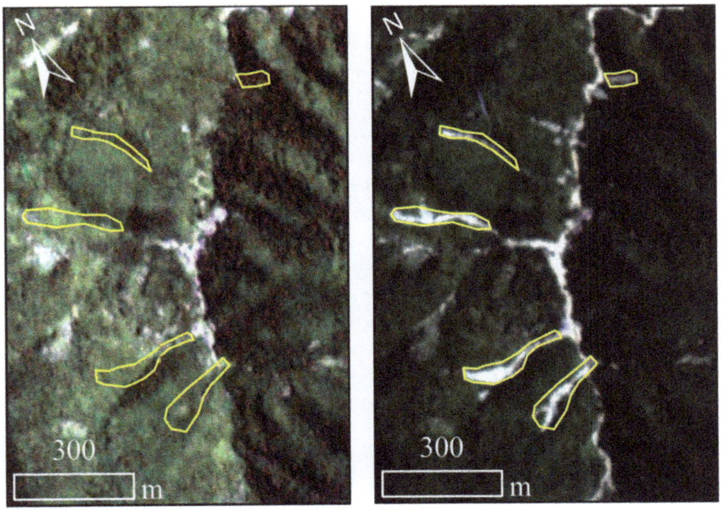

(a) (b)

Figure 4. Typical landslide examples. (**a**) Image acquired before the landslides occurred and (**b**) image acquired after the landslides occurred.

The relationships between the spatial distribution of the landslides and the various influencing factors were analyzed based on the established landslide database. The real probability of landslide occurrence under different rainfall conditions was calculated by combining the Bayesian probability method with a logistic regression model [23]. The sample points were selected to be uniformly distributed throughout the entire study area. The size of the study area was 3002.3 km^2, and 50 points per square kilometer were evenly selected, i.e., 150,100 points in total. To avoid repeated points in each grid, the minimum distance between sample points was set to 30 m. The obtained random points within a landslide area were defined as landslide samples, and those within non-landslide areas were defined as non-landslide samples. Finally, 72 landslide samples and 15,028 non-landslide samples were obtained. The state quantity of landslide occurrence was defined as 1, and the state quantity of non-occurrence was defined as 0.

4. Results and Analysis

4.1. Landslide Database

The landslide database established through artificial visual interpretation contains 414 landslides. The total landslide area was 1.42 km^2, of which the smallest landslide area was 235 m^2 and the largest landslide area was 49,826 m^2. The number and area proportion of the landslides in different zones are presented in Table 1. The largest number of landslides had areas of 0–2000 m^2, accounting for 47.6% (197) of the total number of landslides. Those with areas of 2000–5000 m^2 accounted for 31.7% of the total landslide area, i.e., 0.45 km^2. This indicates that the landslides induced by the typhoon rainstorm were mostly small.

Table 1. Number of landslides and proportion of landslide area.

Landslide Area Zones (m^2)	0–2000	2000–5000	5000–10,000	10,000–20,000	20,000–50,000
Number of landslides	197	149	50	11	7
Proportion of landslides (%)	47.6	36	12.1	2.6	1.7
Landslide area (km^2)	0.25	0.45	0.34	0.14	0.24
Proportion of landslide area (%)	17.6	31.7	23.9	9.9	16.9

4.2. Analysis of Factors Influencing the Landslides

Based on previous studies [24–26] and the actual situation of these typhoon rainstorm-induced landslides, nine influencing factors were selected for the landslide analysis: elevation, slope, aspect, strata, distance to faults, distance to rivers, distance to roads, NDVI, and rainfall (Figure 5). All factor layers were transformed into 30 m × 30 m grid layers, and the landslide surface was converted into a point. The landslide point was extracted into each factor layer. The relationships between the landslide and each factor are shown in Figure 6. To intuitively analyze the relationships between the spatial distribution of the landslides and the influencing factors, the influencing factors were reclassified (Table 2) [24–26]. The slope direction and strata were classified as classification factors, and the other factors were classified as continuous factors.

Table 2. Classification of influencing factors.

Influencing Factors	Classification Standard	Type
Elevation (m)	1. 12–250; 2. 250–500; 3. 500–750; 4. 750–1000; 5. 1000–1250; 6. 1250–1444	Continuous
Slope (°)	1. 0–10; 2. 10–20; 3. 20–30; 4. 30–40; 5. 40–50; 6. 50–80	Continuous
Aspect	1. N (0–22.5, 337.5–360); 2. NE (22.5–67.5); 3. E (67.5–112.5); 4.SE (112.5–157.5); 5. S (157.5–202.5); 6. SW (202.5–247.5); 7. W (247.5–292.5); 8. NW (292.5–337.5); 9. Flat (−1)	Classification
Distance to faults (m)	1. 0–1000; 2. 1000–2000; 3. 2000–3000; 4. 3000–4000; 5. 4000–5000; 6. 5000–10,000	Continuous
Strata	1. unknown strata; 2. Ordovician (O); 3. Cambrian (∈); 4. Triassic (T); 5. Sinian (Z); 6. Silurian (S); 7. Jurassic (J); 8. Cretaceous (K); 9. Quaternary (Qp); 10. Permian (P); 11. Devonian (D); 12. Carboniferous (C); 13. granite porphyry (γπ)	Classification
Distance to rivers (m)	1. 0–1500; 2. 1500–3000; 3. 3000–4500; 4. 4500–6000; 5. 6000–7500; 6. 7500–9000; 7. 9000–31,000	Continuous
Distance to roads (m)	1. 0–2500; 2. 2500–5000; 3. 5000–7500; 4. 7500–10,000; 5. 10,000–12,500; 6. 12,500–15,000; 7. 15,000–36,000	Continuous
NDVI	1. 0.35–0.5; 2. 0.5–0.6; 3. 0.6–0.7; 4. 0.7–0.8; 5. 0.8–0.9; 6. 0.9–0.98	Continuous
3 days rainfall (mm)	1. 175–200; 2. 200–250; 3. 250–300; 4. 300–350; 5. 350–402	Continuous

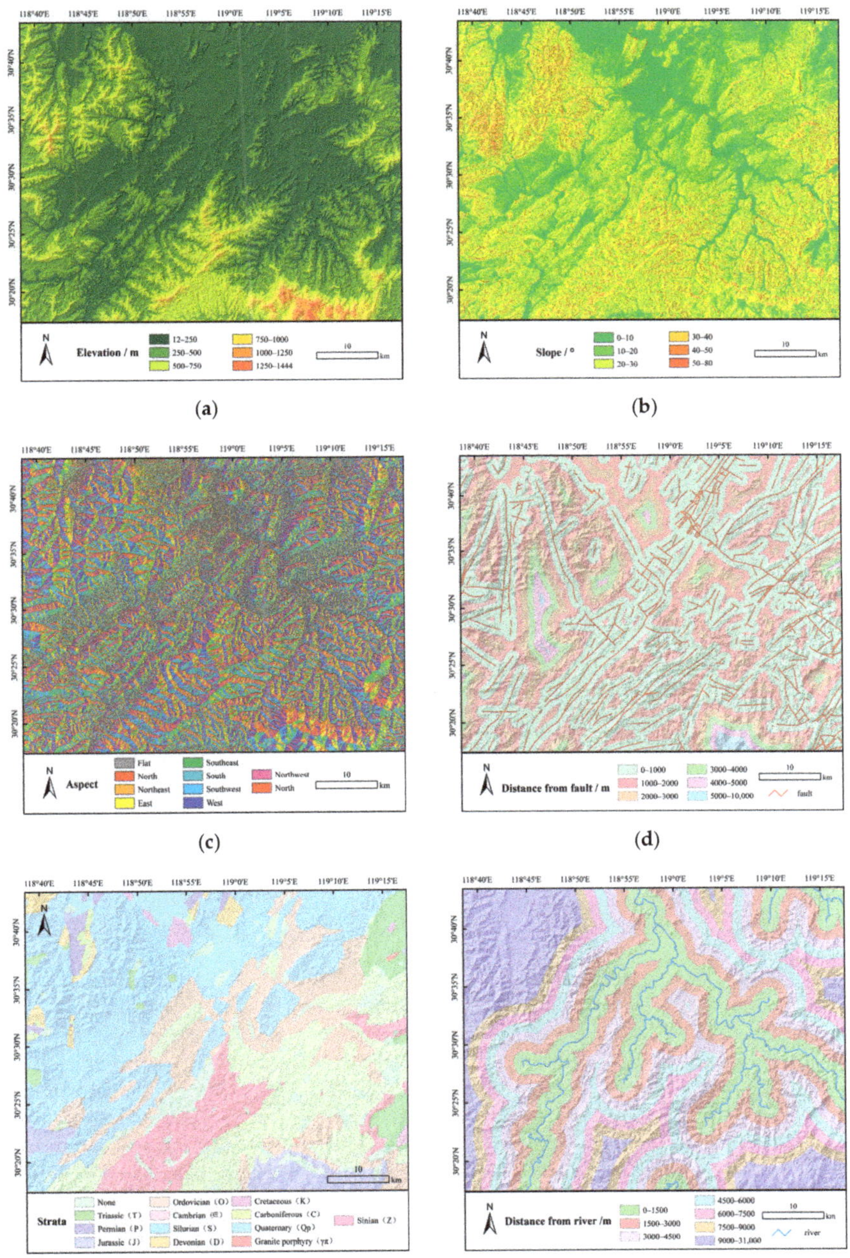

(**a**) (**b**)

(**c**) (**d**)

(**e**) (**f**)

Figure 5. *Cont.*

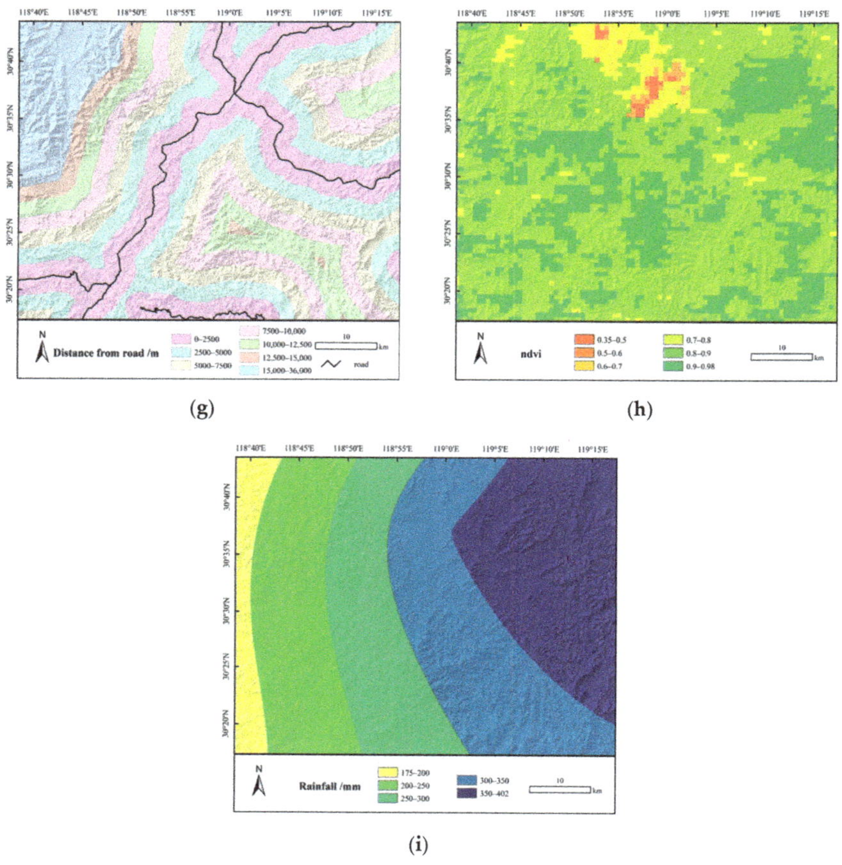

Figure 5. Influencing factor layers. (**a**) Elevation; (**b**) Slope; (**c**) Aspect; (**d**) Distance to faults; (**e**) Strata; (**f**) Distance to rivers; (**g**) Distance to roads; (**h**) NDVI; (**i**) 3 days rainfall.

The classified area and its relationship with LND and LAP are shown in Figure 6. As can be seen in Figure 6, the largest proportion of landslides occurred in the elevation range of 750–1000 m; when the elevation was greater than 250 m, the number of landslides decreased (Figure 6a). In the slope range of 30–40°, the distribution of landslides was more concentrated, and when the slope was greater than 50° or less than 10°, almost no landslides occurred (Figure 6b). The occurrence probability of landslides was the largest in the southeast slope direction (Figure 6c). The landslides were concentrated at distances of 7500–12,500 m from the roads (Figure 6d). The occurrence probability of landslides was the highest in the Sinian strata, and almost no landslides occurred in the Cretaceous, Quaternary, Permian, Devonian, Carboniferous, and granite porphyry strata (Figure 6e). The probability of landslide occurrence was the largest for distances to rivers of 6000–9000 m (Figure 6f). Most of the landslides were concentrated at distances of 3000–4000 m from the faults (Figure 6g). The landslide distribution was concentrated in the NDVI range of 0.8–0.98 (Figure 6h). The landslide distribution was most concentrated in the rainfall range of 300–350 mm (Figure 6i).

4.3. Calculation of Landslide Probability

The landslide samples and non-landslide samples were extracted into each factor layer and substituted into the SPSS software to calculate the regression coefficient of each factor. Finally, an absolute probability index graph was obtained through superposition of each factor layer. By importing the nine factors into the SPSS software for logistic regression analysis and by classifying the factors for classification covariable processing, the regression coefficient of the first range was set to zero. If the regression coefficients of the other ranges are positive, it is more conducive to the occurrence of landslides than the first range. For the continuous factors, if the weight value is positive, the probability of landslide occurrence is positively correlated with the factor, and if the weight value is negative, the probability of landslide occurrence is negatively correlated with the factor. The weights of each continuous variable and the regression coefficients of the different intervals for the classification variables were obtained. The results are presented in Table 3.

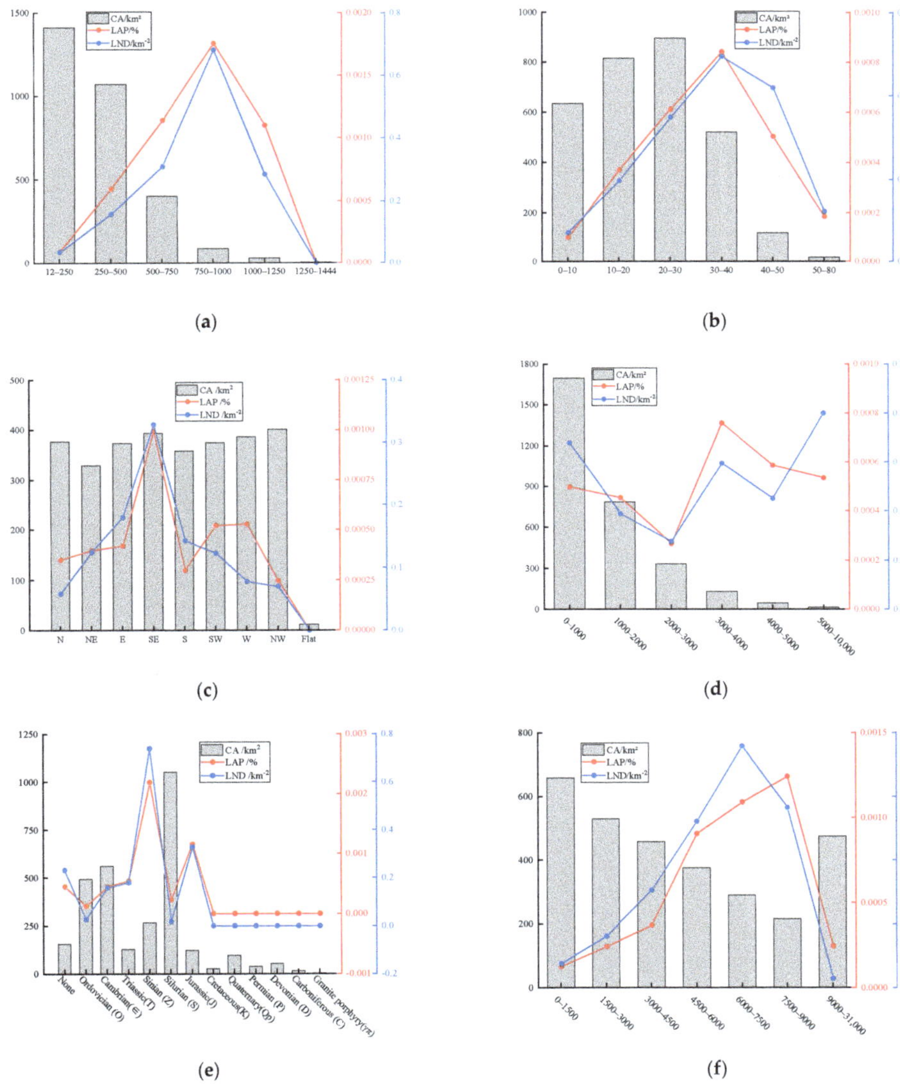

Figure 6. *Cont.*

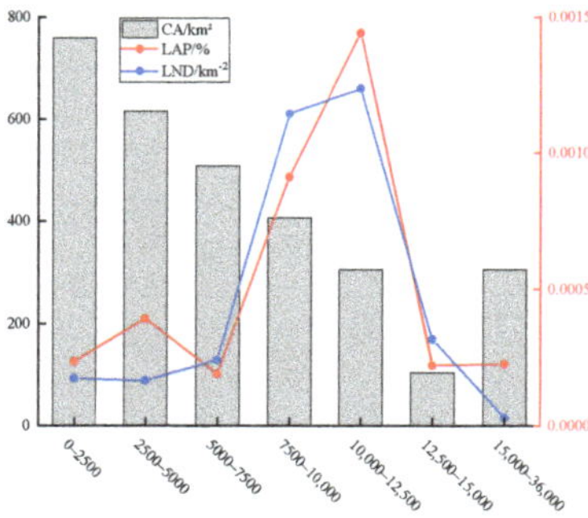

(**g**)

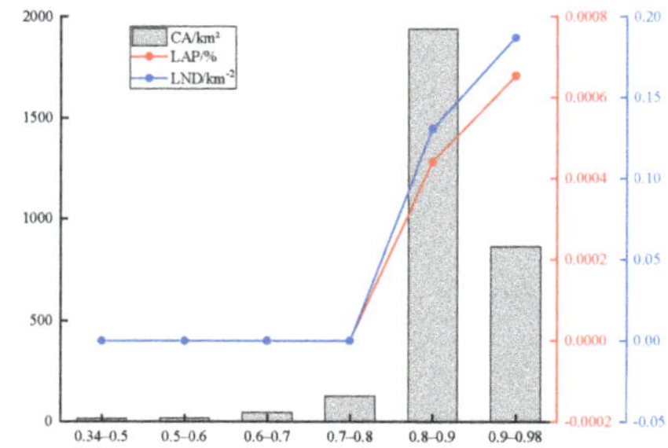

(**h**)

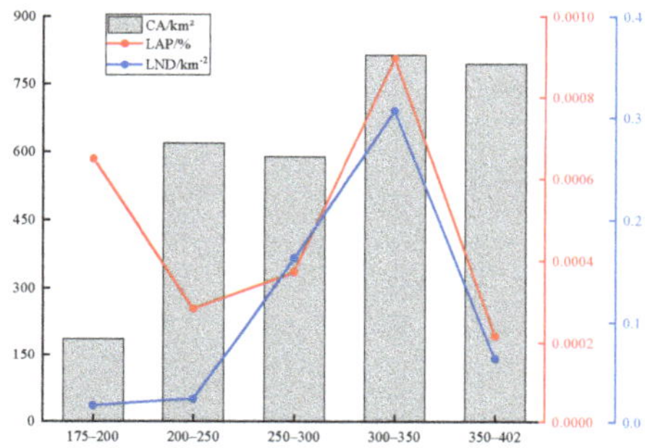

(**i**)

Figure 6. Relationships between landslide distribution and influencing factors. CA: classification area; LND: landslide number density; LAP: landslide area percentage. (**a**) Elevation; (**b**) Slope; (**c**) Aspect; (**d**) Distance to faults; (**e**) Strata; (**f**) Distance to rivers; (**g**) Distance to roads; (**h**) NDVI; (**i**) 3 days rainfall.

It can be seen in Table 3 that the weights of elevation, slope, rainfall, distance to faults, distance to rivers, and distance to roads are positive, so they are positively correlated with the occurrence of landslides. The NDVI weight is negative, so it is negatively correlated with the occurrence of landslides. The weight values of the strata and slope greatly differ in each interval of the classification because there are no landslides in some of the classification ranges; there are many landslides in other classification ranges, and different categories have large differences for rainfall-induced landslides. Based on the logistic regression model, each layer was superimposed in ArcGIS to obtain a landslide occurrence probability model and the accuracy of the model was verified using the receiver operating characteristic (ROC) curve. The area under the curve (AUC) refers to the area under the ROC curve. When the AUC is greater than 0.7, the model has a high accuracy [15,23]. The research results are shown in Figure 7. It can be seen in Figure 7 that the AUC = 0.886, indicating that the accuracy of the model is very high and the results are reliable.

4.4. *Probability of Landslides under Different Rainfall Conditions*

Based on the constructed model, the real probabilities of landslide occurrence in the study area under five different rainfall conditions of 175–200 mm, 200–250 mm, 250–300 mm, 300–350 mm, and 350–402 mm were predicted to be 0.001%, 0.01%, 0.1%, and 1%, which are discontinuous values that divide the study area into five grades (Figure 8). The prediction results can be used to quickly assess the risk of landslide occurrence in the region according to the real rainfall, and they provide a reference for subsequent disaster prevention and mitigation and post-disaster reconstruction. Under different rainfall conditions, the areas with different probabilities of landslide occurrence in the study area are shown in Figure 9. It can be seen in Figure 9 that as the rainfall increases, the areas with a high probability of landslide occurrence become larger, and the areas with a low probability of landslide occurrence become smaller. This demonstrates that rainfall is positively correlated with the occurrence of landslides and landslides are more likely to occur under higher rainfall.

Table 3. Weights of each factor.

Factors	Type	Classification	Weights
Elevation	Continuous		0.166
Slope	Continuous		0.195
Rainfall	Continuous		0.211
NDVI	Continuous		−0.274
Distance to faults	Continuous		0.134
Distance to rivers	Continuous		0.05
Distance to roads	Continuous		0.486
Aspect	Classification	1	0
		2	0.931
		3	1.004
		4	1.402
		5	1.014
		6	1.201
		7	0.926
		8	0.186
		9	−11.167
Strata	Classification	1	0
		2	−13.156
		3	−0.141
		4	−0.232
		5	1.448
		6	−2.043
		7	0.41
		8	−13.589
		9	−13.821
		10	−14.688
		11	−14.631
		12	−14.861
		13	−14.328

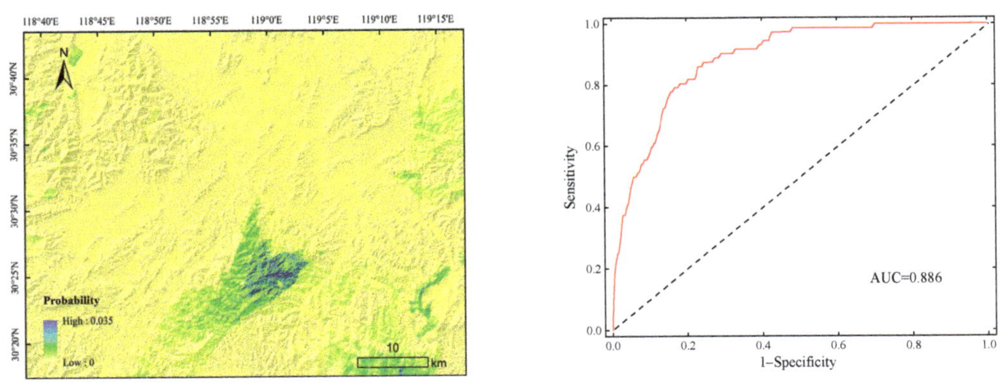

(a) (b)

Figure 7. Landslide occurrence probability model. (**a**) Landslide occurrence probability distribution and (**b**) ROC curve of the model.

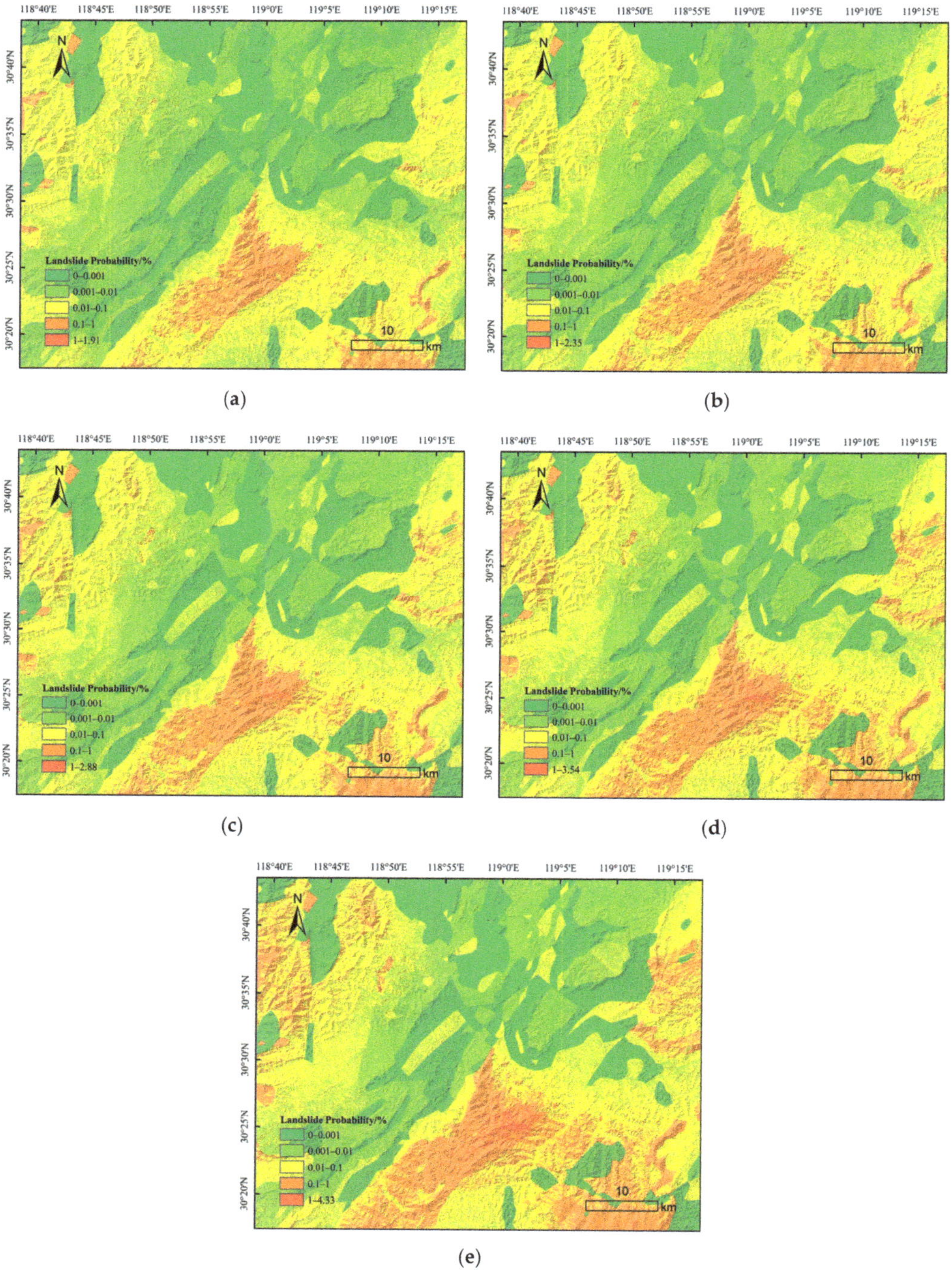

Figure 8. Landslide probability distribution map for different amounts of 3 days of rainfall: (**a**) 175–200 mm; (**b**) 200–250 mm; (**c**) 250–300 mm; (**d**) 300–350 mm; (**e**) 350–402 mm.

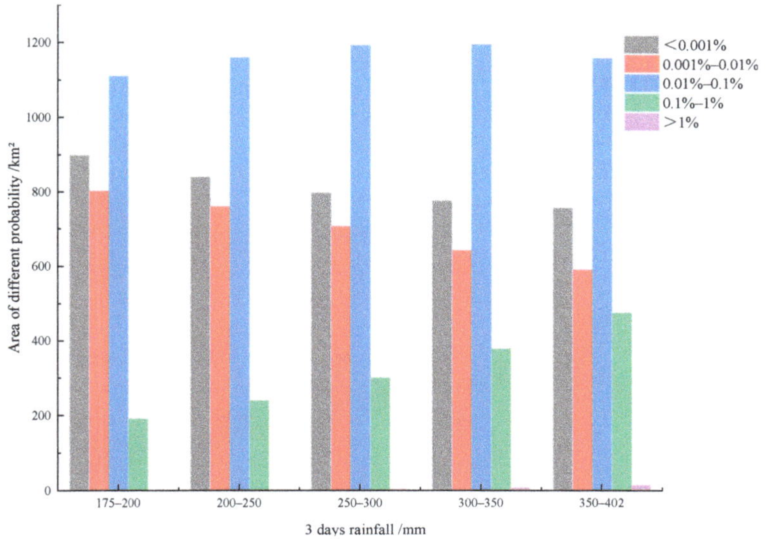

Figure 9. Areas of different probabilities.

5. Discussion

5.1. Landslide Interpretation

In this study, a complete database of shallow landslides in Ningguo City caused by Typhoon Lekima was obtained using visual interpretation of remote sensing images. Then, the Bayesian probability method was combined with a logistic regression model to establish a landslide probability map for the study area. The rationality test using ROC curve shows that the accuracy of the prediction model is very high and the results are reliable. Based on the above results, the real probabilities of landslide occurrence in the study area under five different rainfall conditions of 175–200 mm, 200–250 mm, 250–300 mm, 300–350 mm, and 350–402 mm were predicted.

In recent years, with the popularization of high-precision images and the development of remote sensing technology, new progress has been made in the interpretation and research of landslides caused by extreme rainfall. For example: from 19 to 20 October 2004, Shikoku experienced extreme events of Typhoon Tokage rainfall and 201 small-scale slides in the Moriyuki catchment and 142 in the Monnyu catchment were noticed [27]; from 14 to 16 July 2006, Typhoon Bilis swept over Southern China and, in analyzing the pre- and post-event images of QuickBird and CBERS, a total of 2407 landslide sites in the area around the Dongjiang Reservoir in Hunan Province were inventoried [28]; on 16 October 2013, Typhoon Wipha struck the Izuo-Oshima Volcanic Island, causing torrential rainfall there in a short time, and 44 landslides based on aerial images were interpreted [17]. In this study, satellite images were acquired before and after the rainfall and identified 414 shallow landslides in Ningguo City. Compared with other rainfall landslide sites, Ningguo City is a mountainous and hilly area in southeastern Anhui with complex topography and landforms. Heavy rains have caused floods and triggered massive landslides. In addition, the local lack of adequate defense measures and lack of rescue experience are important reasons for the severe disaster.

5.2. Research on Real Probability Based on the Bayesian Method

Due to the influence of the sampling ratio, the landslide probability obtained using a logistic regression model deviates greatly from the actual landslide probability. In the past, the ratio of landslide sample to non-landslide sample was 1:1 [15,29] and the results obtained by this sampling method are often much larger than the actual landslide probability. In this study, the Bayesian probability method was combined with a logistic regression model. The sample points were selected to be uniformly distributed throughout the entire study area. The size of the study area was 3002.3 km^2, and the total landslide area was 1.42 km^2. A total of 150,100 points were selected. The obtained random points within a landslide area were defined as landslide samples, and those within non-landslide areas were defined as non-landslide samples. Finally, 72 landslide samples and 15,028 non-landslide samples were obtained.

Based on the Bayesian method, the predicted landslide area A_P using the logistic regression model is:

$$A_P = \sum_{i=1}^{m} \sum_{j=1}^{n} P_{i,j} A \tag{1}$$

where $P_{i,j}$ is the landslide probability of row i and column j of the grid, m is the number of rows, n is the number of columns, and A is the area of the unit grid (900 m^2).

According to the results, the predicted landslide area A_P is 1.45 km^2.

The relative difference between the predicted landslide area A_P and the actual landslide area A_a (1.42 km^2) is expressed as:

$$Different = \left| \frac{A_p - A_a}{A_a} \times 100\% \right| \tag{2}$$

The relative difference between the predicted landslide area and the actual landslide area is 2.1%, the error is small, the predicted landslide area is close to the actual landslide area, and the prediction result is reliable.

The landslide prediction probability is not only affected by the sampling ratio, but also by various aspects such as grid resolution. The spatial distribution of predicted probability at different resolutions is different [29]. Ways to improve the accuracy of landslide prediction should be further discussed in future research.

6. Conclusions

Based on the landslides in Ningguo City induced by Typhoon Lekima, in this study, 414 landslides were identified via artificial visual interpretation and a rainfall landslide database was established. To analyze the relationships between the factors and the landslide distribution, nine influencing factors were selected: elevation, slope, aspect, distance to faults, strata, distance to rivers, distance to roads, NDVI, and rainfall. The Bayesian probability method and a logistic regression model were used to establish a landslide occurrence probability model for the study area and to predict the probabilities of landslide occurrence under five different rainfall conditions, i.e., 175–200 mm, 200–250 mm, 250–300 mm, 300–350 mm, and 350–402 mm. The results show that elevation, slope, rainfall, distance to faults, distance to rivers, and distance to roads positively correlate with the occurrence of rainfall-induced landslides, and the NDVI negatively correlates with the occurrence of rainfall-induced landslides.

Landslides were caused by Typhoon Lekima's rainstorm, and many geological disasters were induced by the typhoon's landfall. Ningguo City was only one of the areas that experienced serious geological disasters. If images of the entire typhoon transit area were collected and a large-scale landslide interpretation was performed, a complete database of landslides caused by Typhoon Lekima's rainstorm in the southeast coastal area could be constructed; in addition, similar analysis and landslide prediction could be conducted, which would be of great significance to the disaster prevention and reduction of typhoon rainstorm-induced landslides in the southeast coastal area. This work will be carried out in a future study.

Author Contributions: Conceptualization, Q.H.; methodology, Y.C.; software, J.J.; validation, K.Y.; investigation, C.X.; writing—original draft preparation, Y.C.; writing—review and editing, C.X.; funding acquisition, Y.C. All authors have read and agreed to the published version of the manuscript.

Funding: This research was funded by the Fundamental Research Funds for the Central Universities, CHD, grant number 300102261503, the Natural Science Research Project of the Colleges and Universities in Anhui Province, grant number KJ2020ZD34, and the Postdoctoral Fund in Anhui Province, grant number 2021B545.

Data Availability Statement: The data used to support the findings of this study are included within the article.

Acknowledgments: We thank LetPub (www.letpub.com, accessed on 20 February 2022) for its linguistic assistance during the preparation of this manuscript.

Conflicts of Interest: The authors declare no conflict of interest.

References

1. Jiao, F. Analysis of Energy Evolution Characteristics and Transformation Mechanism of Typhoon "Mujigae" (2015) in the Process of Offshore and Landing. Ph.D. Thesis, Nanjing University of Information Science & Technology, Nanjing, China, 2021.
2. Wang, L.; Luo, Y.; Xu, L.Y.; Zhang, Q.; Chen, Y.; Ye, D.X. Review of Typhoon and Its Related Natural Disasters over the Past 35 Years in China. *Sci. Technol. Rev.* **2006**, *24*, 23–25. (In Chinese)
3. Liu, D.; Liang, P.; Xie, B. Typhoon disaster in China: Prediction, prevention, and mitigation. *Nat. Hazards* **2009**, *49*, 421–436. [CrossRef]
4. Borghuis, A.; Chang, K.; Lee, H. Comparison between automated and manual mapping of typhoon-triggered landslides from SPOT-5 imagery. *Int. J. Remote Sens.* **2007**, *28*, 1843–1856. [CrossRef]
5. Yin, K.; Chen, L.C.; Zhang, G. Regional landslide hazard warning and risk assessment. *Earth Sci. Front.* **2007**, *14*, 85–93. [CrossRef]
6. Xu, J.; Li, W. Characteristics of Rainfall for Typhoon Rainstorm Induced Geological Hazards. *Meteorol. Mon.* **2009**, *35*, 42–48. (In Chinese)
7. Cui, Y.; Hu, J.; Xu, C.; Zheng, J.; Wei, J. A catastrophic natural disaster chain of typhoon-rainstorm-landslide-barrier lake-flooding in Zhejiang Province, China. *J. Mt. Sci.* **2021**, *18*, 2108–2119. [CrossRef]
8. Xiang, C.; Zhao, H.; Liu, Q.; Dong, L. The distribution of torrential rain after landfalling of Typhoon "Lekima" (1909). *J. Meteorol. Sci.* **2020**, *40*, 294–302. (In Chinese)
9. Dahal, R.K.; Hasegawa, S.; Nonomura, A.; Yamanaka, M.; Masuda, T.; Nishino, K. Failure characteristics of rainfall-induced shallow landslides in granitic terrains of Shikoku Island of Japan. *Environ. Geol.* **2009**, *56*, 1295–1310. [CrossRef]
10. Greco, R.; Giorgio, G.; Capparelli, G.; Versace, P. Early warning of rainfall-induced landslides based on empirical mobility function predictor. *Eng. Geol.* **2013**, *153*, 68–79. [CrossRef]
11. Gutierrez-Martin, A. A GIS-physically-based emergency methodology for predicting rainfall-induced shallow landslide zonation. *Geomorphology* **2020**, *359*, 107121. [CrossRef]
12. Althuwaynee, O.F.; Pradhan, B.; Ahmad, N. Estimation of rainfall threshold and its use in landslide hazard mapping of Kuala Lumpur metropolitan and surrounding areas. *Landslides* **2015**, *12*, 861–875. [CrossRef]
13. Segoni, S.; Piciullo, L.; Gariano, S.L. A review of the recent literature on rainfall thresholds for landslide occurrence. *Landslides* **2018**, *15*, 1483–1501. [CrossRef]
14. Teja, T.S.; Dikshit, A.; Satyam, N. Determination of rainfall thresholds for landslide prediction using an algorithm-based approach: Case study in the Darjeeling Himalayas, India. *Geosciences* **2019**, *9*, 302. [CrossRef]
15. Cui, Y.; Bao, P.; Xu, C.; Ma, S.; Zheng, J.; Fu, G. Landslides triggered by the 6 September 2018 Mw 6.6 Hokkaido, Japan: An updated inventory and retrospective hazard assessment. *Earth Sci. Inform.* **2020**, *14*, 247–258. [CrossRef]
16. Cui, Y.; Hu, J.; Zheng, J.; Fu, G.; Xu, C. Susceptibility assessment of landslides caused by snowmelt in a typical loess area in the Yining County, Xinjiang, China. *Q. J. Eng. Geol. Hydrogeol.* **2022**, *55*, 1–10. [CrossRef]
17. Dou, J.; Yunus, A.P.; Tien Bui, D.; Sahana, M.; Chen, C.W.; Zhu, Z.; Wang, W.; Thai Pham, B. Evaluating GIS-based multiple statistical models and data mining for earthquake and rainfall-induced landslide susceptibility using the LiDAR DEM. *Remote Sens.* **2019**, *11*, 638. [CrossRef]
18. Raia, S.; Alvioli, M.; Rossi, M.; Baum, R.; Godt, J.; Guzzetti, F. Improving predictive power of physically based rainfall-induced shallow landslide models: A probabilistic approach. *Geosci. Model Dev.* **2014**, *7*, 495–514. [CrossRef]
19. Capparelli, G.; Versace, P. FLaIR and SUSHI: Two mathematical models for early warning of landslides induced by rainfall. *Landslides* **2011**, *8*, 67–79. [CrossRef]
20. Dou, J.; Yunus, A.P.; Bui, D.T.; Merghadi, A.; Sahana, M.; Zhu, Z.; Chen, C.W.; Khosravi, K.; Yang, Y.; Pham, B.T. Assessment of advanced random forest and decision tree algorithms for modeling rainfall-induced landslide susceptibility in the Izu-Oshima Volcanic Island, Japan. *Sci. Total Environ.* **2019**, *662*, 332–346. [CrossRef]
21. Vasu, N.N.; Lee, S.R.; Pradhan, A.M.S.; Kim, Y.T.; Kang, S.H.; Lee, D.H. A new approach to temporal modelling for landslide hazard assessment using an extreme rainfall induced-landslide index. *Eng. Geol.* **2016**, *215*, 36–49. [CrossRef]

22. Zheng, J.; Zou, Z. Environmental geological characteristics of the mountainarea of southern Anhui Province and treatment of geological disasters due to landslide. *J. Hefei Univ. Technol.* **2006**, *29*, 102–105. (In Chinese)
23. Shao, X.; Xu, C.; Ma, S.; Xu, X.; Shyu, J.B.H.; Zhou, Q. Calculation of landslide occurrence probability in Taiwan region under different ground motion conditions. *J. Mt. Sci.* **2021**, *18*, 1003–1012. [CrossRef]
24. Dai, F.; Xu, C.; Yao, X.; Xu, L.; Tu, X.; Gong, Q. Spatial distribution of landslides triggered by the 2008 Ms 8.0 Wenchuan earthquake, China. *J. Asian Earth Sci.* **2011**, *40*, 883–895. [CrossRef]
25. Shao, X.; Ma, S.; Xu, C.; Zhou, Q. Effects of sampling intensity and non-slide/slide sample ratio on the occurrence probability of coseismic landslides. *Geomorphology* **2020**, *363*, 107222. [CrossRef]
26. Xu, C.; Shen, L.; Wang, G. Soft computing in assesment of earthquake-triggered landslide susceptibility. *Environ. Earth Sci.* **2016**, *75*, 767. [CrossRef]
27. Dhakal, A.S.; Sidle, R.C. Distributed simulations of landslides for different rainfall conditions. *Hydrol. Processes* **2004**, *18*, 757–776. [CrossRef]
28. Dou, J.; Yamagishi, H.; Xu, Y.; Zhu, Z.; Yunus, A.P. Characteristics of the torrential rainfall-induced shallow landslides by Typhoon Bilis, in July 2006, using remote sensing and GIS. In *GIS Landslide*; Springer: Tokyo, Japan, 2017; pp. 221–230.
29. Shao, X.; Ma, S.; Xu, C.; Xu, X. Effects of raster resolution on real probability of landslides. *Remote Sens. Appl. Soc. Environ.* **2020**, *19*, 100364. [CrossRef]

forests

MDPI

Article

A Hybrid Landslide Warning Model Coupling Susceptibility Zoning and Precipitation

Deliang Sun [1], Qingyu Gu [1], Haijia Wen [2,*], Shuxian Shi [3], Changlin Mi [4] and Fengtai Zhang [5]

1 Key Laboratory of GIS Application Research, Chongqing Normal University, Chongqing 401331, China; 20130831@cqnu.edu.cn (D.S.); guqingy00@gmail.com (Q.G.)
2 Key Laboratory of New Technology for Construction of Cities in Mountain Area, Ministry of Education, Chongqing University, Chongqing 400045, China
3 Key Laboratory of Geographic Information Science, Ministry of Education, East China Normal University, Shanghai 200241, China; shelly-shi@outlook.com
4 Natural Resources Development Service Center of Linyi, Linyi 276000, China; michanglin@ly.shandong.cn
5 School of Management, Chongqing Technology and Business University, Chongqing 400054, China; zhfthero45@cqut.edu.cn
* Correspondence: jhw@cqu.edu.cn

Abstract: Landslides are one of the most severe and common geological hazards in the world. The purpose of this research is to establish a coupled landslide warning model based on random forest susceptibility zoning and precipitation. The 1520 landslide events in Fengjie County, Chongqing, China, before 2016 are taken as research cases. We adapt the random forest model to build a landslide susceptibility model. The antecedent effective precipitation model, based on the fractal relationship, is used to calculate the antecedent effective precipitation in the 10 days before the landslide event. Based on different susceptibility zones, the effective precipitation corresponding to different cumulative frequencies is counted as the threshold, and the threshold is adjusted according to the fitted curve. Finally, according to the daily precipitation, the rain warning levels in susceptibility zones are further adjusted, and the final prewarning model of the susceptibility zoning and precipitation coupling is obtained. The results show that the random forest model has good prediction ability for landslide susceptibility zoning, and the precipitation warning model that couples landslide susceptibility, antecedent effective precipitation, and the daily precipitation threshold has high early warning ability. At the same time, it was found that the precipitation warning model coupled with antecedent effective precipitation and the daily precipitation threshold has more accurate precipitation warning ability than the precipitation warning model coupled with the antecedent effective precipitation only; the coupling of the two can complement each other to better characterize the occurrence of landslides triggered by rainfall. The proposed coupled landslide early warning model based on random forest susceptibility and rainfall inducing factors can provide scientific guidance for landslide early warning and prediction, and improve the manageability of landslide risk.

Keywords: landslide susceptibility; antecedent effective precipitation; daily precipitation; hybrid landslide warning model

Citation: Sun, D.; Gu, Q.; Wen, H.; Shi, S.; Mi, C.; Zhang, F. A Hybrid Landslide Warning Model Coupling Susceptibility Zoning and Precipitation. *Forests* **2022**, *13*, 827. https://doi.org/10.3390/f13060827

Academic Editor: Filippo Giadrossich

Received: 27 March 2022
Accepted: 23 May 2022
Published: 25 May 2022

1. Introduction

A landslide is a geological disaster with further impact, and threatens the safety and property of human life to a certain extent [1]. According to the China Statistical Yearbook, from 2000 to 2019, a total of 320,000 natural disasters occurred, causing a total economic loss of 9.894 billion dollars. Of these, 220,000 landslide disasters occurred, accounting for 70.2% of the total. Chongqing is one of the four major geological disaster areas in China. Geological disasters cause about 40–60 deaths and direct economic losses of about 300–400 million yuan per year, accounting for more than 20% of the city's natural disaster losses. Among them, the Longjing landslide in Shizhu County, Chongqing, had a scale of about 142.83 × 104 m^3, directly threatening 2864 residents and 1630 pupils in primary schools, as well as roads, pipelines, and other urban infrastructure. According to the World Health Organization (WHO) report, between 1998 and 2017, the population affected by landslides was 4.8 million, and they caused more than 18,000 deaths. The main impacts of landslides include the destruction of houses and roads, damage to water storage facilities, and the blockage of rivers, causing huge casualties and economic losses [2]. Therefore, it is necessary to conduct a geological disaster investigation and an assessment of potential landslide occurrence areas, and take corresponding preventive measures to reduce the economic losses and casualties caused by landslides.

Landslide susceptibility is an effective method to prevent landslide disasters. It is based on local geological environment factors to evaluate the spatial distribution of landslide occurrence probability in specific areas; it is also an important means to study the occurrence of landslides [3]. The evaluation of landslide susceptibility began in the mid-1970s. Carrara [4] used a large number of data on landslides, topography, landforms, field surveys, etc., based on expert experience and other methods, to evaluate landslide susceptibility in the Columbia area. Many scholars divide landslide susceptibility modeling methods into two categories, qualitative and quantitative. Qualitative methods use prior knowledge to assess the occurrence of landslides [5,6]. This method is based on expert experience, and has the disadvantage of being subjective, which makes the differences between the models more significant. Quantitative methods use mathematical–statistical models to evaluate landslides [7–9]. The accuracy and completeness of the input landslide data have a greater impact on the statistical model. This model has higher requirements for data collection. With the development of GIS and machine learning, many scholars use machine learning methods to conduct landslide susceptibility evaluation studies in different research areas [10–14]. Sun et al. (2020) used an optimized random forest model to establish a landslide susceptibility model in Fengjie County, Chongqing City, and applied the model to Wushan County, Chongqing City, to discuss the generalization ability of the model based on random forest. The results show that the model had high susceptibility simulation ability and a good generalization effect [15]. At the same time, they compared the accuracy of different machine learning methods. Many scholars chose multiple machine learning algorithms—for example, logical model trees, random forest, classification regression trees, XGBoost, etc.—to build a landslide susceptibility model. Moayedi et al. used the artificial neural network model optimized by particle swarm to predict the landslide susceptibility in the Layleh valley area, and achieved good prediction results [16]. Pourghasemi et al. used 10 machine learning algorithms, such as artificial neural network (ANN), boosted regression trees (BRT), and random forest (RF), to model landslide susceptibility in the Ghaemshahr region of Iran, and found that RF (AUC = 83.7%) had the best prediction performance [17]. Zhou et al. developed a novel interpretable model based on SHAP and XGBoost, which provided 0.75 accuracy and 0.83 AUC value for the test sets [18]. Research has shown that the predictive ability of the random forest model is more significant in general [19–22].

Rainfall is the most important factor inducing landslides. At present, there are two main rainfall threshold models, a physical rainfall threshold and an empirical rainfall threshold [23–25]; the empirical threshold is based on the relationship between a large number of landslides and rainfall data, and is currently the most commonly used rainfall threshold model. Rainfall intensity and accumulation are the two main aspects that affect the rainfall threshold. There are many studies on this [26–28]. At present, there are two main types of precipitation threshold statistical models for studying landslides at home and abroad: daily precipitation [29] and antecedent effective precipitation [30,31]. However, research on the impact of precipitation on landslides is lagging [32,33], and threshold research cannot be conducted solely from the daily precipitation or antecedent effective precipitation accumulation. Therefore, it is necessary to adjust the existing precipitation threshold model according to the situation of the study area.

The occurrence of landslides is the result of the combined effects of natural conditions and human activities, in which rainfall triggers the occurrence of landslides. In high-susceptibility areas, relatively low precipitation can trigger the occurrence of landslides. The occurrence of landslides is related to the susceptibility of the area where the landslide is located, and also to the precipitation in the area. The study of landslide-inducing factors mainly considers the daily precipitation or the antecedent effective precipitation. There are few studies based on the susceptibility of landslides that consider multiple precipitation-inducing factors [34,35]. In this paper, the random forest model is used to evaluate the susceptible zones in the study area, and a landslide space–time joint early warning and forecast model is constructed according to the antecedent effective precipitation and the daily precipitation.

2. Materials and Methods

2.1. Methodology

Based on 16 evaluation factors and the random forest model, the landslide susceptibility zoning was established. Subsequently, based on the previous effective rainfall and the rainfall on the day of the landslide, rainfall-induced landslide early warning and forecasting was carried out. Based on the rainfall in the first 10 days of the landslide, the attenuation coefficient of the rainfall in the early stage of the landslide was calculated, and the threshold of the effective rainfall in the early stage of the landslide was constructed according to the frequency of landslide occurrence in each susceptibility zone. At the same time, the landslide warning and forecasting model was adjusted based on the rainfall that day. The specific steps are shown in Figure 1.

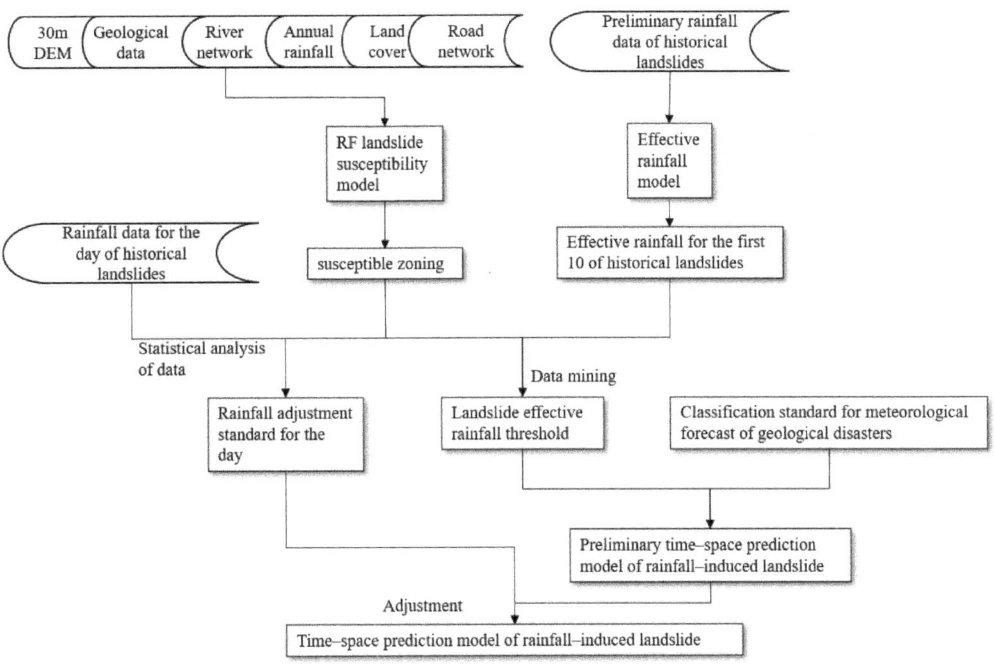

Figure 1. Overall technical route.

2.2. Study Area

Fengjie County is in the north of Chongqing (Figure 2). Fengjie County is located at the junction of the Daba Mountains Arc fold fault zone, the Eastern Sichuan Arc concave fold zone, and the Sichuan, Hunan, and Guizhou Uplift fold zones. Figure 3 shows the distribution of faults. The tectonic stress field is mainly characterized by near north–south compression, and the tectonic form is mainly folded. The direction of the tectonic line is northeast–east, and the anticline and syncline are parallel. In the whole study area, the main development consists of the Damuya anticline, Qumahe syncline, Soling syncline, Catch-up syncline, Qiyaoshan anticline, Wushan syncline, Hengshi anticline, and the Guandu syncline (Figure 3). It has mountainous landforms. The northeast and southeast regions of Fengjie are higher, and the central and western parts are relatively low-altitude. Its strata are mainly Quaternary (Q), Jurassic (J), Triassic (T), Permian (P), Carboniferous (C), Devonian (D), and Silurian (S). The geological conditions of Fengjie County are complex; most of the areas are mountainous areas, and geological disasters are frequent, large in scale, diverse in variety, and wide in scope, causing great loss of life and property. According to the Köppen–Geiger climate classification [36], the study area has a subtropical humid climate with abundant rainfall and long sunshine duration. The average annual temperature is 18.1 °C, the average annual relative humidity is 71.2%, the average annual precipitation is 1132 mm, and the annual sunshine hours are 1639 h.

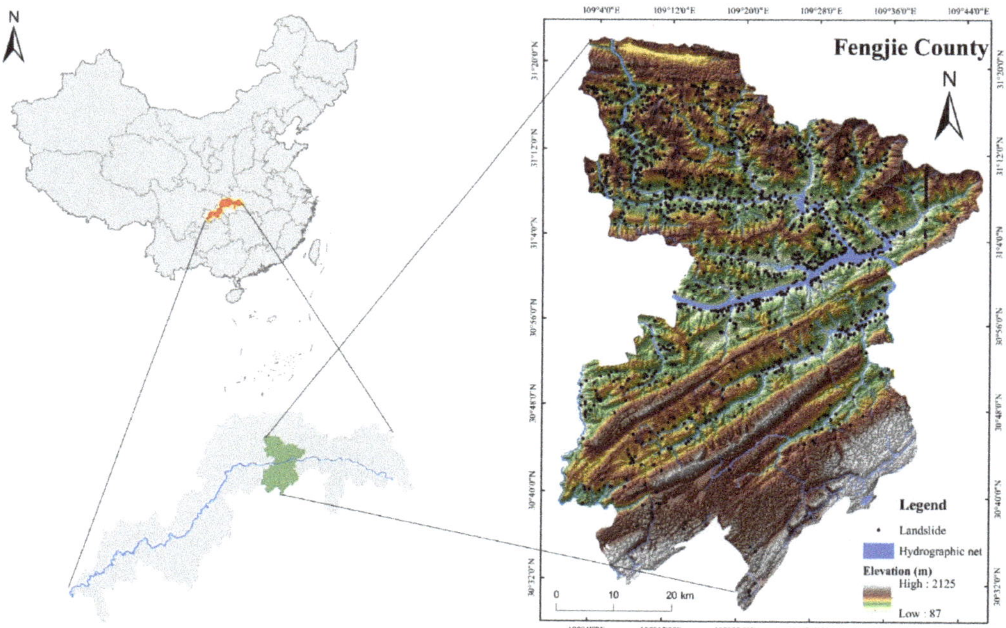

Figure 2. The location of the study area.

Fengjie County has a subtropical monsoon climate with an average annual rainfall of 1132 mm. Most of the landslides in Fengjie County occur from May to October, and the flood season, especially periods of concentrated rainfall, is a time of high incidences of landslide disasters in Fengjie County. Fengjie belongs to the Three Gorges Reservoir Area. The occurrence of landslides is closely related to the reservoir area. The periodic rise and fall of the water level in the reservoir area will cause landslide activity near the reservoir area to show unstable periodicity. The periodical change in reservoir water level will lead to instability of the slope around the reservoir area. In the normal or dry season, the slope is stable or primarily stable. In the flood season or rainstorm period, the slope is saturated, the stability is poor, and the area is prone to landslides [37]. The types and main triggering factors of 1520 historical landslides in Fengjie County from 2001 to 2016 were statistically analyzed (Figure 4). In terms of type, small/shallow/soil landslides account for 82% of the total number of landslides, and large/deep/clay landslides account for only

18%. From the perspective of main triggering factors, rainfall-induced landslides accounted for the majority (75.88%), whereas groundwater (pore water) and human activity-induced landslides accounted for 10.05% and 2.01%, respectively.

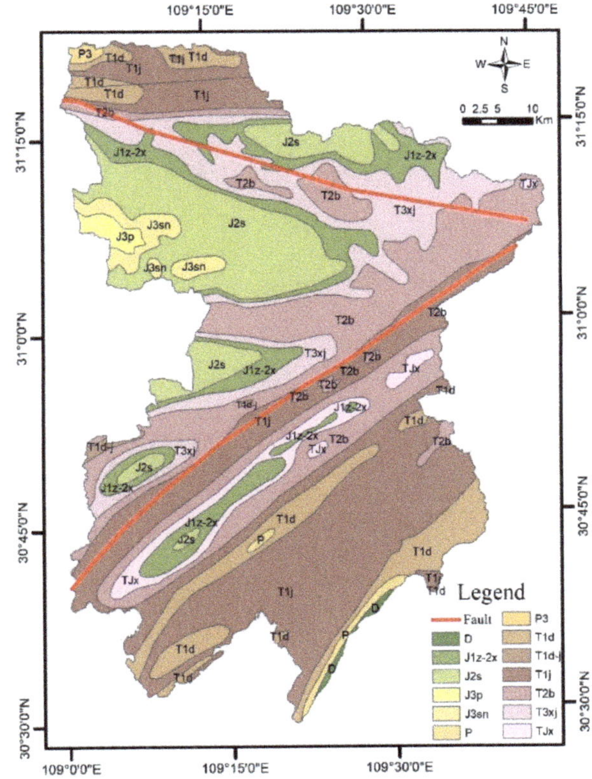

Figure 3. Geological map of the study area.

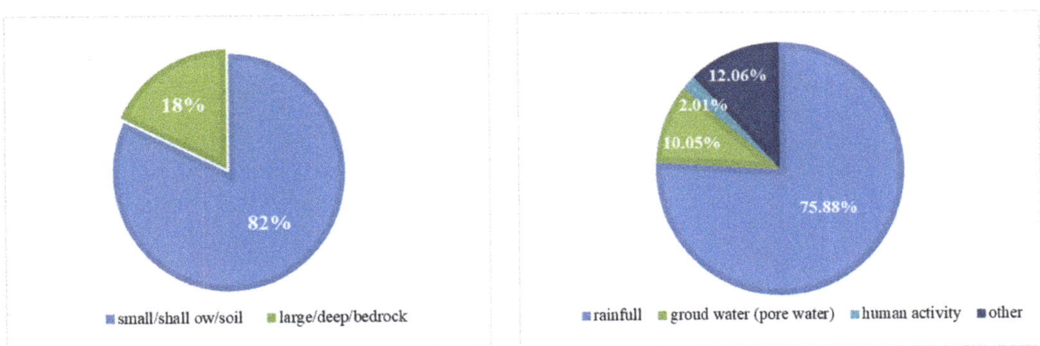

(a) (b)

Figure 4. Scale of landslide and trigger: (**a**) landslide type, (**b**) trigger.

As shown in Figure 5, where the average annual rainfall is high (annual average rainfall over 1060 mm), the historical landslides are also densely distributed. According to statistics on historical rainfall-type landslides in Fengjie, the proportion of landslides occurring in the rainy season (May to October) is more than 90%. This is in good agreement with the concentrated rainfall in Fengjie, and shows that the landslides in this area are positively correlated to precipitation.

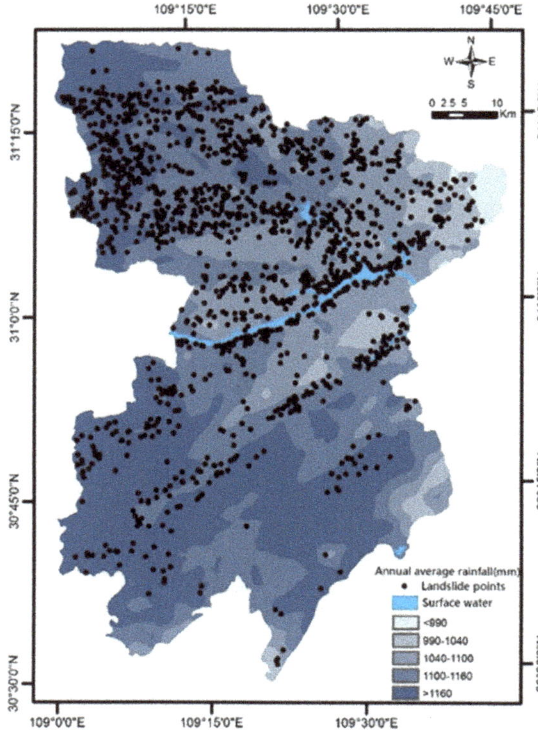

Figure 5. Average rainfall over the years and historical landslides.

The rainfall is extremely uneven and tends to be concentrated in one or several peaks of rainfall intensity, while the rainfall intensity in most other periods is very low or zero. Rainfall pattern analysis was performed on the first four days (96 h) of the occurrence of some rainfall-type landslides in the study area. The 96 h was divided into eight time periods, each of 12 h, and the rainfall in each period was accumulated. Figure 6 shows four typical rainfall patterns in the statistical rainfall-induced landslide data. The following criteria were used to distinguish rainfall patterns. When the peak rainfall occurs in the period 1–2, it is determined to be of the decreasing type; when the peak rainfall occurs in the period 3–6, it is considered to be of the mid-peak type; when it occurs within the 7–8 period, it is considered to be of the increasing type; when there are three or more peak rainfalls, and the difference between them is within 10 mm, it is considered to be of the average type.

According to this standard, of the 284 cases of rainfall-induced landslides, there were 186 cases of increasing rainfall, accounting for 65.5% of the total; 64 cases of mid-peak rainfall, accounting for 22.5% of the total; 19 cases of decreasing rainfall, accounting for 6.7% of the total; and 15 cases of average rainfall, accounting for 5.3% of the total (Figure 7). It can be seen that there was a strong correlation between rainfall and landslides. The impact of rainfall on landslides is not only seen on the day the landslides occur, but also has a certain lag effect on the occurrence of landslides. Therefore, both the rainfall of the day and the previous rainfall have a strong impact on the occurrence of landslides; hence, the threshold for the occurrence of landslides caused by rainfall is divided according to these two aspects.

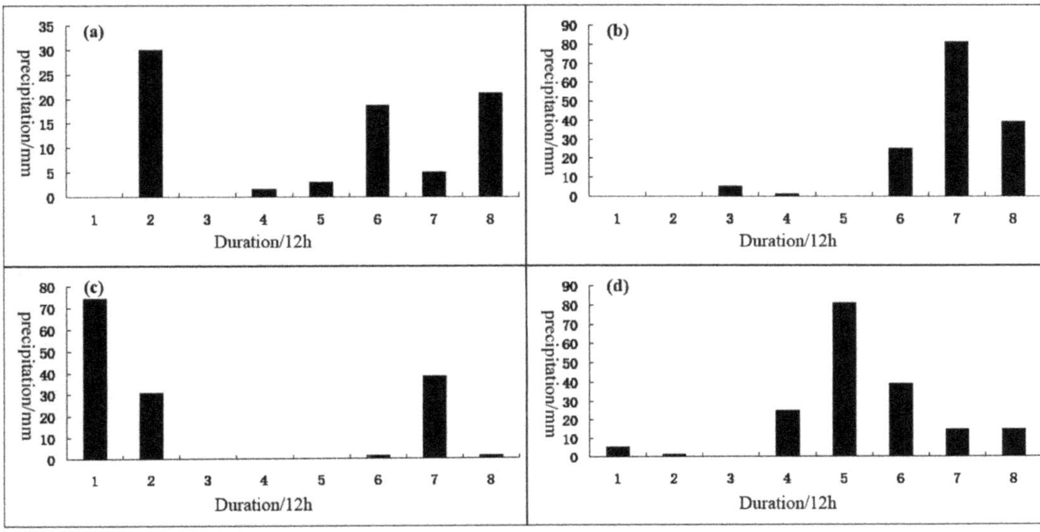

Figure 6. Four typical rainfall patterns in Chongqing: (**a**) average type; (**b**) increasing type; (**c**) diminishing type; (**d**) mid-peak type.

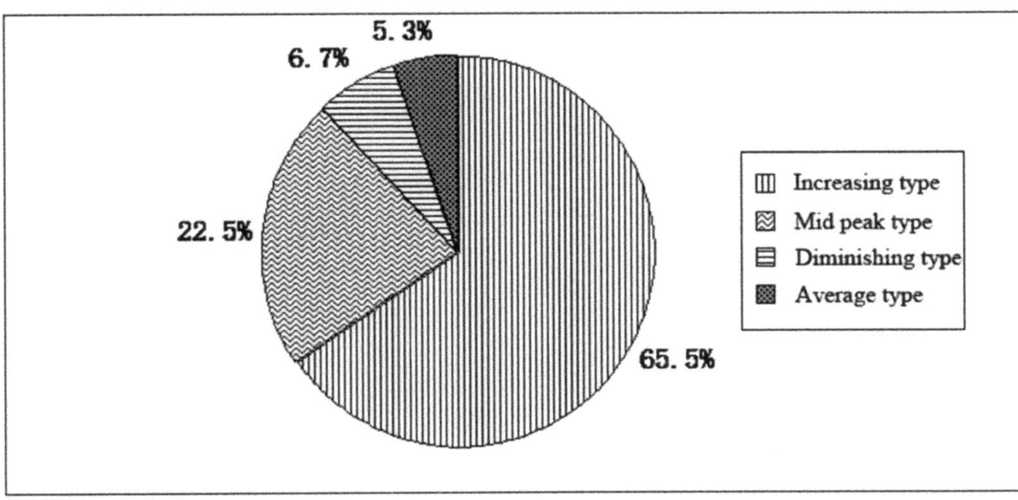

Figure 7. Percentage of four different rainfall types that induce landslides.

2.3. *Susceptibility Zoning*

2.3.1. Selection of Influencing Factors

Landslides are affected by both natural and human conditions. Based on existing research [38,39] and Fengjie County's landslide development characteristics, this paper selects 16 evaluation factors under the influence of four aspects: topography (elevation, slope, aspect, slope position, profile curvature, landforms, topographic wetness index (TWI)), geology (lithology, distance from fault, combination reclassification of stratum dip direction and slope aspect (CRDS)), environmental conditions (normalized vegetation index

(NDVI), distance from rivers, annual average rainfall, land cover), and human activities (the distance from roads and from buildings). The specific meanings as shown in Table 1.

Table 1. Meanings of influencing factors.

Influencing Factor	Meaning
Elevation	The distance from a point along the vertical line to the base surface
Slope	The degree of steepness of the surface unit
Aspect	The direction of the projection of the slope normal on the horizontal plane
Slope position	The landform part of the slope
Landforms	Relatively small-scale landforms, such as hills, valleys, terraces, etc.
Profile curvature	The rate of change of the surface slope at any point on the ground
TWI	The influence of regional topography on runoff flow direction and accumulation
Lithology	Some attributes that reflect the characteristics of the rock
Distance from faults	The distance to the nearest fault
CRDS	The relationship between rock inclination and slope aspect
NDVI	Percentage of vegetation area to the total statistical area
Distance from rivers	Distance to the nearest river
Land cover	Ways the land is used
Distance from roads	Distance to the nearest road
Distance from buildings	Distance to the nearest house
Annual average rainfall	Average annual rainfall over multiple years

2.3.2. Treatment of Influencing Factors

To reduce the influence of different dimensions, the reclassified data are normalized so that the value is between 0 and 1:

$$X = \frac{X' - X_{min}}{X_{max} - X_{min}} \tag{1}$$

where X is the normalized result, X′ is the original data of each factor, X_{min} is the minimum value of each factor, and X_{max} is the maximum value of each factor.

The influencing factors are reclassified, as shown in Table 2.

Table 2. Influencing factor categories of landslides.

Influencing Factor	Grade	Classification Standard
Elevation/(m)	7	1. <340; 2. 340~595; 3. 595~850; 4. 850~1105; 5. 1105~1360; 6. 1360~1615; 7. >1615
Slope/(°)	6	1. <10°; 2. 10~20°; 3. 20~30°; 4. 30~40°; 5. 40~50°; 6. >50°
Aspect/(°)	9	1. Flat; 2. North; 3. Northeast; 4. East; 5. Southeast; 6. South; 7. Southwest; 8. West; 9. Northwest
Slope position	6	1. Valleys; 2. Lower slope; 3. Flat slope; 4. Middle slope; 5. Upper slope; 6. Ridge
Landforms	10	1. Canyons, Deeply incised streams; 2. Mid-slope drainages, shallow valleys; 3. Upland drainages, headwaters; 4. U-shape valleys; 5. Plains; 6. Open slopes; 7. Upper slopes, mesas; 8. Local ridges, hills in valleys; 9. Mid-slope ridges, small hills in plains; 10. Mountain tops, high narrow ridges
Profile curvature	7	1. −1.0; 2. −1~0.5; 3. −0.5~0; 4. 0~0.5; 5. 0.5~1.0; 6. 1.0~1.5; 7. >1.5
TWI	7	1. <10; 2. 10~12; 3. 12~14; 4. 14~16; 5. 16~18; 6. 18~20; 7. >20
Lithology	7	1. TJx; 2. T1j; 3. D; 4. T1d-j; 5. J2s, J1z-2x, J3sn, J3p; 6. T1d, T3xj, T2b; 7. P, P3

Table 2. *Cont.*

Influencing Factor	Grade	Classification Standard
Distance from faults/(m):	11	1. <100; 2. 100~200; 3. 200~300; 4. 300~400; 5. 400~500; 6. 500~600; 7. 600~700; 8. 700~800; 9. 800~900; 10. 900~1000; 11. >1000
CRDS	6	1. Bedding slope; 2. Skewed slope; 3. Inclined slope; 4. Horizontal; 5. Reverse slope; 6. Flat
NDVI	7	1. <0.10; 2. 0.10~0.20; 3. 0.20~0.30; 4. 0.30~0.40; 5. 0.40~0.50; 6. 0.50~0.60; 7. >0.60
Distance from rivers/(m)	7	1. <100; 2. 100~200; 3. 200~300; 4. 300~400; 5. 400~500; 6. 500~600; 7. >600
Land cover	6	1. Cultivated land; 2. Woodland; 3. Meadow; 4. Land used for building; 5. Water area; 6. Unused land
Distance from roads/(m)	7	1. <100; 2. 100~200; 3. 200~300; 4. 300~400; 5. 400~500; 6. 500~600; 7. >600
Distance from buildings/(m)	7	1. <100; 2. 100~200; 3. 200~300; 4. 300~400; 5. 400~500; 6. 500~600; 7. >600
Annual average rainfall/(mm)	5	1. <990; 2. 990~1040; 3. 1040~1100; 4. 1100~1160; 5. >1160

2.3.3. Random Forest

Random forest is a classifier that uses multiple decision trees to train and predict samples. When using data subsets to construct a decision tree, elements of different data subsets can be repeated (that is, sampling with replacement). The random forest can reduce the one-sidedness and inaccuracy problems of single decision tree prediction, and prevent the judgment result of a single decision tree from overfitting, resulting in lower prediction accuracy. The remaining samples can be used as a test set to judge the error rate of the model prediction:

$$Y(x) = \arg \max_{Z} \sum_{i=1}^{k} I(y_i(x) = Z) \tag{2}$$

where Y(x) represents the judgment result of the RF model, $y_i(x)$ represents a single-course DT (decision tree), and Z represents the variable.

2.3.4. Accuracy Verification

The accuracy of the model prediction results can be determined by the receiver operating curve (ROC). The curve is obtained by setting a probability threshold to obtain a series of different binary classification results, and is then compared with the actual results. The closer the ROC is to the upper left, the higher the accuracy of the model. The point of the ROC closest to the upper left corner is the best threshold with the fewest errors, as well as the lowest total number of false positives and false negatives. The area under the curve (AUC) value is the area covered by the ROC, which can quantitatively determine the accuracy of the model. The AUC value is between 0 and 1, and the larger the value, the higher the model accuracy.

The accuracy of the model is verified; with the AUC value of the model training set to 0.95 and the AUC value of the test set to 0.87, the overall accuracy is 0.93 (Figure 8). The three values are all greater than 0.85, indicating the accuracy of the model. Higher accuracy means a better predictive ability, such that it can be used for landslide research.

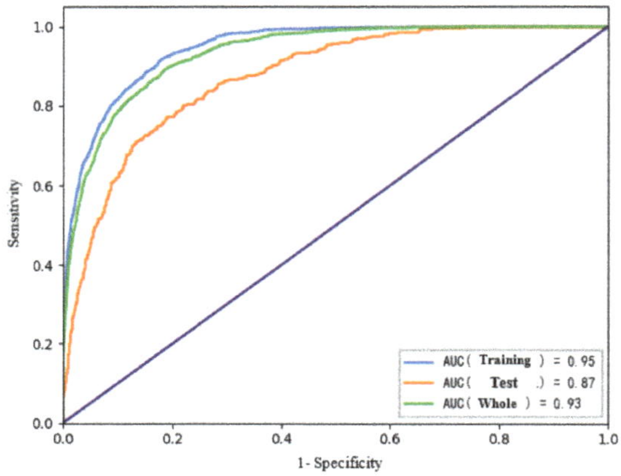

Figure 8. Model ROC curve.

2.4. *Fractal Model of Antecedent Effective Precipitation*

The triggering of landslides is not only related to the precipitation on the day, but also to the effective precipitation of the previous period. Existing studies [40,41] have shown that the cumulative precipitation of the previous period is positively correlated with the frequency of landslides. The greater the accumulated precipitation, the greater the probability of landslides. Due to the effects of interception, evaporation, and seepage, the previous rainfall does not directly act on the slope. In this paper, the precipitation after the attenuation of the previous cumulative rainfall is used as an important parameter for landslide warning and prediction. The effective precipitation of the previous period is the weighted sum of the daily precipitation of the previous period. The weight here is the attenuation coefficient β. The formula for the calculation of the fractal relationship is:

$$\beta = \sum_{n=1}^{m} \left(1 - \left(\frac{n}{n+1}\right)^{\alpha}\right) \tag{3}$$

where n represents the number of days before the occurrence of the landslide; m represents the number of days; and α is the scale index, determined by the formula Rn = C(n + 1)a, where C is the coefficient and the formula is the cumulative precipitation threshold fractal curve, which can be studied by analyzing the regional landslide. The relationship between the occurrence and the cumulative precipitation threshold is derived.

2.5. *Hybrid Model of the Antecedent Effective Precipitation and the Daily Precipitation under the Susceptibility Zoning*

2.5.1. Threshold Model Based on Susceptibility Zoning and Antecedent Effective Precipitation

According to the China Geological Hazard Meteorological Forecast and Early Warning Implementation Plan [42], the geological hazard weather forecast is divided into four levels: blue, yellow, orange, and red. We sorted the effective precipitation in the five different susceptibility regions in the study area, from very low to very high, and formed a first-order matrix. When the cumulative frequency of landslides in each landslide-prone area reached 25%, 40%, and 55%, the corresponding effective precipitation was used as the effective precipitation threshold for the yellow, orange, and red warnings.

$$Yi = \begin{cases} 0, \ (J < Y_{yellow}) \\ 1, \ (Y_{yellow} \leq J < Y_{orange}) \\ 2, \ (Y_{orange} \leq J < Y_{red}) \\ 3, \ (Y_{red} \leq J) \end{cases} \tag{4}$$

where Y_{yellow} is a yellow warning value with a cumulative frequency of 25%, Y_{orange} is an orange warning value with a cumulative frequency of 40%, Y_{red} is a red warning value with a cumulative frequency of 55%, and Y_i is the actual antecedent effective precipitation.

$$\mathrm{Ym} = \mathrm{X}_{\mathrm{fm}} \tag{5}$$

where $\mathrm{Y_m}$ is the early warning threshold; $\mathrm{X_{fm}}$ is the effective precipitation when the cumulative frequency of the landslide reaches a certain amount, which is expressed in matrix form, and the internal size is arranged from small to large; m is yellow, orange, or red; and fm is the cumulative frequency of landslides corresponding to the serial number.

$$\mathrm{fm} = [\mathrm{l}_i * n] \tag{6}$$

where fm is the serial number corresponding to the cumulative frequency of the landslides; i = yellow, orange, or red; $\mathrm{l_{yellow}}$ = 25%, $\mathrm{l_{orange}}$ = 40%, and $\mathrm{l_{red}}$ = 55%; n is the total number of landslides in each group; and [] refers to an integer.

2.5.2. Analysis of Early Warning and Forecast of Landslide Coupled with Daily Precipitation

The effect of the antecedent effective precipitation on the occurrence of landslides is a cumulative process, while daily precipitation has a triggering effect. Therefore, according to the zoning of landslide susceptibility, mathematical statistics and data mining are performed based on the cumulative amount of 24 h precipitation and the frequency of landslides, and the results are adjusted. The specific process is shown in Figure 9.

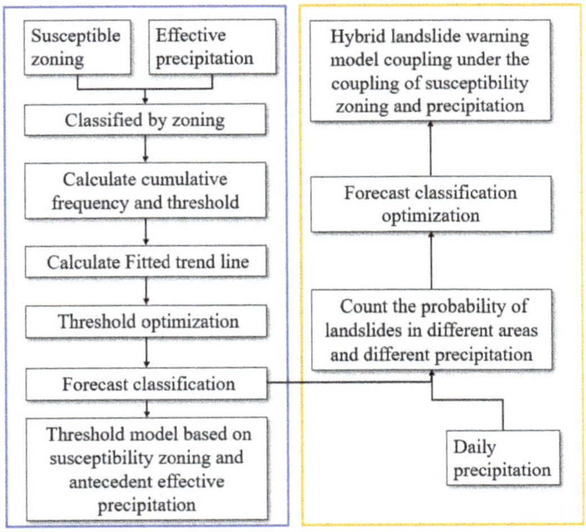

Figure 9. Specific flow chart4. Results and Discussion.

Based on the susceptibility zone and rainfall data, the rainfall data of the 10 days before the landslide in each zone were sorted, the attenuation coefficient of the previous rainfall was calculated, and the rainfall was corrected by this coefficient. We chose yellow, orange, and red as the three warning levels. The rainfall values corresponding to the first 25%, 40%, and 55% of landslide occurrence frequencies in each susceptibility zone were calculated as the thresholds for different grades of previous effective rainfall. When coupling the daily rainfall, we added a blue color to the warning level. The coupling followed certain rules: if the daily rainfall is less than a certain level, no adjustment will be made; when the daily rainfall is greater than a certain level, the warning level needs to be increased; the improvement may be of one level or multiple levels; the daily rainfall thresholds, adjusted by the warning levels of different susceptibility zones, are different.

3. Result

3.1. Random Forest Susceptibility Evaluation

The accuracy of the model prediction results can be determined by the ROC. The closer the curve is to the upper left, the higher the accuracy of the model. The AUC value is the area covered by the ROC curve, which can be quantified to judge the accuracy of the model; the AUC value is between 0 and 1, and the larger the value, the higher the accuracy of the model.

The accuracy of the model is verified; with the AUC value of the model training set to 0.95 and the AUC value of the test set to 0.87, the overall accuracy is 0.93 (Figure 8). These three values are all greater than 0.85, indicating the high accuracy of the model. A model with this level of predictive ability can be used for landslide research.

The random forest model was used to evaluate the landslide susceptibility of the study area, and the range was graded by the results of the natural breakpoint method. After grading, the Landslide Susceptibility Mapping (LSM) result was obtained. As shown in Figure 10, the majority of study areas are in regions with low or very low landslide susceptibility, and are concentrated in the south and southeast; highly prone areas are concentrated on both banks of the Yangtze River and its tributaries, mainly in the north and central regions of Fengjie.

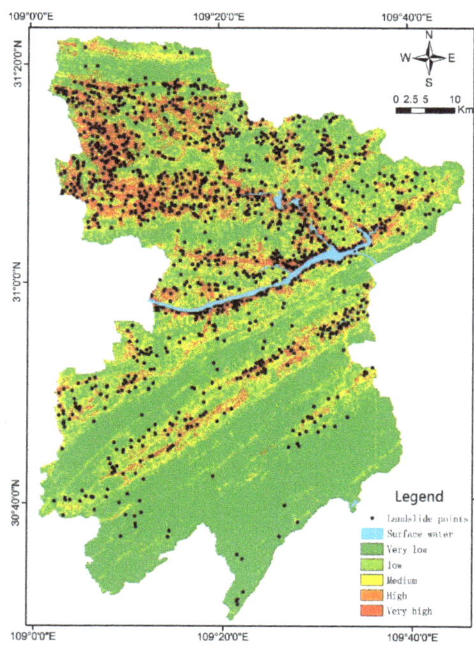

Figure 10. Landslide susceptibility zoning map.

3.2. Antecedent Effective Precipitation Model

Based on 1520 landslides with clear occurrence dates and coordinate records induced by precipitation in the Fengjie area from 2003 to 2016, and the daily precipitation data of the 10 days before the landslide (the day, the day before, the first three, five, and ten days of each landslide, and five other periods of cumulative precipitation), sorted by cumulative precipitation from small to large, we calculated the cumulative precipitation corresponding to 75% and 90% of the cumulative frequency of landslides, and took this as the cumulative precipitation threshold for different landslide occurrence probabilities (Table 3).

Table 3. Influencing factor categories of landslides.

Cumulative Frequency of Landslides/% \ Precipitation/mm	The Landslide Day	From the Day of the Landslide to 3 Days before	From the Day of the Landslide to 5 Days before	From the Day of the Landslide to 10 Days before
75	69.9	146.4	252.9	273.4
90	92.7	233.2	330.7	372.3
Difference	22.8	86.8	77.8	98.9

Figure 11 uses the observation period (days) as the abscissa and the cumulative precipitation threshold as the ordinate to display the cumulative precipitation threshold for each observation period. The accumulated precipitation is the power exponential function of the observation period (n), and values of α can be calculated by fitting, which were found to be 0.609 (75% landslide probability) and 0.603 (90% landslide probability).

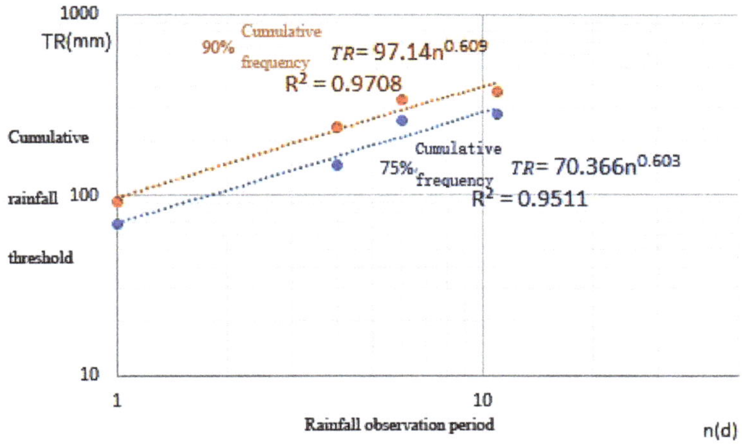

Figure 11. The relationship between the cumulative precipitation threshold and the number of days of the observation period.

Substituting α into Equation (3), we calculated the attenuation coefficient of the daily precipitation in the 10 days prior to the landslide. Table 4 presents the attenuation coefficients of the previous precipitation events, corresponding to 90% of the cumulative frequency of the landslide.

Table 4. Antecedent rain attenuation coefficient.

	1 Day before	2 Days before	3 Days before	4 Days before	5 Days before	6 Days before	7 Days before	8 Days before	9 Days before	10 Days before
90%	0.344	0.219	0.161	0.127	0.105	0.090	0.078	0.069	0.062	0.056

3.3. Calculation and Adjustment of Antecedent Effective Precipitation Threshold Based on Susceptibility Zoning

3.3.1. Threshold Model Based on Susceptibility Zoning and Antecedent Effective Precipitation

According to the antecedent effective precipitation model and susceptibility zoning, the precipitation thresholds of historical landslides were counted. The results are shown in Table 5. The expressions of the three fitting lines of yellow, orange, and red warning levels are: $Y = -6.1687x + 134.03$, $Y = -14.283x + 132.09$, and $Y = -18.954x + 108.83$ ($x = 1, 2, 3, 4,$

or 5, corresponding to very low areas, low areas, moderate areas, high areas, or very high areas, respectively).

Table 5. Effective precipitation thresholds in the first 10 days of different susceptibility zones and different warning levels.

Susceptibility Zoning	Frequency of Landslides	Effective Precipitation in the First 10 Days (mm)	Warning Level
Very low	25%	96	Yellow
	40%	129	Orange
	55%	137	Red
Low	25%	58	Yellow
	40%	87	Orange
	55%	114	Red
Moderate	25%	51	Yellow
	40%	87	Orange
	55%	109	Red
High	25%	49	Yellow
	40%	87	Orange
	55%	109	Red
Very high	25%	6	Yellow
	40%	58	Orange
	55%	109	Red

The original precipitation threshold was adjusted according to the trend line; the optimized precipitation thresholds of each level after calculation are shown in Table 6.

Table 6. Thresholds after adjustment of the effective precipitation in the early stage of different susceptibility zones and different warning levels.

Warning Level	Very Low Areas (mm)	Low Areas (mm)	Moderate Areas (mm)	High Areas (mm)	Very High Areas (mm)
Yellow	90	71	52	33	14
Orange	118	104	89	75	61
Red	128	122	116	109	103

3.3.2. Warning Model Coupled with Daily Precipitation

We performed statistical analysis on the daily precipitation data of 1520 landslides, and obtained the specific adjustment methods of the coupled day's precipitation threshold model, as follows: (1) We counted the cumulative frequency of landslide occurrence, from small to large, unaffected to landslide-prone areas. (2) The cumulative frequency of landslides was less than 0.4. The corresponding daily precipitation type and warning level was not adjusted. For the daily precipitation type corresponding to a cumulative frequency greater than 0.4 and less than 0.7, the warning level was raised by one level; for the daily precipitation type corresponding to a cumulative frequency greater than 0.7, the warning level was raised by 2. (3) We increased the level until the red warning level. Table 7 shows the relationship between the original forecast level and the adjusted level.

Table 7. Forecast level adjustment based on the precipitation of the day.

Original Warning Level	Very Low Areas	Low Areas	Moderate Areas	High Areas	Very High Areas
Blue	Blue	Blue	Blue	Yellow (Light rain)	Orange (Light rain)
Yellow	Yellow	Yellow	Yellow	Orange (Heavy rain)	Orange (Light rain)
Orange	Red (rainstorm)	Red (Heavy rain)	Red (Heavy rain)	Red (Heavy rain)	Red (Heavy rain)
Red	Red	Red	Red	Red	Red

3.4. Instance Verification

3.4.1. Regional Verification Analysis

Figure 12a is a landslide warning map generated based on the antecedent effective precipitation data and landslide susceptibility zoning data of various weather stations on 30 June 2018. Figure 12b is a landslide warning map based on the antecedent effective precipitation data of each weather station on 30 June 2018, and the daily precipitation data, combined with the landslide susceptibility zoning.

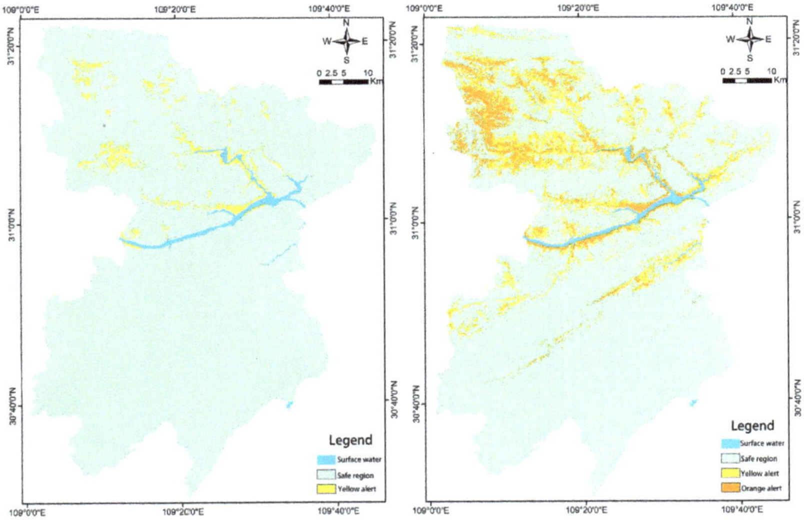

Figure 12. Early warning map of landslide area: (**a**) early warning results based on susceptibility zoning and antecedent effective precipitation; (**b**) early warning results based on a coupled day precipitation.

Most of the study areas were in the warning range, while Fengjie County was mainly in the safe area. Compared with the single early precipitation landslide warning, the coupled precipitation landslide warning model has an orange warning. What is more, in the single landslide warning model, most of the yellow warning areas are replaced by orange warnings, and the yellow warnings in the coupled warning model are attached to the orange warnings.

3.4.2. Monomer Verification Analysis

The rainfall-induced catastrophic deformation of the above five typical landslides can be summarized, as shown in Table 8 below, and their spatial locations are shown in Figure 13. The final analysis results of the five typical cases are yellow to red warnings. All five landslides on the site experienced large deformations due to rainfall, but there was no overall instability damage. Therefore, the overall early warning result is consistent with the actual situation on the ground.

Table 8. Summary table of early warning analysis of typical cases of precipitation-induced landslides in 2017.

Number	Name	Susceptibility Zoning	Antecedent Precipitation (mm)	Daily Precipitation (mm)	Warning Level			Actual Catastrophe
					Antecedent Precipitation	Daily Precipitation	Adjusted Level	
1	Damian	Very high	47.6	85.86	Safe	Yellow	Yellow	Continuous deformation Continuous deformation
2	Hejiawan	Very high	12.32	147.5	Safe	Red	Red	Small area collapse New deformation crack
3	Huoshitan	Very high	88.38	0	Orange	Blue	Orange	Continuous deformation Multiple cracks
4	Zhakou	High	93.8	1.4	Orange	Blue	Orange	Continuous deformation Multiple cracks
5	Xinpu	High	52.26	0	Yellow	Blue	Yellow	Local deformation

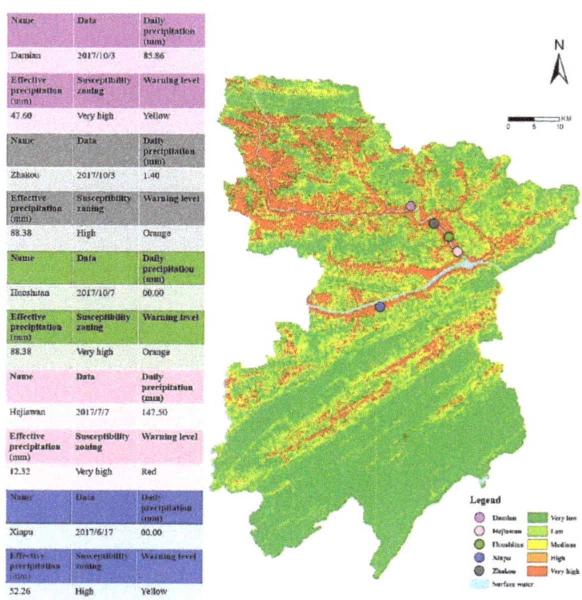

Figure 13. Distribution map of 5 landslide cases in 2017.

The established landslide warning model with the coupling of susceptibility zoning and precipitation was applied to the newly occurring precipitation-induced landslides in the study area and obtained a good verification effect, which is primarily in line with the actual situation.

4. Discussion

4.1. The Importance and Influence of Factors

The contributions of factors to landslides are different. The detection of landslide-inducing factors can provide an important reference for landslide disaster prediction. In this paper, the importance of 16 factors in the model is detected using the reduced Gini index in the random forest model (Figure 14). The highest contributors to landslides are annual average rainfall and elevation.

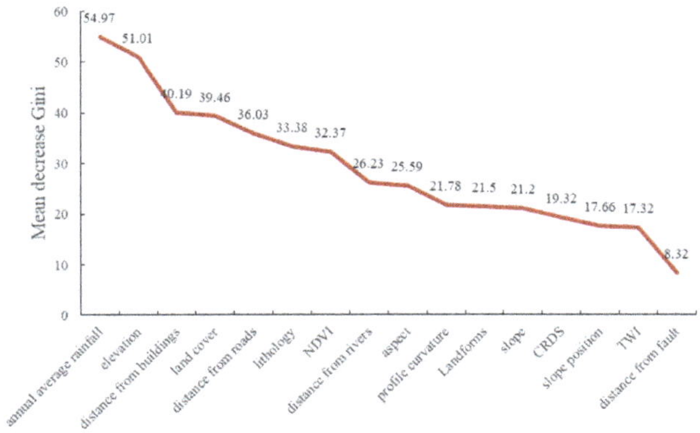

Figure 14. Importance of landslide influencing factors.

The statistical distributions of landslide density based on these two factors are presented in Figure 15. Annual average rainfall is the most important factor. According to the statistics of annual average rainfall and landslide density, the landslide density first increases and then decreases with the increase in annual average rainfall. Because the surface runoff formed by rainfall takes away the unstable soil particles on the slope, the slope is eroded. Rainfall also promotes the growth of local vegetation, thereby inhibiting soil erosion and landslides. Landslide density is negatively correlated with elevation because, usually, in low altitude areas, population agglomeration and human activities easily change the local geological environment, destroy the stability of the slope toe, and lead to a high probability of landslide occurrence.

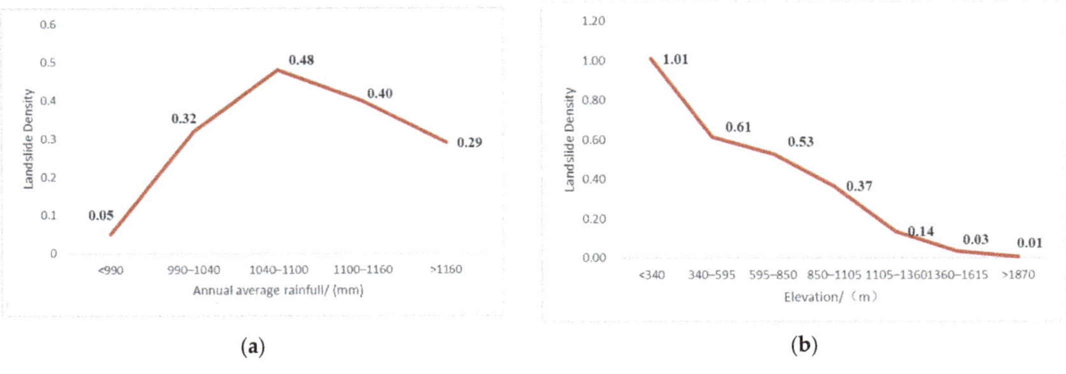

Figure 15. Partial effects on landslide susceptibility of typical conditioning factors: (**a**) annual average rainfall, (**b**) elevation.

4.2. *Effectiveness of Antecedent Rainfall*

Early rainfall can reduce the adsorption capacity, and increase the pore pressure, of soil, thereby reducing the stability of slope, which is an important factor in terms of inducing landslides. The force of early rainfall on the slope is greatly affected by seasonality. Rainfall and evapotranspiration are also different in different seasons. In the rainy season, large amounts of rainfall and low evaporation rates lead to poor slope stability. However, the increase in evapotranspiration from late spring to early autumn resulted in the decrease in soil pressure and the increase in slope stability. Therefore, this paper accounted for the rainfall 10 days before the landslide, referred to as the effective rainfall in the early stage, so as to avoid invalid rainfall, such as evapotranspiration, being included in the model.

4.3. Practical Application of Coupling Model of Susceptibility Zoning and Precipitation

The influence of rainfall on landslide has hysteresis and effectiveness, which cannot only be studied from the antecedent effective precipitation threshold or the daily precipitation threshold. Therefore, the author proposes a model coupling the antecedent effective precipitation and the daily precipitation, in which the antecedent effective precipitation and the daily precipitation complement each other. Based on the landslide susceptibility zoning, we use the coupling method of the antecedent effective precipitation and the daily precipitation to improve the warning ability of the rainfall threshold model. It was found that the warning level of the coupling model is higher than that of the single model, and the warning range is more realistic. Firstly, according to the landslide susceptibility zoning, the antecedent effective precipitation thresholds of three warning levels (yellow, orange, and red) are given. The antecedent effective precipitation alone is not enough to reflect the real-time impact of rainfall on landslides. Therefore, the actual effect of rainfall on landslides is reflected by coupling the daily precipitation with the antecedent effective precipitation threshold.

According to the example verification (Table 8), the susceptibility area of Hejiawan is divided into middle–high susceptibility areas. After the superposition of antecedent effective precipitation of 47.6 mm, the antecedent effective precipitation warning level is 'safe', but combined with the daily precipitation of 85.89 mm, the warning level of Hejiawan landslide is 'red', which is consistent with the actual situation of small-area landslides. The daily precipitation of Huoshitan and Xinpu landslides is 0, and the possibility of landslide occurrence is low. However, combined with the antecedent effective precipitation, the forecast grades of these two places are 'orange' and 'yellow'. The results show that the coupling model of the antecedent effective and the daily precipitation can avoid the occurrence, and reduce the risk, of landslides.

5. Conclusions

In this study, we propose a new precipitation warning model that couples landslide susceptibility, antecedent effective precipitation, and the daily precipitation threshold. We applied the model to actual landslide cases and verified it to explain the practicality and accuracy of the model. Our conclusions are as follows:

(1) The evaluation results of the landslide susceptibility model in Fengjie County based on RF are accurate and reasonable. The AUC value of the test set is 0.87, and the annual average rainfall and elevation are the factors that contribute the most to the model.

(2) The early warning model of landslide susceptibility, the antecedent effective precipitation, and the daily precipitation coupling has higher accuracy than the model of landslide susceptibility and the antecedent effective precipitation coupling, and it can better characterize the mechanism of rainfall-induced landslides.

(3) The landslide warning model based on random forest coupling of rainfall-inducing factors in landslide susceptibility zoning has high warning accuracy, which can provide a reference for areas with the same geological conditions and climatic conditions.

Author Contributions: Conceptualization, H.W. and D.S.; formal analysis, H.W., D.S., Q.G., S.S., C.M. and F.Z.; data curation, Q.G. and S.S.; investigation, H.W., C.M. and F.Z.; writing—original draft, D.S., and Q.G.; writing—review, and editing H.W. and D.S. All authors have read and agreed to the published version of the manuscript.

Funding: This research was funded by the National Natural Science Foundation of China (Grant No.41901214), National Key Research and Development Program of China (Grant No. 2018YFC1505501), the Natural Science Foundation of Chongqing (Grant No. cstc2020jcyjmsxmX0841), and Scientific and Technological Research Program of Chongqing Municipal Education Commission (Grant No. KJQN201800511).

Conflicts of Interest: The authors declare no conflict of interest.

References

1. Huang, F.; Yin, K.; Huang, J.; Gui, L.; Wang, P. Landslide susceptibility mapping based on self-organizing-map network and extreme learning machine. *Eng. Geol.* **2017**, *223*, 11–22. [CrossRef]
2. Chen, L.; Mei, L.; Zeng, B.; Yin, K.; Shrestha, D.P.; Du, J. Failure probability assessment of landslides triggered by earthquakes and rainfall: A case study in Yadong County, Tibet, China. *Sci. Rep.* **2020**, *10*, 16531. [CrossRef] [PubMed]
3. Corominas, J.; van Westen, C.; Frattini, P.; Cascini, L.; Malet, J.P.; Fotopoulou, S.; Catani, F.; Van Den Eeckhaut, M.; Mavrouli, O.; Agliardi, F.; et al. Recommendations for the quantitative analysis of landslide risk. *Bull. Eng. Geol. Environ.* **2013**, *73*, 209–263. [CrossRef]
4. Carrara, A.; Merenda, L. Landslide inventory in northern Calabria, southern Italy. *Geol. Soc. Am. Bull.* **1976**, *87*, 1153–1162. [CrossRef]
5. Fausto, G.; Alberto, C.; Mauro, C.; Reichenbach, P. Landslide hazard evaluation: A review of current techniques and their application in a multi-scale study, Central Italy. *Geomorphology* **1999**, *31*, 181–216.
6. Robert, S.; Westen, C.J.V. Slope instability Recognition, analysis and zonation. *Landslides Investig. Mitig.* **1996**, *247*, 129–177.
7. Di Napoli, M.; Carotenuto, F.; Cevasco, A.; Confuorto, P.; Di Martire, D.; Firpo, M.; Pepe, G.; Raso, E.; Calcaterra, D. Machine learning ensemble modelling as a tool to improve landslide susceptibility mapping reliability. *Landslides* **2020**, *17*, 1897–1914. [CrossRef]
8. Lombardo, L.; Opitz, T.; Ardizzone, F.; Guzzetti, F.; Huser, R. Space-time landslide predictive modelling. *Earth-Sci. Rev.* **2020**, *209*, 103318. [CrossRef]
9. Komori, D.; Rangsiwanichpong, P.; Inoue, N.; Ono, K.; Watanabe, S.; Kazama, S. Distributed probability of slope failure in Thailand under climate change. *Clim. Risk Manag.* **2018**, *20*, 126–137. [CrossRef]
10. Chen, W.; Peng, J.; Hong, H.; Shahabi, H.; Pradhan, B.; Liu, J.; Zhu, A.X.; Pei, X.; Duan, Z. Landslide susceptibility modelling using GIS-based machine learning techniques for Chongren County, Jiangxi Province, China. *Sci. Total Environ.* **2018**, *626*, 1121–1135. [CrossRef]
11. Dou, J.; Yunus, A.P.; Tien Bui, D.; Merghadi, A.; Sahana, M.; Zhu, Z.; Chen, C.W.; Khosravi, K.; Yang, Y.; Pham, B.T. Assessment of advanced random forest and decision tree algorithms for modeling rainfall-induced landslide susceptibility in the Izu-Oshima Volcanic Island, Japan. *Sci. Total Environ.* **2019**, *662*, 332–346. [CrossRef] [PubMed]
12. Kuradusenge, M.; Kumaran, S.; Zennaro, M. Rainfall-Induced Landslide Prediction Using Machine Learning Models: The Case of Ngororero District, Rwanda. *Int. J. Environ. Res. Public. Health* **2020**, *17*, 4147. [CrossRef] [PubMed]
13. Lai, J.S.; Tsai, F. Improving GIS-based Landslide Susceptibility Assessments with Multi-temporal Remote Sensing and Machine Learning. *Sensors* **2019**, *19*, 3717. [CrossRef]
14. Tien Bui, D.; Tuan, T.A.; Klempe, H.; Pradhan, B.; Revhaug, I. Spatial prediction models for shallow landslide hazards: A comparative assessment of the efficacy of support vector machines, artificial neural networks, kernel logistic regression, and logistic model tree. *Landslides* **2016**, *13*, 361–378. [CrossRef]
15. Sun, D.; Xu, J.; Wen, H.; Wang, Y. An Optimized Random Forest Model and Its Generalization Ability in Landslide Susceptibility Mapping: Application in Two Areas of Three Gorges Reservoir, China. *J. Earth Sci.* **2020**, *31*, 1068–1086. [CrossRef]
16. Moayedi, H.; Mehrabi, M.; Mosallanezhad, M.; Rashid, A.S.A.; Pradhan, B. Modification of landslide susceptibility mapping using optimized PSO-ANN technique. *Eng. Comput-Ger.* **2019**, *35*, 967–984. [CrossRef]
17. Pourghasemi, H.R.; Rahmati, O. Prediction of the landslide susceptibility: Which algorithm, which precision? *Catena* **2018**, *162*, 177–192. [CrossRef]
18. Zhou, X.; Wen, H.; Li, Z.; Zhang, H.; Zhang, W. An interpretable model for the susceptibility of rainfall-induced shallow landslides based on SHAP and XGBoost. *Geocarto Int. TGEI* **2022**, 1–27. [CrossRef]
19. Chen, W.; Xie, X.; Wang, J.; Pradhan, B.; Hong, H.; Bui, D.T.; Duan, Z.; Ma, J. A comparative study of logistic model tree, random forest, and classification and regression tree models for spatial prediction of landslide susceptibility. *Catena* **2017**, *151*, 147–160. [CrossRef]
20. Sun, D.; Xu, J.; Wen, H.; Wang, D. Assessment of landslide susceptibility mapping based on Bayesian hyperparameter optimization: A comparison between logistic regression and random forest. *Eng. Geol.* **2021**, *281*, 105972. [CrossRef]
21. Trigila, A.; Iadanza, C.; Esposito, C.; Scarascia-Mugnozza, G. Comparison of Logistic Regression and Random Forests techniques for shallow landslide susceptibility assessment in Giampilieri (NE Sicily, Italy). *Geomorphology* **2015**, *249*, 119–136. [CrossRef]
22. Tsangaratos, P.; Ilia, I.; Hong, H.; Chen, W.; Xu, C. Applying Information Theory and GIS-based quantitative methods to produce landslide susceptibility maps in Nancheng County, China. *Landslides* **2016**, *14*, 1091–1111. [CrossRef]
23. Alvioli, M.; Melillo, M.; Guzzetti, F.; Rossi, M.; Palazzi, E.; von Hardenberg, J.; Brunetti, M.T.; Peruccacci, S. Implications of climate change on landslide hazard in Central Italy. *Sci. Total Environ.* **2018**, *630*, 1528–1543. [CrossRef] [PubMed]
24. Chang, K.-T.; Chiang, S.-H. An integrated model for predicting rainfall-induced landslides. *Geomorphology* **2009**, *105*, 366–373. [CrossRef]
25. Guo, X.; Cui, P.; Li, Y.; Ma, L.; Ge, Y.; Mahoney, W.B. Intensity–duration threshold of rainfall-triggered debris flows in the Wenchuan Earthquake affected area, China. *Geomorphology* **2016**, *253*, 208–216. [CrossRef]
26. Chien-Yuan, C.; Tien-Chien, C.; Fan-Chieh, Y.; Wen-Hui, Y.; Chun-Chieh, T. Rainfall duration and debris-flow initiated studies for real-time monitoring. *Environ. Geol.* **2005**, *47*, 715–724. [CrossRef]

27. Guzzetti, F.; Peruccacci, S.; Rossi, M.; Stark, C.P. Rainfall thresholds for the initiation of landslides in central and southern Europe. *Meteorol. Atmos. Phys.* **2007**, *98*, 239–267. [CrossRef]
28. Vessia, G.; Parise, M.; Brunetti, M.T.; Peruccacci, S.; Rossi, M.; Vennari, C.; Guzzetti, F. Automated reconstruction of rainfall events responsible for shallow landslides. *Nat. Hazards Earth Syst. Sci.* **2014**, *14*, 2399–2408. [CrossRef]
29. Glade, T.; Crozier, M.; Smith, P. Applying Probability Determination to Refine Landslide-triggering Rainfall Thresholds Using an Empirical "Antecedent Daily Rainfall Model". *Pure Appl. Geophys.* **2020**, *157*, 1059–1079. [CrossRef]
30. Mathew, J.; Babu, D.G.; Kundu, S.; Kumar, K.V.; Pant, C.C. Integrating intensity–duration-based rainfall threshold and antecedent rainfall-based probability estimate towards generating early warning for rainfall-induced landslides in parts of the Garhwal Himalaya, India. *Landslides* **2013**, *11*, 575–588. [CrossRef]
31. Sengupta, A.; Gupta, S.; Anbarasu, K. Rainfall thresholds for the initiation of landslide at Lanta Khola in north Sikkim, India. *Nat. Hazards* **2010**, *52*, 31–42. [CrossRef]
32. Jemec, M.; Komac, M. Rainfall patterns for shallow landsliding in perialpine Slovenia. *Nat. Hazards* **2011**, *67*, 1011–1023. [CrossRef]
33. Milne, F.D.; Brown, M.J.; Knappett, J.A.; Davies, M.C.R. Centrifuge modelling of hillslope debris flow initiation. *Catena* **2012**, *92*, 162–171. [CrossRef]
34. Amiri, M.A.; Conoscenti, C. Landslide susceptibility mapping using precipitation data, Mazandaran Province, north of Iran. *Nat. Hazards* **2017**, *89*, 255–273. [CrossRef]
35. Khan, S.; Kirschbaum, D.B.; Stanley, T. Investigating the potential of a global precipitation forecast to inform landslide prediction. *Weather Clim. Extreme* **2021**, *33*, 100364. [CrossRef]
36. Beck, H.E.; Zimmermann, N.E.; McVicar, T.R.; Vergopolan, N.; Berg, A.; Wood, E.F. Present and future Koppen-Geiger climate classification maps at 1-km resolution. *Sci. Data* **2018**, *5*, 180214. [CrossRef]
37. Iadanza, C.; Trigila, A.; Napolitano, F. Identification and characterization of rainfall events responsible for triggering of debris flows and shallow landslides. *J. Hydrol.* **2016**, *541*, 230–245. [CrossRef]
38. Sun, D.; Wen, H.; Wang, D.; Xu, J. A random forest model of landslide susceptibility mapping based on hyperparameter optimization using Bayes algorithm. *Geomorphology* **2020**, *362*, 107201. [CrossRef]
39. Wang, Y.; Sun, D.; Wen, H.; Zhang, H.; Zhang, F. Comparison of Random Forest Model and Frequency Ratio Model for Landslide Susceptibility Mapping (LSM) in Yunyang County (Chongqing, China). *Int. J. Environ. Res. Public. Health* **2020**, *17*, 4206. [CrossRef]
40. Li, C.; Ma, T.; Li, W.; Zheng, A.; Zhu, X. Fractal relation of landslide frequency and rainfall. *Chin. J. Geol. Hazard Control* **2010**, *21*, 87–93.
41. Li, C.; Ma, T.; Sun, L.; Zhu, X. A New Approach to Calibrate Antecedent Precipitation for Rainfall-triggering Landslide Forecast. *Mt. Res. Dev.* **2011**, *29*, 81–86.
42. Liu, C.; Wen, M.; Tang, C. Meteorological early warning of geo-hazards in China based on raining forecast. *Reg. Geol. China* **200[illegible]**, *23*, 303–309.